AGRICULTURAL TECHNOLOGIES AND TROPICAL DEFORESTATION

Agricultural Technologies and Tropical Deforestation

Edited by

A. Angelsen
*Center for International Forestry Research (CIFOR), Bogor, Indonesia
and
Department of Economics and Social Sciences
Agricultural University of Norway
Ås, Norway*

and

D. Kaimowitz
Center for International Forestry Research (CIFOR), San José, Costa Rica

CABI *Publishing*
in association with
Center for International Forestry Research (CIFOR)

333.75137
A278

CABI *Publishing* is a division of CAB *International*

CABI Publishing
CAB International
Wallingford
Oxon OX10 8DE
UK

Tel: +44 (0)1491 832111
Fax: +44 (0)1491 833508
Email: cabi@cabi.org
Web site: http://www.cabi.org

CABI Publishing
10 E 40th Street
Suite 3203
New York, NY 10016
USA

Tel: +1 212 481 7018
Fax: +1 212 686 7993
Email: cabi-nao@cabi.org

© CAB *International* 2001. All rights reserved. No part of this publication may be reproduced in any form or by any means, electronically, mechanically, by photocopying, recording or otherwise, without the prior permission of the copyright owners.

Published in association with:

Center for International Forestry Research (CIFOR)
PO Box 6596 JKPWB
Jakarta 10065
Indonesia

A catalogue record for this book is available from the British Library, London, UK.

Library of Congress Cataloging-in-Publication Data
Agricultural technologies and tropical deforestation / edited by A. Angelsen and D. Kaimowitz.
   p. cm.
   Based on papers presented at a workshop held in Turrialba, Costa Rica, Mar. 11-13, 1999.
   Includes bibliographical references.
   ISBN 0-85199-451-2 (alk. paper)
   1. Deforestation--Tropics--Congresses. 2. Deforestation--Control--Tropics--Congresses. 3. Agriculture--Technology transfer--Tropics--Congresses. I. Angelsen, Arild. II. Kaimowitz, David. III. Center for International Forestry Research.

SD418.3.T76 A47 2001
333.75'137'0913--dc21

00-057960

ISBN 0 85199 451 2

Typeset by AMA DataSet Ltd, UK.
Printed and bound in the UK by Biddles Ltd, Guildford and King's Lynn.

Contents

Contributors ix

Preface xiii

Introduction and Overviews

1 Introduction: the Role of Agricultural Technologies in Tropical Deforestation 1
   Arild Angelsen and David Kaimowitz

2 Technological Change and Deforestation: a Theoretical Overview 19
   Arild Angelsen, Daan van Soest, David Kaimowitz and Erwin Bulte

Developed Countries

3 The Transition from Deforestation to Reforestation in Europe 35
   Alexander Mather

4 Did a Green Revolution Restore the Forests of the American South? 53
   Thomas K. Rudel

Latin America

5 A General Equilibrium Analysis of Technology, Migration and
   Deforestation in the Brazilian Amazon   69
   Andrea Cattaneo

6 Will Intensifying Pasture Management in Latin America Protect
   Forests – or Is It the Other Way Round?   91
   *Douglas White, Federico Holmann, Sam Fujisaka, Keneth Reategui and
   Carlos Lascano*

7 Intensified Small-scale Livestock Systems in the Western Brazilian
   Amazon   113
   *Stephen A. Vosti, Chantal Line Carpentier, Julie Witcover and
   Judson F. Valentim*

8 Technological Progress versus Economic Policy as Tools to Control
   Deforestation: the Atlantic Zone of Costa Rica   135
   Peter Roebeling and Ruerd Ruben

9 Land Use, Agricultural Technology and Deforestation among
   Settlers in the Ecuadorean Amazon   153
   *Francisco Pichon, Catherine Marquette, Laura Murphy and
   Richard Bilsborrow*

10 Ecuador Goes Bananas: Incremental Technological Change and
   Forest Loss   167
   Sven Wunder

11 Soybean Technology and the Loss of Natural Vegetation in Brazil
   and Bolivia   195
   David Kaimowitz and Joyotee Smith

12 Kudzu-improved Fallows in the Peruvian Amazon   213
   David Yanggen and Thomas Reardon

Africa

13 Ambiguous Effects of Policy Reforms on Sustainable Agricultural
   Intensification in Africa   231
   Thomas Reardon and Christopher B. Barrett

14	A Century of Technological Change and Deforestation in the *Miombo* Woodlands of Northern Zambia *Stein Holden*	251
15	Livestock Disease Control and the Changing Landscapes of South-west Ethiopia *Robin S. Reid, Philip K. Thornton and Russell L. Kruska*	271
16	Tree Crops as Deforestation and Reforestation Agents: the Case of Cocoa in Côte d'Ivoire and Sulawesi *François Ruf*	291

Asia

17	Agriculture and Deforestation in Tropical Asia: an Analytical Framework *Sisira Jayasuriya*	317
18	Deforestation, Irrigation, Employment and Cautious Optimism in Southern Palawan, the Philippines *Gerald Shively and Elmer Martinez*	335
19	Agricultural Development Policies and Land Expansion in a Southern Philippine Watershed *Ian Coxhead, Gerald Shively and Xiaobing Shuai*	347
20	The Impact of Rubber on the Forest Landscape in Borneo *Wil de Jong*	367

Summaries

21	Agricultural Technology and Forests: a Recapitulation *Arild Angelsen and David Kaimowitz*	383
22	Policy Recommendations *David Kaimowitz and Arild Angelsen*	403
Index		413

Contributors

Arild Angelsen (arild.angelsen@ios.nlh.no) is an associate professor in development and resource economics at the Agricultural University of Norway and an associate scientist at the Center for International Forestry Research (CIFOR), Bogor, Indonesia.

Christopher B. Barrett (cbb2@cornell.edu) works as an associate professor in the Department of Agricultural, Resource, and Managerial Economics at Cornell University.

Richard Bilsborrow (richard_bilsborrow@unc.edu) forms part of the demography program of the Department of Biostatistics at the University of North Carolina at Chapel Hill.

Erwin Bulte (e.h.bulte@kub.nl) is an associate professor of economics at Tilburg University, the Netherlands.

Chantal Line Carpentier (carpentier@ccemtl.org) prepared her chapter while working at the Wallace Institute for Alternative Agriculture. She now serves as a program manager for the Commission on Environmental Collaboration (CEC) in Montreal.

Andrea Cattaneo (a.cattaneo@cgiar.org) is an environmental economist at the International Food Policy Research Institute (IFPRI) in Washington, DC.

Ian Coxhead (coxhead@facstaff.wisc.edu) is an associate professor in the Department of Agricultural and Applied Economics and a faculty member of the Center for Southeast Asian Studies at the University of Wisconsin-Madison.

Wil A. de Jong (w.de-jong@cgiar.org) is a senior scientist at the Center for International Forestry Research (CIFOR), Bogor, Indonesia.

Sam Fujisaka (s.fujisaka@cgiar.org) is an agricultural anthropologist. His work for this book was done while working for the Centro Internacional de Agricultura Tropical (CIAT).

Stein Holden (stein.holden@ios.nlh.no) is an associate professor in development and resource economics at the Agricultural University of Norway.

Federico Holmann (f.holmann@cgiar.org) is a livestock economist with a joint senior scientist position between Centro Internacional de Agricultura Tropical (CIAT) and the International Livestock Research Institute (ILRI), based in Cali, Colombia.

Sisira Jayasuriya (s.jayasuriya@latrobe.edu.au) is a reader in economics at La Trobe University, Melbourne, Australia.

David Kaimowitz (d.kaimowitz@cgiar.org) is principal economist at the Center for International Forestry Research (CIFOR), based in San José, Costa Rica.

Russell L. Kruska (r.kruska@cgiar.org) is a geographer who works with geographical information systems (GIS) at the International Livestock Research Institute (ILRI) in Nairobi, Kenya.

Carlos Lascano (c.lascano@cgiar.org) is an animal nutritionist and the tropical forages project manager in Centro Internacional de Agricultura Tropical (CIAT), Cali, Colombia.

Catherine Marquette (cmarquette@earthlink.net) is a research associate at the Chr. Michelsen Institute in Bergen, Norway.

Elmer Martinez (elmmarti@indiana.edu) is a recent graduate of the Kelly School of Business Administration at Indiana University.

Alexander Mather (a.mather@abdn.ac.uk) is professor and department head in the Department of Geography, University of Aberdeen, UK.

Laura Murphy (llmurphy@neosoft.com) is adjunct assistant professor in Tulane University's Department of International Health and Development and a postdoctoral researcher at the Carolina Population Center, University of North Carolina, Chapel Hill.

Francisco Pichon (fpichon@worldbank.org) is a rural sociologist at the World Bank, Washington, DC, and a visiting professor at the Institute of Geography, Norwegian School of Economics and Business Administration in Bergen, Norway.

Thomas Reardon (reardon@msu.edu) is an associate professor in the Department of Agricultural Economics at Michigan State University in East Lansing, Michigan, USA.

Keneth Reategui (depam@terra.com.pe) is an agronomist and coordinator of the Amazonian Participatory Research Project (DEPAM) in Pucallpa, Peru.

Robin S. Reid (r.reid@cgiar.org) is a systems ecologist leading research on livestock and the environment at the International Livestock Research Institute (ILRI) in Nairobi, Kenya.

Peter Roebeling (peter.roebeling@alg.oe.wau.nl) is a PhD researcher in the Development Economics Group at the Department of Economics and Management, Wageningen Agricultural University, the Netherlands.

Ruerd Ruben (ruerd.ruben@alg.oe.wau.nl) is an associate professor in the Development Economics Group at the Department of Economics and Management, Wageningen Agricultural University, the Netherlands.

Thomas K. Rudel (rudel@aesop.rutgers.edu) is a rural sociologist and professor in the Department of Human Ecology at Rutgers University in New Brunswick, New Jersey.

François Ruf (ruf@africaonline.co.ci) is an agricultural economist working for Centre de Coopération Internationale en Recherche Agronomique pour le Développement (CIRAD) in Côte d'Ivoire.

Gerald Shively (shively@agecon.purdue.edu) is an assistant professor of agricultural economics at Purdue University.

Xiaobing Shuai (shuai@eudoramail.com) is a former graduate student of the University of Wisconsin-Madison, and currently an analyst with Capital One Financial Corporation.

Joyotee Smith (e.smith@cgiar.org) is an economist with the Center for International Forestry Research (CIFOR), Bogor, Indonesia.

Philip K. Thornton (p.thornton@cgiar.org) is an agricultural systems analyst and project coordinator of the systems analysis and impact assessment team at the International Livestock Research Institute (ILRI) in Nairobi, Kenya.

Judson F. Valentim (judson@cpafac.embrapa.br) was the General Director of the Empresa Brasileira de Pesquisa Agropecuaria (EMBRAPA) Research Centre in Acre while working on his chapter.

Daan van Soest (d.p.vansoest@kub.nl) is an environmental economist and associate professor at Tilburg University, the Netherlands.

Stephen A. Vosti (vosti@primal.ucdavis.edu) was a research fellow at International Food Policy Research Institute (IFPRI) when writing his chapter, and is currently a visiting assistant professor in the Department of Agricultural and Resource Economics, University of California, Davis.

Douglas White (d.white@cgiar.org), an agricultural and environmental economist, is a senior research fellow with Centro Internacional de Agricultura Tropical (CIAT) in Pucallpa, Peru.

Julie Witcover (witcover@primal.ucdavis.edu) was a research analyst research fellow at International Food Policy Research Institute (IFPRI) when writing her chapter, and is currently a PhD student in the Department of Agricultural and Resource Economics, University of California, Davis.

Sven Wunder (s.wunder@cgiar.org) currently works as an economist at the Center for International Forestry Research (CIFOR) in Bogor, Indonesia, and was at the Centre for Development Research in Copenhagen, Denmark, while writing his chapter.

David Yanggen (d.yanggen@cgiar.org) works as an agricultural and natural resource management policy analyst in Quito, Ecuador, with Centro Internacional de la Papa (CIP) and Montana State University.

Preface

As researchers and economists working at the Center for International Forestry Research (CIFOR), we have been struck for some time by an apparent contradiction. On the one hand, many of those in the development, environmental and agricultural research communities firmly believe that better agricultural technologies can save forests by producing more food on the existing land area. On the other hand, basic economic theory suggests that anything that makes agriculture more profitable should stimulate land expansion and deforestation. Which view is correct? This book attempts to answer that question.

In 1998, we completed a review of economic models of deforestation but it failed to provide much insight into how technological change might affect deforestation. Data problems had largely kept researchers from including technological change in their analysis. When they did include it, they got results that pointed in varying directions. The studies' failure to specify what type of technical change they were analysing and in what context it occurred made a meaningful comparison practically impossible.

Therefore looking in more detail at the link between agricultural technology and forest cover seemed like a perfect candidate for follow-up research. It is important. We know surprisingly little about it. And lots of myths surround the issue.

Our first task was to develop a theoretical framework that could help us single out which factors to consider and formulate initial hypotheses. Next we had to put together a number of case-studies and test the hypotheses. As part of that process we held a workshop on the subject at Centro Agronómico Tropical de Investigación y Enseñanza (CATIE) in Turrialba, Costa Rica, on

11–13 March 1999. Most of the case-studies included in this book were presented during that workshop. We subsequently added those by de Jong, Mather and Roebeling and Ruben.

A number of people have contributed along the way. At CIFOR, Jeff Sayer, the Director General, gave his unconditional support to the project. Neil Byron, Joyotee Smith and William Sunderlin participated in many of the discussions leading up to this book. Joyotee, together with Stein Holden and Steve Vosti, participated in laying out the workshop programme and writing the concept paper. Julie Witcover served as a rapporteur during the workshop and provided valuable inputs in the preparation of the introduction and summary chapters. At CATIE, Marta E. Núñez and Miguel Caballero were invaluable in facilitating the workshop. Øystein E. Berg prepared the index of the book. Ambar Liano at CIFOR has provided excellent secretarial and administrative services throughout the process.

We are also thankful to the participants at the workshop; they all contributed substantively to the process and our understanding of the technology–deforestation link. The reviewers of the chapters also deserve credit, although the principle of anonymity prevents us from mentioning their names.

The main contributors to this volume are the authors whose research is presented here. We are grateful for the opportunity to work with all of you. Hopefully both sides have benefited from the cooperation. Each and every chapter has a unique story, and together they make this book the first systematic review of the impact of agricultural technologies on tropical forests.

The Norwegian Ministry of Foreign Affairs sponsored the workshop and the editorial and publication process. We are grateful for their financial assistance. Without it this book would not have materialized.

Arild Angelsen and David Kaimowitz
Ås, Norway and San José, Costa Rica, June 2000

Introduction: the Role of Agricultural Technologies in Tropical Deforestation

Arild Angelsen and David Kaimowitz

1. What Kind of World do we Live in?

Imagine a world where the demand for food and other agricultural products is constant or increases regularly as populations and incomes grow. Land can only be used for agriculture or forest. Then the only ways to keep more land in forest are to increase agricultural yields, reduce population growth or depress incomes. The amount of land devoted to agriculture equals the total demand for agricultural products divided by the average yield (output per hectare). Technological progress resulting in higher yields means less land in agriculture and more in forest.

Now imagine another world. Farmers who live in this second world will do anything they can to increase their profits. They can sell all the produce they want for a fixed price and obtain all the land, labour and credit they need, also for a fixed price. What will these farmers do if a profitable technological change increases their yields or lowers their input costs? They will certainly cultivate more land since farming has become more profitable. If agriculture and forest are still the only possible land uses, forest cover will decline. Unlike in our first world, technological progress leads to forest destruction.

Which world do we live in? Does technological progress in agriculture protect or endanger tropical forests? Do we face a 'win–win' situation between farmer incomes and food production on the one hand and forest conservation on the other? Or is there a trade-off between the two?

This book attempts to answer these questions. The answers depend heavily on the assumptions we make about type of technology, farmer characteristics, market conditions, policy environment and agroecological conditions, among other things. Thus, the real question is when does technological progress lead to greater or lower tropical deforestation? We want to identify technologies and contexts that are likely to produce win–win outcomes and help decision-makers that face serious trade-offs to make hard choices.

The book contains cases from Latin America, sub-Saharan Africa and South-East Asia, in addition to two studies on the historical experience of developed countries in Europe and the USA. They cover a wide range of technological changes (new crops, higher-yielding varieties, mechanization, irrigation, fertilizers, pest control, etc.) in different agricultural systems (shifting cultivation, permanent upland cultivation, irrigated farming or lowland cultivation and cattle ranching). The comparative approach permits us to distil the key conditioning factors in the technology–deforestation link.

2. Policies Based on False Assumptions?

Higher agricultural production and forest conservation are both vital for achieving sustainable development in poor countries. Most people understand and appreciate the importance of higher agricultural production to improve farmers' well-being. For some time researchers have debated about what role agriculture plays in economic development, but it is now widely recognized that good agricultural performance is key for high economic growth (World Bank, 1991). Growing evidence also supports the idea that agriculturally driven growth reduces poverty and improves income distribution more than industrially driven growth (Mellor, 1999).

At the same time, international concern about the adverse consequences of tropical deforestation is also rising. Forest clearing contributes to climate change, biodiversity loss, reduced timber supply, flooding, siltation and soil degradation. This in turn affects economic activity and people's livelihoods. The Food and Agriculture Organization (FAO, 1997) estimates that 12.7 million ha of tropical forest was lost each year during the first half of the 1990s. In some cases deforestation is probably appropriate, in the sense that the benefits are higher than the social costs. However, in many it is not.

Current policies and institutional arrangements often lead to inappropriate deforestation, in part due to false assumptions about the causal relations that link the policies to forest clearing (for an elaboration, see Angelsen and Kaimowitz, 1999). One such dubious assumption is that higher productivity and better agricultural technologies will almost always benefit forest conservation. This 'win–win' assumption has dominated recent policy debates on agricultural technologies and deforestation. It is grounded in various hypotheses, which we critically review below.

2.1. The Borlaug hypothesis

By definition, average yield multiplied by area gives total production. Thus, if we keep global food demand fixed, then higher average yield reduces agricultural area, as in our first world discussed above. With food demand expected to grow steadily over the next decades, one could argue that using new technologies to make agriculture more intensive is the only way to avoid rising pressure on tropical forests. This sort of thinking recently led the former vice-president of the World Bank to state that Central African agriculture needs 4% productivity growth annually to save the region's rain forest (Serageldin, quoted in Gockowski *et al.*, 2000).

This line of reasoning also underlies the position that the Green Revolution has had a positive effect on forest cover. Green-Revolution enthusiasts often stress that new varieties of rice, wheat and maize, combined with greater use of fertilizers, irrigation and pesticides, helped save millions of hectares of tropical forest. They argue that, without a Green Revolution, Asian countries in particular would have had to expand their cropland to feed their population. We refer to this argument as the Borlaug hypothesis, in recognition of the key role that Norman Borlaug, the 'father of the Green Revolution', had in promoting it.

The Borlaug hypothesis probably holds for aggregate food production at the global level, at least as long as one assumes that no land uses exist except forest and agricultural land. However, it is much less clear that it applies to technological changes that affect specific products, particularly at the local and regional levels. Technological change at the forest frontier often has minimal impact on agricultural prices. Therefore, the increased profitability effect may dominate and lead to greater agricultural expansion.

Perhaps more importantly, forest, cropland and pasture are not the only land uses that exist. There are large areas of fallow, savannah, brush and other land uses out there. This means that increases or decreases in cropland and pasture may or may not lead to a corresponding change in forest cover. It may simply be that more fallow gets put back into agricultural use or vice versa.

2.2. The subsistence hypothesis

The micro-level version of the Borlaug hypothesis is what we refer to as the subsistence hypothesis. If one assumes that smallholder farmers: (i) live close to the subsistence level of consumption; (ii) are primarily concerned with meeting that subsistence target; (iii) only use family labour on their farms; and (iv) have no alternative uses for that family labour, then technological progress should reduce deforestation. Higher yields allow farmers to get their subsistence income from a smaller area. In addition, if the new technology is labour-intensive, the farmer will have to reduce the amount of land he or she cultivates to adopt it.

The subsistence hypothesis underlies many integrated conservation and development projects (ICDP). Higher income from agriculture (or other activities) is supposed to reduce farmers' need to encroach upon protected areas. Similarly, the assumption that agroforestry – as a way of intensifying land use – will limit conversion of primary forests to slash-and-burn agriculture has been a key element of the Alternatives to Slash-and-Burn (ASB) programme coordinated by the International Centre for Research on Agroforestry (ICRAF) (ASB, 1994).

One can dispute the subsistence hypothesis on several accounts. Most farmers probably do not exhibit the 'limited wants' or 'full belly' preferences that the hypothesis assumes. They aspire to give their children a proper education, buy a new bicycle or maybe a motorcycle, put a proper roof over their head, etc. Thus, if a new technology presents fresh economic opportunities, farmers are likely to expand their agricultural land unless their labour and/or capital constraints keep them from doing so. Although they are far from perfect, local labour markets exist. Farmers can usually sell some labour off-farm and can hire labour. In addition, technologies that create new economic opportunities can stimulate migration to forest frontiers, increasing forest conversion. As the ASB-Indonesia programme has acknowledged in a recent assessment of the issue:

> It is naïve to expect that productivity increases necessarily slow forest conversion or improve the environment. Indeed quite the opposite is possible, since increased productivity of forest-derived land uses also increases the opportunity costs of conserving natural forests. These increased returns to investment can spur an inflow of migrants or attract large-scale land developers and thereby accelerate deforestation . . . ASB research in Indonesia has shown that land use change normally involves tradeoffs between global environmental concerns and the objectives of poverty alleviation and national development.
>
> (Tomich *et al.*, 2000)

2.3. The economic development hypothesis

The Borlaug hypothesis applies at the international or global (macro) level. The subsistence hypothesis focuses on the household or village (micro) level. We can also identify a third argument that links technological progress in agriculture and forest conservation at the regional or national (meso) level. The argument goes as follows. Higher productivity in agriculture – of which improved technologies are a crucial element – contributes to economic development and growth, which, in turn, is associated with other changes that limit forest conversion. These include reduced poverty and population growth, more and higher-paying off-farm jobs, increased demand for environmental services and products from managed forests and higher government capacity to enforce environmental regulations.

This chain of causation provides the underlying rationale for the so-called environmental Kuznets curve (EKC), which posits the existence of a bell-shaped relation between income and environmental degradation. At early stages of economic development, when per capita incomes are low, growth exacerbates environmental problems, but eventually growth helps reduce these problems. This idea is also linked to the forest transition hypothesis, which suggests that the decline in forest cover will eventually level out as countries develop and forest cover will slowly increase.

Again, we have a plausible positive link between technological progress in agriculture and forest conservation. But does it pass the empirical test? The historical experience of the developed countries provides some support for the forest transition hypothesis. Nevertheless, most tropical forest-rich countries are decades away from the inflection point. Economic growth in these countries provides better infrastructure, which stimulates deforestation. Reduced poverty might relax farmers' labour and capital constraints, which previously had effectively limited deforestation. Higher demand for agricultural products stimulates agricultural encroachment. The political priorities and weak administrative capacity of developing-country governments often impede effective forest protection, which potentially could counterbalance these effects. The limited statistical evidence on the EKC is also inconclusive (Kaimowitz and Angelsen, 1998). For example, one recent study finds no statistically significant relation between deforestation and per capita income (Koop and Tole, 1999).

2.4. The land degradation–deforestation hypothesis

Many tropical farmers practise unsustainable farming methods. After a few years of cultivation, loss of soil fertility and weed problems force them to move on and clear additional forest somewhere else. While such shifting-cultivation systems may be perfectly sustainable as long as population densities remain low, when population rises these systems may degrade the natural resources. New technologies can allow farmers to maintain productivity without degrading their resources. This, in turn, should reduce their need to abandon land and clear additional forests to make new plots. Farmers may not want to use land in an extensive fashion, but with their existing technology they have little choice.

This volume provides several examples of situations where farmers clear land, exploit it for several years and then move on to forest areas they had not cleared previously. Farmers have good reasons for behaving like this. Smallholders often have high discount rates and exhibit short time horizons, which leads them to ignore the long-term effects of land degradation on productivity. The economic context and government policies sometimes make it difficult or costly to intensify their production in a sustainable fashion.

For example, affordable inputs may not be available when farmers need them. Lastly, as long as 'unutilized' potential farmland exists, farmers will generally find it cheaper to expand the area under cultivation than to intensify. This is one of Boserup's (1965) main hypotheses. If given the choice, farmers will expand into new areas before they intensify.

Another key question related to the land degradation–deforestation hypothesis is the following: Does sustainable intensification stop – or at least reduce – expansion and deforestation or will it accelerate deforestation by making farming more profitable? In other words, is it a question of intensification *or* expansion, or is the most likely outcome intensification *and* expansion? Many chapters of this book address that question.

3. The Book's Aims and Scope

3.1. Definitions of technological progress (change)

Technological progress (change) can be defined as an increase (change) in total factor productivity (TFP), which is a key concept in economic theory. It simply implies that farmers can produce more with the same inputs, or the same output with fewer inputs. As long as prices remain constant, an increase in TFP will increase profits.

Technological change should be distinguished from agricultural intensification. The latter can be defined as higher input use (or output) per hectare. Intensification and yield-increasing (land-saving) technological change are related terms. But change in technologies may or may not lead to intensification, and intensification can occur without any change in the underlying technology.

Some types of new technologies are embodied in inputs and capital goods, as in the case of improved seeds and fertilizers. Others are disembodied, which means that they rely entirely on new management practices or information. This volume discusses mostly embodied technological changes.

A crucial aspect of new technologies is their effect on how intensively farmers use different factors of production (mainly labour, capital and land). Do the per-hectare requirements of labour and other inputs increase or decline? Technologies may be labour-saving, capital-intensive, and so on. In Chapter 2 we provide more precise definitions of each type of technological change.

Capital-intensive technological change takes various forms. For our purposes it is critical to distinguish between those that save labour, such as tools and draught animals, and those that save land, such as fertilizers. By definition, the former reduce the amount of labour demanded per hectare. The latter often have the opposite effect. How higher capital input use affects the demand for labour depends on which of these two types of capital farmers adopt.

3.2. The key variables that determine how technological change affects forests

The key question this book seeks to answer is how technological change in agriculture affects tropical forest cover. Economic theory allows us to organize the main arguments into a consistent framework and derive hypotheses that can be empirically tested. Prior to the Costa Rica workshop mentioned in the preface, we presented a list of hypotheses about the key conditioning factors and asked the authors of the case-studies to address them. The main variables that we hypothesized might affect how technological change influences forest cover were the following:

1. *Type of technology*: labour and capital intensity, the type of capital involved and the suitability of the technology for recently cleared forest areas.
2. *Farmer characteristics*: income and asset levels (poverty) and resource constraints.
3. *Output markets*: farmers' access to markets, the size and demand elasticity of those markets and how they function.
4. *Labour market*: wage rates, ease of hiring in and hiring out and feasibility of in- and out-migration.
5. *Credit markets*: availability and conditions (interest rate) of loans.
6. *Property regime*: security of property rights and how farmers acquire rights to forest.
7. *Agroecological conditions*: quality of land (slope, soil, rainfall) and accessibility.

In Chapter 2, we use economic theory to derive more explicit hypotheses about how many of these factors can affect the rate of deforestation. In Chapter 21, we summarize the empirical evidence from the case-studies for each of these variables.

3.3. Isolating the technology–deforestation link

In the process of putting together this book, we have tried to stay focused on the link between technology and deforestation. As much as possible, we have avoided entering into a general discussion of the causes of deforestation or of agricultural innovation in poor countries. We felt – and continue to feel – that to say something new we had to maintain a narrow focus. There are nevertheless several caveats. One cannot understand the technology–deforestation link without understanding the wider context. Indeed, it is precisely the interaction between technology type, farmer characteristics and context that produces particular forest outcomes.

Many factors influence the rate of deforestation. From an empirical perspective, it is hard to separate out the marginal effect of technological change.

For example, an increase in the price of a crop suitable for frontier agriculture will directly stimulate the crop's expansion but may also indirectly promote the use of new technologies for that crop. Conversely, new technologies might induce changes in population patterns, infrastructure and policies, which all influence deforestation.

For the most part, we have tried to take technological change and adoption as exogenous and discuss what they imply for forest clearing. But it is not easy to separate adoption from the effects of technological change. Farmers must first adopt a technology before it can have an impact on forest. The theory of induced technological innovation (Boserup, 1965; Hayami and Ruttan, 1985) tells us that researchers develop and farmers adopt technologies that reflect the scarcity (price) of different factors. Forest frontiers tend to have abundant land and scarce labour and capital. Thus farmers will generally prefer technologies that save labour and capital rather than land. Labour-saving technologies are more likely to augment the pressure on forest because they free labour for expanding agriculture. Unfortunately, this means that the type of technology frontier farmers are mostly likely to adopt is the one most likely to increase forest clearing. If we think about it in these terms, we might say that one of this book's central themes is to explore under what circumstances Boserup might be wrong. In other words, when might farmers be willing to intensify even though they still have the option of expanding extensively?[1]

3.4. Sustainable agricultural intensification

The issues this book deals with form part of a broader agenda related to tropical agriculture and sustainable development. That agenda is concerned with finding ways to combine several objectives: (i) increased food production and farmer incomes; (ii) equitable distribution of the resulting benefits; (iii) minimal degradation of existing farmland; and (iv) minimal expansion of agricultural land into natural forests.

The book focuses on (iv), although it pays attention to the trade-offs and synergies between (iv) and the other objectives, particularly (i). While analysts normally think of the negative environmental effects of agriculture in terms of land degradation, they should not lose sight of the negative consequences of forest clearing and forest degradation. There may be a trade-off between the two types of effects. Extensive tree-based systems have low impacts on soil erosion and fertility, but may have large impacts on primary forest cover.

The simple forest–non-forest dichotomy tends to sweep a lot of these important issues under the carpet. As noted earlier, the real world includes secondary forest (fallows) in shifting-cultivation systems, tree crops, agroforestry systems and other land use, all of which provide different levels of environmental services. A number of chapters in this book touch on this issue.

Our focus on deforestation does not imply that this should be the sole – or even dominant – criterion for assessing agricultural technologies. The question is not whether to promote technological change in tropical agriculture, but what type of change to promote. We firmly believe technological progress in tropical agriculture is critical to increasing rural income, improving food security and contributing in general to economic growth and development. But we also believe the current rate of tropical deforestation is too high.

4. Key Conclusions

Below we present the main conclusions and policy lessons that emerge from the studies in this book. Chapters 21 and 22 elaborate these main ideas in greater detail.

1. Trade-offs and win–lose between forest conservation and technological progress in agriculture in areas near forests appear to be the rule rather than the exception. However, win–win opportunities exist. By promoting appropriate technologies and modifying the economic and political environment in which farmers operate, policy-makers and other stakeholders can foster them.
2. New technologies are more likely to encourage deforestation when they involve products with elastic demand (supply increases do not depress prices much). This typically applies to export commodities. The stories of commodity booms and deforestation are almost always about export crops. On the contrary, higher supplies typically depress the price of products sold only in local or regionalized markets rather rapidly. That dampens the expansionary impact of the technological change and may even override it. But it also dampens the growth in farmers' income.
3. New technologies often create economic opportunities, which tend to attract migrants. Otherwise agricultural expansion would inevitably bid up local wages, which would choke off further expansion. Commodity booms can only be sustained if there is a large pool of abundant cheap labour or the technology involved is very capital-intensive. Elastic product demand combined with an elastic supply of labour provides optimal conditions for the introduction of new crops, leading to massive deforestation. On the other hand, when productivity improvements in agriculture coincide with growing employment opportunities in other sectors, the former may not stimulate forest conversion, as demonstrated by the historical experience of the developed countries.
4. Most farmers operating at the forest frontier are capital- and labour-constrained. Thus, the factor intensities of the new technology matter a lot. Technologies that free labour may allow farmers to expand the area they cultivate or release labour to migrate to the agricultural frontier. On the other hand, labour-intensive technologies should limit the amount of family labour available for land expansion and bid up local wages, therefore discouraging deforestation. Since farmers are labour-constrained, we can – as a rule – expect them to prefer labour-saving technologies. Thus, with some important exceptions, we are not likely to get the type of technological change that would save

the forests. Even labour-constrained farmers may adopt labour-intensive technologies if they are the only alternative available to produce certain profitable or less risky crops or to achieve some other household objective.

5. Agricultural land expansion often requires capital to buy cattle or planting material, hire labour or purchase other goods. Capital (credit) constraints can therefore limit expansion. Technological progress should increase farmers' ability to save and thus to invest in activities associated with deforestation. Similarly, higher off-farm wages can provide farmers with the capital they need to expand their operation, even though they increase the opportunity costs of labour.

6. Technological progress in the more labour- and/or capital-intensive sectors of agriculture, which are normally not close to the forest frontier, is usually good for forest conservation. Technological progress in these more intensive sectors shifts resources away from the frontier by bidding up wages and/or lowering agricultural prices. There are exceptions. For example, the new technology may displace labour and push it towards the agricultural frontier or it may generate the funds farmers use to invest in forest conversion.

7. Smallholders normally maintain several production systems. Technological progress in the more intensive systems may shift scarce resources away from the extensive ones, thus reducing the overall demand for agricultural land. But the increased surplus can also be used to invest further in the expansive system (typically cattle), increasing land demand.

Some people may find the overall tone of this book overly pessimistic about the feasibility of achieving win–win solutions. But we are convinced certain trade-offs do exist and policy-makers must sometimes make hard choices. Many policies that are good for agricultural development frequently promote deforestation, including improving access to markets, credit, transportation infrastructure and technologies (Kaimowitz and Angelsen, 1998; Angelsen and Kaimowitz, 1999). Policy-makers can make better choices if they explicitly consider the existing trade-offs and alternatives. Sometimes, they can also identify win–win solutions. In either case, decision-makers need to anticipate the possible effects of promoting different types of technologies in various contexts and cannot assume from the outset that the outcome will be win–win. It is not a matter of slowing down agricultural intensification to save forests, but rather of identifying technologies and intensification strategies that come as near to win–win as possible.

5. The Contributions

Chapter 2 provides a theoretical overview. After that, we have arranged the chapters geographically. We start with two studies of the historical experience in developed countries (Chapters 3 and 4), followed by eight chapters on Latin America (5–12), four on Africa (13–16) and four on Asia (17–20). Then come a summary and a set of policy recommendations (Chapters 21 and 22).

Arild Angelsen, Daan van Soest, David Kaimowitz and Erwin Bulte spell out the theoretical framework in Chapter 2. First, they provide precise definitions of technological change and classify technological change into different types, based on their factor intensities. The theory discussion starts off with a single farm household. Two key concepts for understanding how that household will respond to technological changes are economic incentives and constraints. The former relate to how new technologies influence the economic return of different activities. The latter have to do with how the technologies modify the labour and capital constraints that farmers face. Then, the chapter shifts to the macro level and discusses how aggregate changes in output supply and input demand affect prices, wages, migration and investment.

In Chapter 3, Alexander Mather examines the historical role of technological change in agriculture in Denmark, France and Switzerland. During the 19th and 20th centuries, many European countries underwent a forest transition: forest cover stopped declining and began to rise. New agricultural technologies contributed to this transition. Together with improvements in the transport network, they helped break the link between local population size and agricultural area. Marginal land went back to forest. Nevertheless, technological progress was only one of several radical societal changes that took place and it is difficult to assess its specific contribution. Industrialization created new urban jobs and stimulated a rural exodus. Coal replaced fuel wood as the main source of energy supply. The state emerged as a legislative and technical agent for environmental management.

Thomas Rudel provides a related story from the American South during the period from 1935 to 1975 in Chapter 4. Yield increases in the more fertile areas put farmers in more marginal areas out of business and their lands reverted to forests. The type of technological change influenced the increase in forest cover, since fertilizers and mechanization were both more suited to the more productive lands. Even though mechanization displaced labour, it did not promote deforestation because the expelled labour moved to the cities. In a context of rising opportunity costs for labour, land degradation led to the reforestation of marginal lands, which could no longer compete in agriculture.

In the first chapter on Latin America, Chapter 5, Andrea Cattaneo presents a general equilibrium analysis of a wide range of technological options for the Brazilian Amazon. An increase in TFP increases deforestation nearly always in the short run and always in the long run. Labour-intensive technologies for perennials reduce deforestation sharply. In annuals, this occurs in the short run, but in the long run labour and capital migrate to the Amazon to take advantage of the profits offered by the new technology and the net result is more deforestation. Capital-intensive technological change involving livestock and perennials lowers deforestation in the short run, since farmers are capital-constrained. But in the long run deforestation greatly increases. Cattaneo concludes that there are trade-offs between income generation, food security, equity and deforestation. Technological change in perennials is good for deforestation and equity, while livestock innovations are good for income and food security. Improvements in annual crops are not a preferred choice for any of the objectives.

Livestock researchers in Latin America have argued for some time that intensifying pasture systems can help reduce deforestation. Douglas White, Federico Holmann, Sam Fujisaka, Keneth Reategui and Carlos Lascano critically examine this claim in Chapter 6. Based on evidence from three research sites in Peru, Colombia and Costa Rica, they conclude that it is not so much that pasture technologies reduce deforestation but rather that forest scarcity resulting from past deforestation encourages ranchers to adopt more intensive pasture technologies. Forest scarcity drives up land prices, which make intensive growth more attractive than extensive growth. The authors conclude that research should focus less on how intensification affects deforestation and more on finding ways to make deforestation and extensive land use less attractive for farmers.

In Chapter 7, Stephen A. Vosti, Chantal Line Carpentier, Julie Witcover and Judson F. Valentim provide a detailed study of farmers' options for pasture and cattle production systems in the western Brazilian Amazon. Using a linear programming farm model, they find that many of the more intensive production systems are attractive to farmers and they adopt them. However, these more intensive systems will increase the pressure on remaining forest on farmers' land. The intensive systems are more profitable and the extra profits help relax farmers' capital constraints. Although the authors conclude that improved pasture technologies are a win–lose rather than a win–win alternative, they note that policy-makers may be able to offer ranchers more profitable livestock alternatives in return for a commitment to conserve their forest.

Peter Roebeling and Ruerd Ruben use a methodology similar to the previous chapter in their study from the Atlantic zone of Costa Rica, presented in Chapter 8. They compare the effectiveness of technological progress and price policies in improving agricultural incomes and reducing deforestation. Technological progress generally generates larger income effects than economic policies and leads to similar levels of deforestation. Better pasture technologies stimulate deforestation on large farms, again suggesting a win–lose situation. The authors are optimistic, however, that, with an appropriate mix of policies, policy-makers should be able to simultaneously increase incomes and reduce deforestation.

The next chapter (9), by Francisco Pichon, Catherine Marquette, Laura Murphy and Richard Bilsborrow, describes the results from detailed household surveys of smallholder settlers in the Ecuadorean Amazon. The adoption of a labour-intensive crop, coffee, in a context where households are labour-constrained has limited deforestation on most farms. Farmers grow coffee even though it is labour-intensive and does not provide the highest immediate income. Coffee has, however, a guaranteed market and low transportation costs and is important for farmers' long-term income security. Some farmers have gone for systems involving greater forest clearing, usually based on cattle raising. Farmers who obtain more capital as a result of productivity increases or improved access to credit usually invest it in cattle, which uses a lot of land but little labour, or coffee, using hired labour. This implies a win–lose-type situation, where the same factors that restrict farmers' forest clearing also limit their incomes.

Still in Ecuador but in a different context, Sven Wunder analyses the banana booms in Chapter 10. The initial production systems farmers adopted shortly after the Second World War used land in an extensive fashion and required the farmers to frequently change locations. The technologies were labour-intensive. But, rather than reducing deforestation, the expansion of banana production stimulated massive in-migration, which was associated with much greater forest clearing. Roads built for bananas opened new areas to cultivation. During a second period, the introduction of the 'Cavendish' variety and mechanization made banana-growers demand less land and labour. The fragility of the 'Cavendish' variety made frontier regions with poor transportation infrastructure less suited for bananas. Stagnant banana markets combined with higher yield reduced banana-related deforestation, although the decline in employment on the banana plantations provoked some forest loss, as unemployed banana workers began clearing forest to grow other crops. The population shifts and infrastructure developed during the initial boom had lasting effects, which carried on into later periods. From a comparative perspective, the deforestation resulting from Ecuador's banana boom was probably less than would have occurred with similar booms of other agricultural products, since bananas are comparatively higher-value crops that require lots of labour and capital per hectare.

Soybeans in Brazil and Bolivia present us with a more recent commodity-boom story. Over the past three decades, the new crop has had a profound impact on land use, as David Kaimowitz and Joyotee Smith document in Chapter 11. Brazilian farmers now plant almost 13 million ha of soybean, a crop virtually unknown in that country 50 years ago. The production system is very capital-intensive and uses much less labour than most alternative land uses. In the Brazilian Cerrado and Santa Cruz, Bolivia, soybean cultivation directly replaced the natural vegetation. In the Brazilian South, where it mainly replaced other crops, soybean expansion displaced large numbers of agricultural labourers and small farmers, who could not afford the high capital costs. This induced a great push-migration to the frontier regions of the Amazon and Cerrado. Kaimowitz and Smith also note that new soybean technologies and policies favouring soybean expansion reinforced each other and lifted production levels high enough to justify the creation of a massive infrastructure of roads, processing facilities and input distribution outlets. They also favoured the emergence of a powerful soybean lobby, which was able to ensure long-term government support for the crop.

Shifting cultivators are the focus of many controversies. One relates to their share of tropical deforestation. Another concerns how getting shifting cultivators to adopt more intensive technologies might affect their land-use patterns. In Chapter 12, David Yanggen and Thomas Reardon analyse how the introduction of kudzu-improved fallows affected the demand for forest by shifting cultivators in Peru. Kudzu is a leguminous vine that speeds up soil recuperation. This allows farmers to use shorter fallow periods, which in principle should reduce their need for agricultural land. But kudzu also saves labour and increases productivity, which pull in the opposite direction. On balance, the authors' household data show that kudzu induced a shift from primary to secondary forest clearing, with a modest increase in total forest clearing.

Beginning with Chapter 13, we move to Africa. Thomas Reardon and Christopher Barrett discuss the challenge of sustainable agricultural intensification (SAI), broadly defined as production systems that allow greater production without depleting soil nutrients or otherwise degrading the natural resources. Most farmers on the continent are intensifying without investing enough in maintaining soil fertility. Such soil mining eventually leads them to expand their production on to fragile lands. Reardon and Barrett argue that economic liberalization in a context of poorly functioning markets has made it more difficult for farmers to adopt an SAI path. In particular, reductions in fertilizer subsidies and government credit programmes have induced farmers to mine their soils and adopt more extensive agricultural systems.

The following chapter (14) looks at many of the same issues within the specific context of northern Zambia. Stein Holden gives a historical treatment of two major technological changes during the 20th century: the introduction of cassava in the *chitemene* shifting-cultivation system and the adoption of a more capital-intensive maize cultivation system. The *chitemene* system required each household to clear significant amounts of forest, but market imperfections limited total deforestation. The introduction of cassava improved yields and increased the number of people agriculture could support in the region. It also reduced labour requirements and made production less risky. In the short run it reduced deforestation. But in the long run and in areas with better market access it had the opposite effect, since it permitted higher population densities and a surplus to sell to markets. In the 1970s, government credit and price policies encouraged the adoption of hybrid maize, grown with fertilizers. In the short run this reduced deforestation, as farmers cut back their *chitemene* area. Holden notes, however, that the long-run outcome might not be so favourable, since the fertilizers acidify the soils. Many farmers have abandoned the maize–fertilizer system in response to structural adjustment policies and gone back to *chitemene* systems, and deforestation has increased as a result.

In Chapter 15, Robin Reid, Philip Thornton and Russell Kruska review how trypanosomosis, a major livestock disease, affects the African landscape and how efforts to control it might change that landscape. Disease control can encourage the use of animal traction, which permits farmers to cultivate about twice as much land as cultivating with hand-hoes. Based on remote-sensing data and other spatial data, the authors conclude that disease control encouraged agricultural expansion in the Ghibe valley in south-west Ethiopia, their study area. People moved toward lower-elevation areas and cleared land for cultivation near the rivers. But they also point out that many areas that have trypanosomosis lack the conditions that might lead disease control to induce significant deforestation.

Over the past four centuries, cocoa has moved from country to country, constantly bringing deforestation in its path. In Chapter 16, François Ruf reviews the two most recent touchdowns of the cocoa cyclone, Côte d'Ivoire and Sulawesi in Indonesia. It costs less to grow cocoa in recently cleared forest than in old cocoa plantations. This and the ageing of the cocoa farmers after several decades of cocoa boom provide the main driving forces behind the continuous shifts in location. Farmers are only likely to find it worthwhile to replant and intensify cocoa production once forest has become scarce. Thus, like White *et al.*, Ruf concludes that deforestation triggers technological change. It is not just the other way around. The cocoa-boom story resembles the banana-boom case, presented by Wunder, in several aspects: abundant and accessible forest, a large reservoir of potential migrants and (expectations about) rising prices. Ruf reviews several technological changes in cocoa cultivation in the two countries and shows how, in most cases, they encouraged deforestation. He also argues that the adoption of green-revolution technologies in the lowlands of Sulawesi stimulated deforestation in the uplands by displacing labour and providing investment capital for cocoa expansion.

This upland–lowland dichotomy is central in Asian agriculture. In Chapter 17, Sisira Jayasuriya uses a trade-theoretic analysis to discuss how the two sectors interact. He systematically reviews what impact various technological changes in either of the sectors will have on upland deforestation in situations with fixed and endogenous prices, with and without migration, with capital- and labour-intensive technologies, with distinct types of property rights and with different upland-crop income elasticities. Jayasuriya argues that improving the productivity of crops like rubber, tea, oil-palm and coffee, which compete for land with forest, will aggravate deforestation. The Green Revolution in wet-rice agriculture, which depressed real food prices and increased agricultural employment, may have had a significant pro-forestry effect. However, one cannot assume that low lowland food prices will always have a benign effect on forests. Lower food prices raise incomes and that can stimulate demand for upland products, such as vegetables, and actually increase deforestation.

Chapter 18 provides a concrete example of favourable lowland–upland interactions. Gerald Shively and Elmer Martinez use farm-level data to document how technological change in lowland agriculture in Palawan, Philippines, gave a win–win outcome. Irrigation investments reduced the amount of labour required per hectare during each cropping season but increased the number of crops per year, leading to higher overall labour demand. This resulted in more job opportunities and higher wages for upland households, who responded by reducing the amount of land they cleared by almost half. The households cut back mostly on cash-crop (maize) production, rather than subsistence-crop (rice) cultivation, which remained practically constant.

In Chapter 19, Ian Coxhead, Gerald Shively and Xiaobing Shuai analyse the implications of technology changes in maize and vegetable production in Mindanao, the Philippines, in the broader context of agricultural and macroeconomic policies. The authors discuss how changes in the level and variability of yields and prices determine cultivated area. Reducing the variability of maize yields reduces total area, presumably because farmers no longer have to cultivate so much maize to guarantee food security (a kind of 'full belly' effect). Reducing vegetable-crop yields has no effect on total areas or may even have the opposite effect. Improvements in technology do not induce farmers to increase their area in vegetable areas, in part because they are credit- and labour-constrained. They cannot hire outside labour for vegetables, because the crop requires special skills and high-quality care.

The last case-study in the volume is by Wil de Jong. In Chapter 20, he deals with the impact of rubber on the forest landscape in Borneo (Indonesia and Malaysia). Although many associate rubber with deforestation, de Jong finds that the crop contributed little to encroachment into primary forest in his study areas. In fact, it encouraged farmers to restore agroforests in certain areas, since the typical rubber production system combines planted rubber with natural regeneration. The fact that the study areas were isolated areas with low migration contributed to this outcome. In addition, farmers had a reservoir of old fallow land where they could plant rubber, and local authorities and the national government restricted forest conversion. In locations with other characteristics, introducing rubber might have led to a rather different outcome.

Chapter 21 summarizes the key insights from the above case-studies. First, it discusses the technology–deforestation link in six different types of cases: developed countries, commodity booms, shifting cultivation, permanent upland (rain-fed) agriculture, irrigated (lowland) agriculture and cattle production. Next, it returns to the hypotheses presented in section 3.2 and Chapter 2, and discusses the key conditioning factors in the technology–deforestation link. A number of factors determine the outcome. Among these, labour-market effects and migration are critical in a majority of the cases. Another critical effect is that new technologies can help relax farmers' capital constraints, which may lead to higher or lower deforestation, depending on how they invest their additional funds.

Chapter 22 offers policy recommendations. It presents some typical win–win and win–lose situations. It also relates the issues this volume discusses with the current trend towards greater economic liberalization and globalization and with the overall policy objectives of poverty reduction and economic growth.

Note

1 To be fair, Boserup (1965, 1981) acknowledged that population growth (land scarcity) is not the only factor that drives technological progress and intensification.

References

Angelsen, A. and Kaimowitz, D. (1999) Rethinking the causes of deforestation: lessons from economic models. *World Bank Research Observer* 14(1), 73–98.

ASB (1994) *Alternatives to Slash-and-Burn: a Global Strategy*. ICRAF, Nairobi.

Boserup, E. (1965) *The Conditions for Agricultural Growth: the Economics of Agrarian Change under Population Pressure*. George Allen & Unwin, London, and Aldine, Chicago.

Boserup, E. (1981) *Population and Technological Change: a Study of Long-Term Trends*. University of Chicago Press, Chicago, and Basil Blackwell, Oxford.

Food and Agriculture Organization (FAO) (1997) *State of the World's Forest, 1997*. FAO, Rome.

Gockowski, J., Blaise Nkamleu, G. and Wendt, J. (2000) Implications of resource use intensification for the environment and sustainable technology systems in the Central African rain forest. In: Lee, D. and Barrett, C. (eds) *Tradeoffs and Synergies: Agricultural Intensification, Economic Development and the Environment*. CAB International, Wallingford, UK, pp. 197–220.

Hayami, Y. and Ruttan, V.W. (1985) *Agricultural Development: an International Perspective*. Johns Hopkins University Press, Baltimore.

Kaimowitz, D. and Angelsen, A. (1998) *Economic Models of Tropical Deforestation: a Review*. Center for International Forestry Research (CIFOR), Bogor, Indonesia.

Koop, G. and Tole, L. (1999) Is there an environmental Kuznets curve for deforestation? *Journal of Development Economics* 58, 231–244.

Mellor, J. (1999) Pro-poor growth – the relation between growth in agriculture and poverty reduction. Unpublished paper prepared for USAID.

Tomich, T., van Noordwijk, M., Budidarsono, S., Gillison, A., Kusumanto, T., Murdiyarso, D., Stolle, F. and Fagi, A.M. (2000) Agricultural intensification, deforestation, and the environment: assessing tradeoffs in Sumatra, Indonesia. In: Lee, D. and Barrett, C. (eds) *Tradeoffs and Synergies: Agricultural Intensification, Economic Development and the Environment*. CAB International, Wallingford, UK, pp. 221–244.

World Bank (1991) *World Development Report 1991: the Challenge of Development*. Oxford University Press, Oxford.

Technological Change and Deforestation: a Theoretical Overview

2

Arild Angelsen, Daan van Soest, David Kaimowitz and Erwin Bulte

1. Introduction[1]

This book seeks to answer the question: 'Does technological progress in tropical agriculture boost or limit deforestation?' Theory alone provides few unambiguous answers. But it can sort out the main arguments, structure the discussion and provide testable hypotheses. This chapter sets out basic economic theories relevant to the book's central question.

The first step in any scientific discussion is to define the key terms. In our case, the terms 'technological change', 'technological progress' and 'intensification' and the terms used to describe various technologies lend themselves to a certain degree of confusion. Thus, before going into the theories themselves, section 2 provides basic definitions and classifies technologies based on factor intensities. Factor intensities are critical to determining how new technologies affect forest clearing when farmers are labour- and/or capital-constrained, as is often the case.

Section 3 explores how farm households make decisions about clearing forest and how technological change affects those decisions. This constitutes the microeconomic part of our story and our discussion draws from the rich literature on agricultural household models. This literature suggests that whether or not technological progress reduces deforestation will depend on the constraints farmers face, the market conditions and the type of technology involved.

The next step is to look at the aggregate effects of all farmers' decisions, sometimes referred to as the general equilibrium effects. This provides the macroeconomic part of the story, which we present in section 4. We take into account that technological change might alter prices and wages and induce migration. In other words, certain variables that are fixed at the household level are endogenous when we consider the agricultural sector as a whole and the interaction between this and other sectors.

Other chapters in this book deepen and extend our discussion of the relevant theories. Reardon and Barrett, Shively and Martinez, and Yanggen and Reardon use models similar to those presented in this chapter. Holden, Roebeling and Ruben, and Vosti *et al.* use farm programming models to test empirically some of the hypotheses we derive. Cattaneo and Jayasuriya extensively discuss general equilibrium effects. Coxhead *et al.* deal with the issue of risk, which this chapter does not deal with much. Kaimowitz and Smith discuss another issue that we ignore here, namely the economics of scale.

2. Defining Technological Change[2]

One can approach technological change or technological progress in agriculture from different angles. Economic theory normally defines technological progress as an increase in total factor productivity (TFP). This implies that farmers produce more physical output with the same amount of physical inputs or, conversely, the same output with fewer inputs. Others define technological progress as any change in the production process that increases net profit. This definition partly overlaps with the previous one. As long as prices remain constant, an increase in TFP will increase profits.

New technologies take various forms. They may be embodied in inputs and capital goods, as in the case of improved seeds and fertilizers, or they can be disembodied, which means that they rely entirely on new management practices or information. Analysts often describe the impact of new technologies, embodied or otherwise, in terms of how intensely they use various inputs (mainly labour, capital and land). Thus, technologies may be labour-saving, capital-intensive, and so on. At times, the exact meaning of these terms appears convoluted, so we shall explain how we use them in this text.

The most intuitive approach is to start with a situation where farmers must use a fixed proportion of inputs to produce their output. They cannot substitute labour for land, capital for labour, or any other factor for another factor. Economists refer to such situations by saying that farmers have Leontief-type technologies. Equation (1) gives the amount of output, Y, they can produce using the inputs labour (L), land (H) and capital (K):[3]

$$Y = \min[L, H, K] \qquad (1)$$

This functional form rules out substitution between inputs. One can think about a Leontief situation as a recipe. To produce a cake, you need a fixed amount of eggs, flour, milk and appliances. Two ovens cannot make up for a lack of milk, you cannot substitute flour for eggs, etc. The Leontief production function undoubtedly oversimplifies the situation. In real life, farmers can, to a certain extent, substitute between inputs. For example, they can apply herbicides or do the weeding by hand. Even so, the insights and definitions from the simple Leontief case apply to more general formulations.

One can define factor (input) intensities in relation to output or other factors of production. Most chapters in this book use definitions that refer to the amount of labour or capital per unit of land (H). Hence, y, l and k denote output, labour and capital per hectare, respectively, and by dividing by H we can write equation (1) as:

$$y = \min[l, k] \tag{2}$$

Table 2.1 classifies technologies based on the change in physical yield and factor intensities. A labour-intensive technology increases labour input per hectare, whereas a labour-saving technology has the opposite effect. Similarly, a capital-intensive technology increases capital inputs per hectare and a capital-saving technology reduces them. A labour-saving technology may increase or decrease yield, but farmers will only adopt it if it increases their profits (recall the definition of TFP). Falling labour costs should more than compensate for any possible fall in output or revenues. New technologies can be both labour- and capital-intensive. One such example would be a fertilizer technology that increases both the use of the capital input (fertilizer) and the need for weeding (labour). Pure yield-increasing technologies raise yields without altering the labour and capital requirements per hectare. Economists also call these labour- and capital-neutral technologies, also referred to as Hicks neutral technologies.

Some analysts use the term land-saving to describe technologies. But, once we measure inputs per unit of land, the concept becomes meaningless. The term yield-increasing technologies captures much of what these analysts are referring to. Unlike pure yield-increasing technologies, simply calling something a yield-increasing technology may or may not imply higher input use.[4]

The most widely used alternative to measuring factor intensities in terms of units of land is to measure them in terms of output.[5] Dividing through by Y in equation (1), we get:

Table 2.1. Classification of technologies based on change in yield and factor intensities.

Type of technology	Yield (y)	Labour per ha (l)	Capital per ha (k)
Labour-intensive	+	+	?
Labour-saving	?	–	?
Capital-intensive	+	?	+
Capital-saving	?	?	–
Pure yield-increasing (Hicks neutral)	+	0	0
Yield-increasing and input-intensive ('land-saving')	+	+	+

$$1 = \min[l^Y, h^Y, k^Y] \tag{3}$$

The coefficients inside the brackets refer to the minimum amount of each input needed to produce one unit of the output. The term 'land-saving technology' now makes perfect sense. It is a technology that reduces h^Y. Similarly, the new meaning of labour and capital intensity becomes l^Y and k^Y. The term yield-increasing technological progress loses its meaning. In the remainder of this chapter, we define labour and capital intensities in terms of input per hectare (we use equation (2) rather than equation (3)).

Capital-intensive technological change can take various forms. For our purposes, it is critical to distinguish between two categories of capital-intensive technological change, those that save labour, such as tools and draught animals, and those that save land, such as fertilizers. By definition, the former reduce the amount of labour demanded per hectare. The latter often have the opposite effect. Thus, how higher capital input use affects the demand for labour depends on which type of capital farmers adopt. Below we show that this greatly influences how technological change will affect deforestation.

Many people associate the concept of technological change with that of agricultural intensification, which they understand to mean higher input use (or output) per hectare.[6] Intensification therefore relates to the terms yield-increasing and land-saving technological change. None the less, technological change and agricultural intensification are not synonymous. Change in technologies may or may not imply intensification. Intensification can occur without any change in the underlying technology (in economic jargon: without any change in the production function).

Where do new technologies originate? In some cases, outside development or extension agencies generate and introduce new technologies. In other cases, the technologies arise 'spontaneously' within the rural communities themselves. Such 'spontaneous' technological changes often respond to changes in the context. For example, changes in population density may trigger the search for land-saving technologies, as initially hypothesized by Boserup (1965). Similarly, changes in relative prices may induce technological change, as farmers find ways to switch from expensive inputs to cheaper ones or to introduce more valuable crops.

So far, the discussion may seem to suggest that farmers produce only a single output. In reality, farmers often produce multiple outputs, including annual crops, tree crops, livestock products and processed goods, and they use more than one production or land-use system to produce these outputs. This has implications for our definition of technological progress. When we define technological progress and TFP at the farm level, this will include: (i) technological progress for a particular crop and/or production system; (ii) the introduction of a new crop and/or production system (technology) with higher TFP; and (iii) a shift in farm inputs towards crops/systems with higher TFP. In all three cases TFP at the farm level increases, and therefore they qualify as technological progress.

3. Farm-level Effects

Farmers respond to economic opportunities. They allocate their scarce resources (land, labour and capital) to meet their objectives. These objectives include things like ensuring family survival, maximizing income or minimizing risk. Available technology, assets, market conditions, land tenure and other factors constrain the choices farmers have available. Technological change may modify these constraints and provide incentives that encourage farmers to allocate their resources in a different manner. Two key concepts to understand farmers' response to technological change are therefore constraints and economic incentives.

To analyse how farmers may change their land use in response to technological changes, we start with the analytically simplest case, where farmers are integrated into perfect output and input markets.[7] Even though this is rather unrealistic at the forest margin, it is a useful starting-point. After analysing this simple case, we then introduce market imperfections, labour and capital constraints, farms with multiple outputs and dynamic wealth and investment effects that affect the capital constraint.

3.1. The perfect market case

Consider a farm household that produces one commodity with a fixed-input (Leontief) technology. Labour, capital and output per hectare are all fixed. Land is abundant (i.e. its price equals zero) and agricultural expansion takes place in 'empty' forest. When farmers move inputs and outputs between a village and the field, they incur transport costs. Thus, land rent diminishes as you move further from the village centre.

Given these assumptions, we can use a von Thünen approach to determine how far the agricultural frontier advances. The frontier will expand until the net profit or land rent is zero.

$$r = py - wl - qk - vd = 0 \qquad (4)$$

Equation (4) shows the variables that influence land rent and hence the limits of the agricultural frontier. Output price (p), yield (y), wage rates (w), per-hectare labour requirements (l), the price of capital (q), per-hectare capital requirements (k), transport costs per km (v) and distance in km (d) determine the land rent per hectare (r). The outer limit of agriculture is the distance d^*, where the land rent is zero. This agricultural frontier determines the total amount of land in production. Figure 2.1 presents land rent as a function of distance (the rent gradient).

This simple model can represent several types of technological change (see Table 2.1). As long as markets are perfect, the type of technological change will not affect the qualitative results. All types of technological progress increase the rent at any given distance (the rent gradient shifts upward) and

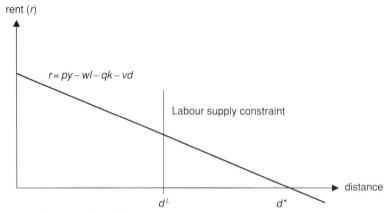

Fig. 2.1. The agricultural frontier.

promote the expansion of the agricultural frontier. Thus, with perfect markets, technological progress unambiguously stimulates deforestation.

The perfect-market model provides an important insight. Technological progress in frontier agriculture makes it more profitable and therefore leads farmers to expand into forests. Although this conclusion is based on stylized and unrealistic assumptions, one should not simply discount it, since, to one degree or another, it also applies to the more realistic models presented below. Several of the cases discussed in this book demonstrate that, even though the real world is much more complex, this simple prediction is often borne out in real life. When technological change makes agriculture at the frontier more profitable, deforestation increases.

3.2. The constrained farm household

In the previous subsection, we assumed that no transaction costs keep farmers from trading freely in any market. In practice, transaction costs may be so high that farmers decide it is not worth their while to participate in certain markets (Sadoulet and de Janvry, 1995). This means that, de facto, those markets do not exist for some households. Peasant households may also not face a complete set of markets for other reasons, such as the inability to share risks.

The absence of certain markets has important consequences for how households are likely to respond to technological changes. One commonly missing market is the labour market. Family labour often has few alternative uses outside the farm and many households cannot afford to hire labour. Thus, they must rely entirely on family labour. In such circumstances, the amount of labour the family has available will limit how much land it can use. Assuming that the maximum labour input the family has access to is L^S, we get:

$$L^D = L(d;l) \leq L^S \tag{5}$$

L^D is the demand for labour on the farm. The greater the distance (i.e. the larger the total cultivated area) and the higher the labour intensity (l), the more labour the farmer will demand. If the labour constraint is binding, equation (5) will determine the boundaries of the agricultural frontier, not equation (4). The vertical line in Fig. 2.1 illustrates this case. Under these circumstances, the forest frontier will be at d^L, instead of d^*.

Capital constraints can also affect the relation between technological change and deforestation. The availability of capital (K^S) can constrain the expansion of the agricultural frontier, and can be modelled as follows:

$$K^D = K(d;k) \leq K^S \qquad (6)$$

When farmers' limited access to labour and/or capital constrains their ability to expand their area, the type of technological change will influence how technological change affects deforestation. For example, when households have a limited amount of capital (K^S) at their disposal, the only way they can adopt a new capital-intensive technology is if they reduce their cultivated area. More generally, technological changes that allow farmers to use less of their scarce factor will boost deforestation. Innovations that are intensive in the scarce factor will reduce deforestation.

Adding more realistic features to the model may modify these results. For example, we have assumed that farmers cannot substitute between different inputs. However, in reality, farmers may find ways to relax their capital constraint by substituting labour for capital. If they do, the new technology will not necessarily reduce deforestation. The new capital-intensive technology may also help the farmers become eligible for credit or persuade them to request more credit, thus removing their capital constraint and allowing them to expand their area.

Equally important is the fact that the profits farmers obtain in previous periods largely determine their access to capital in the current period. We would expect any technological progress that improves farmers' profits to relax their future cash constraints. Technological changes may provide the funds farmers need to expand. Hence, thanks to technological change, farmers who initially behaved as if they were credit-constrained may accumulate capital over time and start behaving more like unconstrained profit-maximizers (Holden, 1998).

The utility-maximizing household

Households' well-being does not only depend on how much food and other goods they consume. People also need leisure. Households choose the number of hours they work based on the returns to labour and the pleasure they derive from pursuing other activities. Therefore, labour supply is not fixed, although the total amount of available time is. In mathematical terms, the household's time constraint is:

$$L^D = L(d;l) \leq L^S = L^T - c^L \tag{7}$$

where L^T is the total amount of time the household has available and c^L is the time it dedicates to non-production activities. In such settings, the situation portrayed in Fig. 2.1 is no longer straightforward. Consider the case of pure yield-increasing technological change. If y goes up, it will have (contradictory) effects on the amount of time farmers spend working in their fields. On the one hand, technological progress increases the returns to labour. This encourages households to work more and take less leisure. In other words, if the rent function in Fig. 2.1 shifts outwards, the household has an incentive to supply more labour, thus shifting d^L to the right. This is the so-called substitution effect. On the other hand, technological progress makes our household richer. We can expect it to use some of its additional income to take more leisure time. As long as the household cannot hire labour, to consume more leisure it must work fewer hours. Technological progress may thus decrease the labour supply, shifting d^L to the left. This is the income effect. Depending on which of the two effects dominates, technological progress may increase or decrease deforestation.

The opposite of the perfect-market model is the subsistence (or full-belly) model, based on what we called the subsistence hypothesis in Chapter 1. Here the crucial assumption is that people seek a predefined fixed level of material well-being and have little interest in going beyond that level. As soon as a household achieves this level, the household will turn to leisure or other non-production activities. Any yield-increasing technological progress will then unambiguously benefit forest conservation. As the rent function in Fig. 2.1 shifts upward, the household will be able to achieve the same amount of income using less labour, capital and land. Thus, the supply of labour simply decreases in response to technological progress. In this case, there is no conflict between the welfare and conservation objectives. Although the subsistence model may accurately describe the individual farmer's response to technological change in certain circumstances, there is little evidence to suggest that the model applies at the aggregate level (Holden *et al.*, 1998; Angelsen, 1999a).

In summary, if farmers face a set of perfect markets, technological change will spur deforestation. When farmers are labour (capital)-constrained, as is often the case at the forest margin, labour (capital)-saving technological progress will probably lead to more deforestation. Labour- and/or capital-intensive technological progress will lead to less deforestation, unless the constraints are 'soft' and/or there is a large 'investment' effect (i.e. higher profits relax future capital constraints). Technological change affects household income and this may affect the amount of labour they supply.

3.3. Intensive and extensive production systems at the household level

In this section, we extend our discussion to situations where farms maintain two production systems: one intensive and one extensive. The former has higher yield and labour and capital intensities than the latter. This allows us to capture how shifts between intensive and extensive systems provoked by technological change determine the overall demand for agricultural land. Farmers choose to engage in more than one production system for several reasons. These include: risk spreading, distributing seasonal labour requirements, the gender division of labour, the desire for self-sufficiency, the presence of multiple soil types, production systems that correspond to various stages in a land-use cycle and distinct transport costs, depending on the location of the crop or pasture. Below we use the transport-cost argument to explain the coexistence of intensive and extensive farming systems, although we could have used some of the other factors.

Figure 2.2 illustrates the land rents of the two production systems, again inspired by von Thünen. Our farmer will locate the intensive system closer to the centre (village) and the extensive system between the intensive system and the forest.[8]

We can now distinguish between the intensive (d^i) and extensive (d^e) frontiers. To those interested in conserving natural forests, the extensive frontier is the most relevant. It is worth emphasizing, though, that in real life many extensive systems are based on tree crops and provide some of the same environmental services as natural forests.

As long as we have perfect markets, technological change in the intensive sector will not affect the extensive frontier. Farmers treat the two systems as separate activities and make their decisions about how to maximize their profits in each system without taking into account the other system. Perfect markets imply that the two systems do not compete with each other for inputs. Farmers use each input up to the point where marginal revenues equal marginal cost. In the case of the extensive sector, the results from section 3.1 directly apply. Technological change will promote deforestation.

More interesting results emerge when farmers face constraints and have to allocate a fixed amount of labour and/or capital between the two systems. Consider first technological change within the extensive production system.

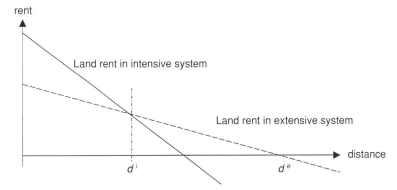

Fig. 2.2. Intensive and extensive frontiers.

Capital- and labour-saving technological change will still increase demand for land, thus spurring deforestation. But the effect will be even stronger because farmers can shift labour and capital from intensive to extensive cultivation. The fact that farmers can shift resources between the intensive and extensive systems implies that labour- or capital-neutral technological progress will also encourage deforestation. Capital- and/or labour-intensive technological changes have ambiguous effect, but – unlike in the case of one production system – they may lead to more deforestation. The net effect depends on the initial size of the two sectors, the difference in capital and/or labour requirements and the increase in capital and/or labour requirements following the technological change.

As long as farmers are resource-constrained, labour (or capital)-intensive or neutral technological progress in the intensive system will contract the extensive frontier. Farmers will divert their scarce labour and capital away from the extensive system and into the intensive system. Labour (or capital)-saving technological progress has two contradictory effects on the extensive frontier. It will shift resources to the intensive sector, but it also frees labour. In one analytical model, assessed by Angelsen (1999b), the first effect always dominates. Labour-saving technological progress in the intensive sector reduces the overall demand for land. To what extent one can generalize these results, however, remains uncertain.

If one takes into account the dynamic interactions between the two sectors, one can obtain rather different results. Technological progress in the intensive sector can serve as a source of capital that farmers use to expand the extensive sector. In other words, increased profits in intensive agriculture can relax the capital constraint and allow farmers to invest more in activities involving forest clearing (see Ruf, Chapter 16, this volume).

Including these dynamic interactions also leads to ambiguous results with regard to the impact of off-farm income opportunities. In the unconstrained world, off-farm opportunities increase the opportunity cost of labour. That makes land expansion more expensive and causes the agricultural frontier to contract. But farmers can also use increased wage earnings to invest more in hiring labour to clear forest, purchasing more cattle and similar activities (see Vosti *et al.*, Chapter 7, this volume).

At least four important lessons emerge from this brief discussion. First, the effect of technological progress on deforestation greatly depends on which agricultural subsector the technological progress occurs in. Secondly, if farmers can switch between different systems, technological change will affect overall land demand much more than in situations with only one production system. Thirdly, in multiple production-system contexts, even labour- and/or capital-intensive technological progress in the extensive system may lead to more deforestation, because of the opportunity to shift resources to the frontier. Fourthly, dynamic investment effects resulting from higher farm income due to technological change in any system (or due to off-farm income increases) can increase the pressure on forests.

4. Macroanalysis: General Equilibrium Effects

The previous section focused on the individual household's response to changes in technological parameters and prices. However, technological progress is unlikely to involve only one household. And, if a large number of households adopt the new technologies, this will have economic repercussions beyond those envisioned in section 3. These macroeconomic effects can either diminish or enlarge the microeconomic impact discussed in section 3. We can identify two major types of macroeconomic effects. The first operates through changes in the number of households living in the forest area – i.e. through migration to or from the extensive margin. The second works via changes in prices.

4.1. Migration

The impact of technological progress on deforestation depends on the number of agricultural households at the extensive margin, since that will determine to what extent aggregate labour supply constrains agricultural expansion. Typically, people compare the level of well-being that they can expect in different regions and choose to live where they will do best. To analyse this type of decision, we assume there are two regions, uplands and lowlands, and that the expected per capita income in each region declines as the number of people in the region rises.[9] People will migrate from one region to another until each region has the same level of per capita income, as illustrated by point L_1 in Fig. 2.3. The length of the box is total population.[10]

Technologies influence the location of the curves. Consider first a technology that only functions in so-called traditional lowland agricultural

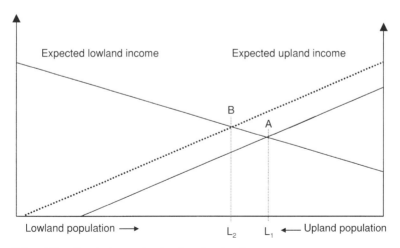

Fig. 2.3. Migration between lowland and upland.

areas, and not at the extensive margin in the uplands, where the forests are. The technology may only apply to the lowlands because it can only be used in certain types of soils or requires good access to markets or other institutions, or for some similar reason. Introducing a technology like this shifts the lowland income curve upward, thus reducing upland population and deforestation. Creating attractive economic opportunities outside the uplands, either in agriculture or elsewhere, is therefore an important tool for securing forest conservation.

Now consider a technological change that applies to upland regions, but not elsewhere. The dotted line in Fig. 2.3, where L_2 is the new equilibrium, reflects this. Since the technological change makes upland agriculture more attractive, compared with activities in the lowlands, people migrate to the uplands (from L_1 to L_2). Possible reasons why the change may occur only in the uplands are that the cultivation of the crop enjoys a forest rent or that the region has specialized in certain crops due to its comparative advantage.

To determine the aggregate effect of technological change on deforestation, we can multiply the per-household effect (section 3) by the number of households living at the margin. Once one takes into account the potential of technological change to attract additional households to the forest margins, the risk that technological changes can increase deforestation generally increases substantially.

The shape of the two curves in Fig. 2.3 strongly influences the magnitude of the impact of technological change. The level of lowland incomes directly determines upland labour supply, so the lowland income curve and the upland labour supply curve are the same thing. If the curve is flat (i.e. upland labour supply is elastic), new upland technologies will have large effects and many potential migrants will move to the forest in response to the new economic opportunities. Similarly, we can consider the upland income curve to be the labour demand curve. A flat curve implies that the uplands can absorb a lot of migrants without exhausting the economic opportunities – in part, perhaps, because forests are abundant. Thus, when both curves are flat (migration keeps labour constraints from emerging and forests abound), the conditions are ideal for technological change in the uplands to provoke massive forest clearing. The commodity booms discussed in this book by Wunder (Chapter 10) and Ruf (Chapter 16) provide good examples of such situations.

4.2. Endogenous prices

The second main macroeconomic feedback mechanism operates via price changes. These include both output and input prices (including wages, although the previous section indirectly dealt with wage changes). If innovations substantially increase the supply of agricultural output (and possibly greatly increase the demand for labour), output prices may go down while wages and other input prices may rise.

We can decompose the price effect into two elements: (i) how sensitive market prices are to changes in supply (the demand elasticity); and (ii) how much supply increases in relation to the size of the market. The relative increase in production in the region affected by the technological change and the region's market share determine the latter.

Based on this, we can distinguish between agricultural outputs destined mostly for domestic markets, such as subsistence food crops, e.g. maize and cassava, and products sold in international markets, such as banana, rubber, coffee and cocoa. With respect to the latter, in many cases, no matter how much technological change increases yields in a country, the aggregate effect will not be large enough to influence world prices. Although there are exceptions, most individual countries face a horizontal demand curve for export crops. Hence, the assumption of fixed prices for agricultural crops, underlying the micromodels in section 3, still largely holds when it comes to export crops. However, substantial increases in the supply of crops produced for the domestic market will exert strong downward pressure on prices, since the demand curve in these cases can be quite steep (in other words, demand is inelastic).

Depending on whether the increase in agricultural productivity outweighs the price decline induced by the rise in aggregate supply, revenues may go up or down in the individual households. If technological progress affects crops whose price is not very sensitive to changes in supply (as is the case for most export crops), the increase in productivity will generally exceed the price decrease, so agricultural activities will expand at the expense of forests. On the other hand, if agricultural prices are quite sensitive to changes in supply, the price decrease may outweigh the productivity increase. The literature refers to this latter situation as a treadmill. The more farmers produce, the less they earn, and hence the less incentive they have to clear additional forest (at least as long as the income effect is not dominant (see section 3.2)).

Technological progress may be region-specific, benefiting some producers but not others. If agricultural productivity rises outside the forest region and farmers both in and out of the forest region produce the same crop and sell it in the same market, which has downward-sloping demand, deforestation should decrease. Frontier farmers will receive lower prices, even though they did not benefit from the innovation, which makes them worse off. This will induce households to move away from the frontier and, as long as they face perfect markets, households will produce less than they otherwise would. Households in imperfect market situations may produce more or less, depending on the magnitude of the income and substitution effects.

Thus far, we have ignored the role of factor prices. Given that developing-country agriculture tends to be rather labour-intensive, wages may play an important role. As long as the labour supply is not perfectly elastic (for example, because the labour force is fixed), increasing the demand for labour will bid up wages. Thus, if technological progress generates additional employment, wages will go up and this may discourage forest clearing. In alternative specifications, land and labour can be substitutes. If the price of labour increases, farmers may use more land instead of labour, which implies greater deforestation.

5. Conclusions

In this chapter, we have discussed how economic theory predicts that technological change will affect deforestation. We conclude that the impact will depend on: (i) the type of technical change; (ii) the presence of market imperfections; (iii) the extent to which farmers can substitute between factors; (iv) the way households balance work against leisure; (v) whether the technology affects the intensive or extensive production systems; (vi) how much people migrate in response to regional income differentials; and (vii) how steep the demand and supply curves for outputs and inputs are. Dynamic wealth effects may play a role if innovations allow farmers to accumulate resources that they then use to finance investments in activities associated with forest conversion.

Taking all this into account, we would like to stress two central results and mention one caveat.

First, if both the input and output markets are well developed and 'perfect', we can expect technological progress to promote deforestation. However, high transaction costs at the forest frontier may limit farmers' access to certain markets. Without well-functioning labour and capital markets, technological change will have ambiguous effects, depending on whether it relaxes binding constraints or makes them bind even tighter. If farmers have several production systems and can divert inputs from one to another as their relative profitability changes, this may magnify the micro-level effects.

Secondly, if technological change affects the production possibilities of many farmers, general equilibrium effects arise. In general, the price effects tend to 'dampen' the micro-level effects. For example, if supply increases as a result of new technologies, this may depress prices and effectively counteract the initial incentive to deforest. For the migration effects, however, it matters a great deal whether the innovations perform better on the forest frontier or in traditional agricultural areas, since a greater impact in one of the two areas may trigger migration to or from the frontier.

Finally, it is worth pointing out that all of the previous discussion largely ignored the complex relation between technological change, land degradation and deforestation. Households at the extensive margin often deplete their soils and then move on and deforest a new parcel. Technologies that reduce land degradation reduce the incentive to 'cut, crop and run', thereby lessening the pressure on natural forests. On the other hand, sedentary agriculture generally retains fewer characteristics of natural ecosystems than fallow systems or extensive agricultural land uses. In addition, for all the reasons discussed in section 3, any technology that increases profits can potentially result in greater land clearing. This further illustrates the complexity of the relation between technological progress in agriculture and forest conservation, and the difficulty of reaching unambiguous general conclusions.

Notes

1   This chapter is based on two papers presented at the Costa Rica workshop: Angelsen, A., Kaimowitz, D., Holden, S., Smith, J. and Vosti, S., Technological change in agriculture and tropical deforestation: definitions, theories and hypotheses; and Bulte, E., van Soest, D. and van Kooten, G.C., Opening Pandora's box? Technological change and tropical deforestation. We would like to thank our co-authors for their inputs in the process of writing the present chapter.

2   Initially, we made a distinction between the concepts of 'technological progress' and 'technological change', the latter being broader. Since these terms are used more or less interchangeably in the debate, we have not maintained the distinction in this book, except that technological change also includes technological progress in reverse (technological regress). The same goes for the terms 'technological' and 'technical', although we prefer the former.

3   Note that we have defined the inputs L, H and K such that exactly one unit is required to produce an output Y.

4   Some people use the term 'land-intensive' interchangeably with 'land-saving' or 'yield-increasing'. It is, however, by no means self-evident whether land-intensive means that farmers use a lot of land per unit of labour and capital (also referred to in the literature as extensive land use) or the opposite.

5   One could argue that agricultural intensification should be defined in terms of output per unit of the scarce factor. Since labour is often the scarce factor at the forest frontier, a third option would be to divide through by labour in equation (1), and focus on labour productivity (output per worker) and the land and capital requirements per worker.

6   To measure input intensity with more than one input, we need a common yardstick, normally a monetary unit. This raises several issues. Should we use farm-gate or social prices? What prices should we use to value non-market output and inputs? Are the relevant prices those that existed before the technological change or after?

7   A 'perfect' market implies that farmers can take prices as given, can buy or sell as much as they want at that price and have perfect information, and that the input or output involved is homogeneous, e.g. family and hired labour are perfect substitutes.

8   This model and the corresponding results are taken from Angelsen (1999b). See also Randall and Castle (1985) for a more general treatment of the von Thünen model with two production systems.

9   The latter assumption may be realistic for agriculture, but is unlikely to hold for urban areas, because most industrial activities exhibit increasing returns to scale (see Murphy et al., 1989).

10   This analysis makes a few simplifying assumptions. There are no migration costs. Marginal income equals average income. No one is unemployed, nor do they prefer to live in a particular region for non-monetary reasons. The figure is a simplified version of the Harris–Todaro migration model (e.g. Stark, 1991).

References

Angelsen, A. (1999a) Agricultural expansion and deforestation: modelling the impact of population, market forces and property rights. *Journal of Development Economics* 58(1), 185–218.

Angelsen, A. (1999b) Technological change and deforestation: a farm household model with intensive and extensive production. Draft, Center for International Forestry Research, Bogor, Indonesia

Boserup, E. (1965) *The Conditions for Agricultural Growth: the Economics of Agrarian Change under Population Pressure.* George Allen & Unwin, London, and Aldine, Chicago.

Holden, S.T. (1998) The role of capital markets in the process of deforestation. Mimeo, Center for International Forestry Research, Bogor, and Norwegian Agricultural University, Ås.

Holden, S.T., Pagiola, S. and Angelsen, A. (1998) *Deforestation and Intensification Decisions of Small Farmers: a Two-period Farm Household Model.* Discussion Paper No. D-26/1998, Department of Economics and Social Sciences, Agricultural University of Norway, Ås.

Murphy, K.M., Schleifer, A. and Vishny, R.V. (1989) Industrialization and the big push. *Journal of Political Economy* 97, 1003–1026.

Randall, A. and Castle, E.N. (1985) Land markets and land resources. In: Kneese, A.V. and Sweeney, J.L. (eds) *Handbook of Natural Resources and Energy Economics,* Vol. II. Elsevier Science, Amsterdam, pp. 571–620.

Sadoulet, E. and de Janvry, A. (1995) *Quantitative Development Policy Analysis.* Johns Hopkins University Press, Baltimore.

Stark, O. (1991) *The Migration of Labour.* Blackwell, Cambridge, Massachusetts.

The Transition from Deforestation to Reforestation in Europe

Alexander Mather

1. Introduction[1]

During the 19th and 20th centuries, many European countries underwent a 'forest transition'. Net national forest cover stopped declining and began to increase (Mather, 1992). This has led some to speculate that developing countries currently experiencing deforestation may eventually undergo a similar transition.

Data deficiencies and the fact that technological changes in agriculture coincided with other major social, political, economic, technological and cultural changes prevent a rigorous analysis of agricultural technology's role in Europe's forest transition. Among the most important confounding variables are the rural exodus, industrialization, improved transport systems, forest regulation and political control and the shift from fuel wood to coal.

Nevertheless, the evidence suggests that technological change in agriculture contributed significantly to the forest transition. This implies that it might also influence deforestation and its control in present-day developing countries. Thus, examining Europe's experience with forest-cover change over the last several centuries can provide a broader historical perspective for understanding current forest trends in developing countries.

Throughout history, farmers have responded to the need to produce more food to satisfy a growing population and rising per capita consumption by expanding agricultural area and/or managing their existing area more intensively. Since forest constitutes the natural vegetation in most areas capable of producing crops, extensive agricultural expansion is likely to reduce forest

©CAB *International* 2001. *Agricultural Technologies and Tropical Deforestation*
(eds A. Angelsen and D. Kaimowitz)

area. Intensification, on the other hand, does not directly affect forests but often requires farmers to adopt new cultivation techniques.

Technical changes in agriculture, particularly changes that increase yields, permit farmers to supply more food from smaller areas. This can relieve the pressure to clear additional forest for crops and livestock. Eventually, farmers may even abandon areas, paving the way for reforestation (by natural regeneration or plantation). Similarly, transportation improvements can facilitate the concentration of agricultural production in more fertile areas and allow people in other regions to purchase their food from elsewhere and stop growing crops on marginal lands.

How population pressure and systemic stress affect resource management has long been debated. Although one can identify Malthusian trends in a number of European countries in the periods leading up to the forest transition, it is hard to explain the transition itself within a Malthusian framework. Contrary to the Malthusians' expectations, population growth and forest expansion have gone hand in hand in these countries for several hundred years. This suggests that, as Boserup predicted, farmers responded to increased population pressure by intensifying their agricultural systems. As the following case-studies make clear, various types of stress coincided with the forest transition and favoured the emergence of new paradigms of resource management, including new technological paradigms for agriculture.

Sections 2, 3 and 4 of this chapter review the transition in three countries – Denmark, France and Switzerland. Despite geographical differences and environments ranging from coastal lowland to alpine, certain similarities characterize the three cases. Section 5 discusses the role of agricultural technology within the broader context of socio-economic and political change. The final section concludes and draws out some points that may be of relevance for today's developing countries.

2. The Forest Transition in Denmark

By 1800, Denmark had lost all but some 4% of its forest cover. Then the decline stopped and forest cover continuously expanded from the mid-19th century, although with certain fluctuations. Today, Denmark has nearly three times more forest than in 1800. The transition coincided with major changes in land and forest tenure, forest management and the political context. There was more to it than technological change.

Rural restructuring began in the 1780s and proceeded rapidly. A group of 'improvers', inspired by Enlightenment ideas, in effect 'captured the machinery of State' and set out to modernize the country (Smout, 1987: 87). C.D.F. Reventlow, an influential landowner, and the young Prince Regent were key figures in this group.

In 1786, the government formed a special agricultural commission, charged with reforming landlord–tenant relations and the enclosure of open field and commons (Tönesson, 1981). Raaschou-Nielsen (1990) estimates that only one-tenth of Danish peasant farms were enclosed in 1790. By 1830, this had risen to 99%. Enclosure and the privatization of the commons initially reduced total forest area by as much as one-third (Fritzbøger, 1994). The government compensated for the loss of grazing rights in newly privatized forests by allowing farmers to clear common woodlands with light tree cover and convert them to cropland (Sabroe, 1954).

New laws required livestock owners to maintain fences and keep their cattle off other people's property (Friedmann, 1984). Between 1781 and 1788, the state demarcated its royal forests from the adjoining agricultural fields and excluded all cattle (Sabroe, 1954). This process was then repeated on private land. Thus, forest and agricultural land became increasingly separate. That facilitated a rise in tree planting. As a result, the number of private estates with tree plantations grew from 19 in 1791 to 53 in 1805 and 101 in 1830 (Jensen, 1993).

In 1805, the government passed the Forest Preservation Act, which granted 'overwood' (woods composed of tall trees, as opposed to scrub) to landowners but required that they maintain them as forest. Reventlow helped draw up the act, with the idea that it was a temporary measure to prevent further forest depletion until forest owners fully accepted that managing their forests 'scientifically' was profitable (Grøn, 1960). The act stipulated that all forest had to be enclosed by 1810 and that landowners had to replant all cleared land (Fritzbøger, 1994). Thus, the preindustrial system of managing forests for multiple use, including cattle grazing, slowly gave way to forests managed mainly or solely for timber production.

Denmark's early adoption of 'scientific' forest management systems from Germany facilitated the implementation of the 1805 act. As early as 1763, the government adopted a sustained-yield management system, promoted by German forester von Langen, in some of its royal forests. The forests were fenced to keep out cattle and subdivided into blocks intended for annual felling. Reventlow and other influential landowners promoted the new 'scientific' outlook, which was disseminated through an evolving system of forestry education. The first forestry training schools opened in 1786 and the first university programme began in 1800 (Sabroe, 1954).

A series of crises triggered the 1805 act. According to Kjærgaard (1994), Denmark suffered a chronic multidimensional ecological crisis during the 18th century. One aspect of this was a loss of forest cover, due to rising population and limited agricultural resources. Apart from that, the Napoleonic Wars disrupted Copenhagen's firewood imports from Holstein. This made firewood acutely scarce and prices doubled between 1780 and 1800 (Friis and Glamann, 1958). The chronic shortage of wood helped provide a climate of opinion conducive to the adoption of new forest management regimes, while the acute shortage triggered their implementation.

The 1805 act helped stem further deforestation. But significant reforestation did not begin until 1860. The loss of secure timber supplies from Norway after its independence in 1814 was not enough to persuade people to reforest, even after the postwar agricultural recession released land that could have been used for that purpose. Moreover, the state went bankrupt in 1813 and had no resources to promote reforestation.

Ultimately, it was the loss of part of the national territory, in Schleswig-Holstein, that really got reforestation off the ground. This blow to national pride apparently provoked a strong national sentiment and a desire to use the country's land resources as fully as possible (Jensen, 1993). One manifestation of this was the creation of the Danish Heath Society, which worked on converting the 'wasteland' of the Jutland heaths to arable land and forest. Earlier attempts at the afforestation of the moors had achieved little. But, by the 1860s, the combination of technical advances, political-economic climate and national mood had provided the conditions for more sustained forest expansion.

During this whole period, agriculture and forestry expanded concurrently. Forest expansion was not linked to agricultural retrenchment. Cultivation and tree planting went together, both spatially and functionally. In Jutland, the farmers used the wages they received for planting trees to expand their agricultural holdings. Farmers more generally continued to bring previously uncultivated areas into production and combined extensive with intensive growth (Nielsen, 1988). In the two decades after 1788, crop yields increased 25% and agricultural production doubled (Friedmann, 1984). During the last third of the 19th century, the national livestock herd also increased greatly. While some attempts to convert heathland to cropland proved over-optimistic and the land was subsequently abandoned and afforested, this process did not really take off until the 1890s, decades after large-scale afforestation had begun (Jensen, 1976).

The case of Denmark suggests that the relation between forest trends and technological change in agriculture is complex. To suggest that the latter 'caused' deforestation to stop or reforestation to take off is an oversimplification. Both stemmed from modernization and had their roots in political and philosophical changes. Factors such as the spirit of the Enlightenment and the national mood of the 1860s proved more important than technological change in agriculture *per se*, though the latter certainly reduced the pressure to encroach on the areas of forest that remained at the end of the 18th century. Within this context, there is little doubt that technological improvements in agriculture helped stabilize the forest area, but in a wider sense than some recent work on tropical deforestation and technological change might imply.

3. The Forest Transition in Switzerland

Switzerland and Denmark have quite different histories and geographies. However, forest area in both countries has expanded substantially since the

19th century and they share common features in terms of perception of crisis, legislation and agricultural change.

Data are simply too sparse and unreliable to provide a long-term forest curve, but analysts generally agree that Switzerland's forest area has almost doubled since the mid-19th century. Prior to that, growing population and demand for wood had attenuated the forest for several centuries and the problem probably worsened during the 18th century. As in Denmark, forest and farmland initially formed part of the same system and forests provided fodder as well as wood. Crop yields were poor, not least because only a small fraction of the animal dung was available to use as fertilizer (Pfister, 1990).

The introduction and widespread adoption of the potato was one factor accelerating population growth. Its higher productivity per unit area, compared with cereal crops, effectively increased the carrying capacity of the land in terms of food production. To some extent, emigration provided a safety valve. Nevertheless, local population continued to grow. In the absence of a commensurate change in land management, this resulted in environmental stress. Pasture productivity and the capacity to make hay rose more slowly (at least in the upland areas) and farmers partly replaced cows with goats (the 'poor man's cow'). The goats' activities, combined with the growing fuel demands of the expanding population, seriously degraded mountain forests (Pfister, 1983). By the early 19th century, highland population growth had led to both environmental degradation and pauperization (Pfister and Messerli, 1990). Thus the introduction of potatoes in Switzerland suggests that the technological change might have had a negative effect on the forest.

Existing land management systems could not cope with population growth and rising wood demand. Most of the forest was communally owned. Traditionally, a series of complex communal mechanisms had strictly limited the cutting of wood for fuel and construction. Many villages had elected councils responsible for such controls. The emphasis was on not allowing resource use to outpace forest growth, maintaining forests' protective functions and providing each household with an equitable share of the annual cut (Netting, 1972). These systems proved effective when population was relatively stable. But, once population began to grow rapidly, the demand for forest resources created more stress than this type of communal regulation could handle. By the mid-19th century, many areas were experiencing fuel shortages (Marek, 1994).

Besides this chronic stress, the country also suffered specific crises. Particularly damaging floods occurred in the 1830s and again in the 1850s, and the Swiss Forestry Society helped convince authorities that forest problems were partially responsible for the floods. In response to a petition from the society's president, Elias Landolt, in 1856, the federal government commissioned a major survey of forest condition under Landolt's supervision. The report concluded that forests were being depleted and that deforestation had made river discharge more irregular and increased the risk of avalanches and falling rocks (Landolt, 1862). The wider conclusion was that alpine deforestation was not just a local problem; it affected the whole nation. Further floods in 1868, which caused 50 deaths, served to emphasize the apparent link between deforestation and floods.

In response to these perceived threats, over the next few years, the federal government initiated reforestation efforts and strengthened regulation of existing forests. These efforts culminated in the Forest Police Law of 1876 (IFF, 1976). This law, which applied to Alpine and pre-Alpine forests, prohibited any reduction in forest area. It required farmers to obtain permits to fell forests and to either replant felled areas or compensate for them by reforesting some other land in the vicinity. It also regulated traditional forest-use rights. The cantons or federal government could require farmers to afforest bare lands to create protective forest and could appropriate private land for that purpose. In short, the state began intervening in forest management.

Whether or not the frequency of flooding really had anything to do with deforestation has little bearing on our discussion. The perceived links provided a basis for a 'crisis narrative', which helped legitimize Landolt's attempts to set up a federal forest regulatory system. The dominant social construction relating forests and floods made it possible for the forester-scientist to mobilize favourable public opinion and political support and steer the state towards stricter regulation and reforestation. The new widespread belief that deforestation in the mountains could endanger the lowlands justified federal intervention (Schuler, 1983). The passing of the first federal forestry law in 1876 was a milestone in this respect and had strong parallels with the Danish law of 1805. The new regulations changed the use of the forests and weakened their links to farming through grazing and fodder collection. As in Denmark, forest and farmland became increasingly separate.

To attribute the forest transition solely to Landolt and the 1876 act would be wrong. The passage to a regulated forest economy became possible only after agriculture and the economy in general began to modernize (Schuler, 1984). The establishment of the Swiss Confederation in the mid-19th century marked a milestone in the evolution of the modern nation-state. In Switzerland, as in neighbouring countries, the new state assumed the right to intervene in the management of the forest resource. Individuals such as Landolt had access to the state apparatus, and the corollary was that they were able to privilege 'scientific-rational' constructions of the forest and the forest–flood relationship above others. These actors did not operate solely as individuals. In 1843, they founded the Swiss Forestry Society, which was influential in promoting forest management and conservation., Shortly after the Swiss Confederation was established, a Department of Forestry was created and, in 1855, the Federal Polytechnic School began providing training in forest management. The following year, the Swiss Forestry Society obtained funding for forest research, and work began on the causes of flooding. In short, several institutional developments had occurred before the 1860s, involving both the state and civil society, which helped satisfy the preconditions for moving towards more sustainable forest resource management.

As in other countries, agriculture became more market-orientated in the second half of the 19th century and this led to changes in where production was located. Although there was some urban growth, much of the industrialization was in rural areas. This meant that the rural exodus and abandonment of agricultural land were slower than in France and other countries. Land abandonment did occur, however, and it became more widespread in the early 20th century (Hauser, 1975). By the second half of the 19th century, the agricultural labour force (as opposed to the rural population in general) was clearly in decline. Both the absolute and relative numbers of people who depended on agriculture fell, and this favoured the reforestation of certain lands previously used for agriculture.

In areas such as Emmental, technological change in agriculture contributed to significant forest expansion (Gerber, 1989). During the 19th century, a shift in dairy farming from the alpine regions to the valleys, combined with a broader trend towards less intensive farming in marginal areas, allowed forests to expand through natural regeneration. This, along with the decline in the agricultural labour force, may help explain the muted tone of resistance to new regulatory measures, such as the Forest Law of 1876.

The Landolt Report concluded that annual wood removals exceeded increments by around 30%. If this pattern persisted over decades, as seemed likely, forests clearly would have suffered. The removals were largely for domestic and industrial fuel. Socio-economic trends and industrial growth increased the demand for firewood, and lower temperatures over the previous century may have aggravated the problem (Pfister, 1990, 1994; Pfister and Messerli, 1990).

During the second half of the 19th century, however, the demand for local fuel wood and timber waned. The Swiss began to substitute coal for fuel wood and the expansion of the railway system made it easier to import both fuel wood and coal from abroad. The first Swiss railway opened in 1844. Six years later there were still only 24 km of track. But by 1860 that figure had risen to over 1000 km and by the end of the century it was over 3000 km (Statistischen Bureau, 1900). Partly as a result, fuel imports doubled between 1860 and 1870, and had trebled 2 years later (Société Suisse des Forestiers, 1874). The growth of the transport network also favoured the rise of market-orientated agriculture and the concentration of agricultural production in the more fertile areas. These trends made it easier to expand the area in forest.

By the end of the 19th century, Switzerland had moved from the wood age to the fossil-fuel age (Table 3.1.) Fuel-wood consumption only declined a modest 9% between 1850 and 1910. By itself, that was probably not enough to induce a forest transition, although it may have facilitated it. More importantly, the new sources of energy made possible new forms of employment and lifestyles that were less dependent on local resources. In other words, the trends in energy use shown in Table 3.1 reflected broader economic, social, political and technological changes. As people generally came to depend less on local natural resources for their food, fuel and livelihoods, it became easier

Table 3.1. Primary energy balance in Switzerland 1851–1910 (based on Marek, 1994).

	1851		1910	
	TJ	%	TJ	%
Wood	18,920	88	17,190	16
Peat	2,050	9	0	0
Coal	664	3	83,570	78
Petroleum	0	0	740	1
Water power	90	<1	5,270	5
Total	21,724	100	106,770	100
Per capita	9.03*		28.45*	

*Gigajoule.
TJ, terajoule.

to stop depleting the forest resources in more marginal areas. This was particularly true in the Alpine regions.

It is hard to escape the conclusion that 'development' in general, including its economic and political dimensions as well as technological change in agriculture, silviculture and transport, helped reverse the decline in Switzerland's forests. It is noteworthy, however, that technical change in agriculture had both negative and positive effects on forest area. The introduction of the potato made it possible to support a larger population. (Indeed, the mounting pressure on the available food supply as a result of population growth may partly explain the potato's rapid adoption.) This probably exacerbated forest degradation, since cultivating potatoes obviously could not alleviate fuel shortages. Conversely, the introduction and increasing use of sown grasses and the rise of commercial dairy farming influenced forest cover more favourably. As in the neighbouring areas of France, these changes helped concentrate farming in the more productive lowlands and valleys and gradually lessened pressure on the higher areas. This, in turn, facilitated reforestation. It is important to emphasize, however, that changes in transport and in attitudes about forests and their management accompanied these changes in agriculture.

4. The Forest Transition in France

The period between the late 18th and early 19th centuries was decisive in French forest history. Following a long forest decline, sometime in the mid-19th century forest cover started to grow again, possibly as early as 1830. During the second half of the century, the trend accelerated. Since the early 19th century, the area has more or less doubled. The forest has now recovered all the area it lost since the 14th century, although its character and spatial distribution are quite distinct.

After a period of expansion following the Black Death, the French forest contracted almost continuously until the early 19th century. Between 1750 and 1800, the population increased from 24.5 million to 29.1 million. Farmers met the associated growth in food demand by expanding agricultural area, rather than intensifying. Some commentators contend that agricultural yields changed little between *c.* 1750 or even earlier and the early 19th century (Morineau, 1970), although a revisionist view has it that yields on large farms near cities began increasing around 1750 (Moriceau, 1994). Even if the revisionist view is correct, this does not alter that fact that the productivity of most poor farmers in more remote areas was relatively stagnant.

Overall, the arable area increased from 19 million ha in 1751–1760 to 23.9 million ha in 1781–1790 (Abel, 1980). Much of this area may have come from heathland, but Bourgenot (1977) estimates that farmers cleared more than 500,000 ha of forest between 1760 and 1780. Cultivation extended into difficult environments. For example, in the high Auvergne and in the Ardèche, farmers cleared fresh plots to grow rye (Jones, 1990). The cultivators' grazing animals and growing demands from industry added further pressure on the forests (de Planhol with Claval, 1988). According to Clout (1983), four times as much woodland was cleared for rough pasture as for crops. Iron-making and charcoal production developed in the Pyrenean valleys and other areas. Theoretically, coppicing allowed producers to obtain wood for these industries sustainably, without having to degrade the forest. But that potential was not always achieved (Bonhote and Fruhauf, 1990).

Deforestation proceeded apace in some Alpine areas. Between 1791 and 1840, Basses-Alpes lost 71% of its forest area and Var 44%. Corvol (1987) estimates the annual rate of deforestation for the country as a whole at between 0.8 and 1.4% per annum.

Agricultural expansion on to poor land, especially in the mountains, soon proved unsustainable and led to soil erosion and other forms of degradation (Sclafert, 1933). By the early 19th century, the Causses and other southern land had become 'landscapes of desolation' and, in Provence, woodlands 'were becoming rarer every day' (Clout, 1983: 124–125). In 1819, Prefect Dugied of Hautes-Alpes asserted that deforestation and erosion had rendered large areas in the department unproductive. He urged the government to prohibit further clearing and to promote the conversion of cleared land to artificial grassland and the reforestation of extensive areas (Ponchelet, 1995). Environmental stresses apparently also affected other parts of France. Blaikie and Brookfield (1987: 135), for example, describe soil erosion in Champagne and Lorraine during the 1790s and early 1800s as 'catastrophic'.

Technological change in agriculture almost certainly facilitated the forest transition. Cereal yields increased gradually during the 19th century, and more rapidly thereafter. By the second half of the 19th century, bare fallows, previously the largest 'agricultural' land use, were being phased out (Clout, 1983). With less idle land, farmers could produce the same amount in a smaller area (Sutton, 1977).

Rotation grasses allowed landowners to concentrate grazing in 'artificial meadows'. Braudel (1990: 277) has called these grasses the 'the motor of a powerful and necessary agricultural revolution'. Already established in some areas, such as the Paris basin, by the 1760s, they slowly expanded for several decades and then took off after the 1830s (Jones, 1990). Cattle numbers rose, largely on the more productive improved pastures, but the sheep herd declined progressively from around mid-century. This took pressure off the commons and the unimproved pastures and forest, which facilitated tree regeneration or at least reduced the pressure on the remaining forest and scrub.

Intensive agriculture gradually replaced extensive agriculture, although each was concentrated in different areas. Agriculture continued to encroach upon the 'marginal' lands on the borders of heath and forest, while most intensification was in the 'better' areas of the lowlands or valley floors.

By the second half of the 19th century, the agricultural frontier was stagnating or even retreating. Farmers abandoned certain areas and the forest eventually returned (Bourgenot, 1993). This retreat was linked to a 'rural exodus', which accelerated during the period, thanks to urban and industrial growth in the lowlands. The effects of the rural exodus and of agricultural intensification intertwine and cannot be meaningfully separated. Both could result in abandoning of agricultural land, thus making it available for natural forest regeneration, plantations or other purposes. The growth of the market economy and of transport links facilitated the concentration of crop production in the more productive areas and weakened the bonds of local subsistence.

Despite the dearth of strong statistical evidence, the hypothesis that the technological transformation of agricultural and the rural exodus led the forest area to increase is credible, especially for upland and marginal regions. Both the methods and the economic orientation of the two forces came together to reduce pressure on forests (Rinaudo, 1980). And, as agriculture became increasingly market-orientated, traditional peasant use of the forest waned.

But agricultural change was certainly not the only factor underlying the forest transition. The source of energy also fundamentally changed, as coal replaced fuel wood both in industry and more generally. This did not take pressure off the forest overnight, but it gradually reduced it. From 1837 onwards, it became cheaper to use coal to produce iron than to use charcoal and by mid-century the per-unit energy cost of the former had fallen to only one-sixth that of the latter. Partly as a result, coal consumption grew 15-fold between 1815 and 1860 (Braudel, 1990; Table 3.2). As late as 1852, more than one-quarter of all fuel wood went to the furnaces, but consumption dwindled rapidly thereafter (Brosselin, 1977). Domestic fuel-wood consumption also declined quickly, at least in the cities. In Paris, per capita fuel-wood consumption was 1.80 stère (m^3) in 1815, but only 0.45 in 1865 and 0.20 in 1900. After about 1900, 'the production of firewood had become an anachronism' (Brosselin, 1977: 105 tr.).

Table 3.2. Trends in energy resources in France 1809–1855 (million tonnes carbon equivalent) (based on Benoit, 1990).

	1809	1835–1844	1855
Consumption by iron-making			
Wood	1.700	1.225	1.322
Coal	–	0.610	2.510
Total (balance from water energy)	1.800	2.000	4.000
French energy balance			
Wood	11.40	10.70	9.00
Coal	0.80	4.49	12.00
Total (balance from water and wind energy)	13	17	23

The introduction of a new forest law (Code Forestier) in 1827 and other new forest policies was also significant (Baudrillard, 1827). Three separate but interrelated factors combined to bring this about: the perception of a crisis, the rise of the state and the emergence of forest science. State, commune and other public forests were to be managed according to a prescribed regime. Forest clearing could be prohibited under certain conditions and, as time passed, the range of conditions subject to such prohibitions increased. State forestry officials were instructed to demarcate forest limits and enforce regulations on livestock grazing, the taking of wood and other activities. Initially, only modest provision was made for reforestation, involving tax exemptions for forests established in certain mountain settings.

Communal forests now came under the jurisdiction of the state forest service, and local rural areas increasingly lost control over how their forests were managed. The Code and its implementation reflected the 'official' view – that deforestation and forest depletion should be halted, especially in peripheral areas, such as the Alps and the Pyrenees (Clarenc, 1965). State officials used crisis narratives to legitimize both the 1827 Code and direct state intervention in reforestation of mountain terrain. From the peasant viewpoint, however, the Code represented an unjust interference in their traditional use of the forest. In practice, the Code focused on industrial timber production (rather than other forest products), prohibited peasants from continuing their customary practices and failed to address their needs adequately. For example, some communes were authorized to cut only one-sixth of their fuel-wood requirements (Sahlins, 1994). This alienated local peasant users of the forest, notably in the districts suffering from population pressures, such as the Pyrenees, Alps and Jura.

The state used coercion to impose the new order and, perhaps not surprisingly, the peasants resisted. The clearest case of this was 'La Guerre des Desmoiselles', where conflicts between peasants dressed as women and forest guards led to the deployment of thousands of troops. Resistance died down, however, as the century progressed and people increasingly moved away from the rural areas. Similarly, the reforestation programme introduced in the mountains in the 1860s initially met with resistance, but opposition faded as the population declined.

Beginning in the mid-19th century, as the rural population began to decline, in some areas peasants switched from clearing mountain slopes to intensively cultivating the irrigated lowlands and from grazing sheep and goats to raising cattle (Freeman, 1994). Decreases in cropland were especially sharp in the higher regions, where severe demographic pressure had driven expansion in previous decades, and population now began to fall. The population in Alpes-de-Haute-Provence, for example, dropped from 154,000 in 1870 to 118,000 in 1900 (Devèze, 1979). Similar trends were apparent in the Pyrenees. In one canton, the population fell by one-third between 1836 and 1906 (Fruhauf, 1980). Following the rural exodus of the late 19th century, areas deforested in the 17th century to grow crops and raise livestock reverted to forests. In the (translated) words of Fel and Bouet (1983: 222): 'as a general rule, the greater the fall in population, the more the forest extends'. With reduced grazing and browsing, prospects for regeneration improved and resistance to the programme of reforestation weakened.

The combination of technological change in agriculture and in transport, along with the development of a market system, allowed agriculture to concentrate in the more productive areas (at a variety of scales, ranging from the local to the international). The corollary was that abandoned land was available for forest expansion, through either regeneration or planting. As technology and the development of market relations gradually decoupled the historical link between population growth and agricultural expansion, population growth now no longer (necessarily) meant encroachment on the forest.

The conclusion from the French case is similar to that of Denmark and Switzerland. Technological change in agriculture and agriculture's increasing market orientation significantly helped to stabilize and eventually expand forest area. Technological change accelerated the concentration of agricultural production on higher-quality land and the development of the transport network allowed the decoupling of local population size and agricultural production. But the French case also resembles those of Denmark and Switzerland in that it is difficult, if not impossible, to separate out the specific contribution of technological change in agriculture. Political, social and economic change coincided with technological change, and all occurred in a time of philosophical change. This coincidence in time was not a chance event: the various dimensions of change were interrelated. In relation to forest trends, some of the changes may have functioned as immediate or proximate factors, while others (such as philosophical and political change) were more fundamental and underlying.

5. The Role of Agriculture in the Forest Transition

The agriculture of Denmark, Switzerland and France underwent radical change during the period in which the forest transition was taking place. Agriculture modernized and become more market-orientated. The economic context in which agriculture operated, its land organization and tenure and agricultural technology all experienced profound change. The dramatic expansion of the transport network, and especially of railways, made this possible and the rapid growth of urban population further accelerated it.

The rewards to technological change varied, depending on the environmental conditions. New methods often had more success on fertile land than in more marginal areas. This held at a variety of spatial scales. The better land in the northern half of France, and in the Paris basin in particular, was more suited for the new methods of wheat production than land in the mountain valleys. Similarly, the use of sown grass in Alpine valleys allowed livestock production to intensify and reduced pressure on mountain pastures from grazing. Uneven development characterized agricultural change and one facet of that was the abandonment of marginal land, or a least a reduction of agricultural pressures on it.

Abandonment, however, was not the result of agricultural change alone. The growth of opportunities for industrial employment in the cities greatly encouraged the move away from a semi-subsistence agricultural system in the mountains or other marginal areas. The 'rural exodus' from such areas decreased the pressure on forests from agriculture, grazing and fuel-wood collecting and made it possible for some forests to regenerate naturally in some areas. Technological change in others sectors, particularly transport, led to the substitution of fuel wood by fossil fuels. That also took pressure off the forest and meant that population growth was no longer closely associated with increasing fuel-wood consumption.

Even without technological change, market forces and learning processes can lead to the spatial reorganization of agricultural production and the concentration of agriculture in more favourable environments, but technological change is likely to accelerate that process (Mather and Needle, 1998). Such adjustment may lead landowners to abandon certain areas and allow forests to regenerate there.

In each of the three countries, the emerging modern state employed a 'crisis narrative' to legitimize state intervention in environmental management (in the form of forest codes and/or reforestation). The alleged crises involved wood shortages, erosion, flooding and various other resource and environmental problems. Technological change in silviculture was involved, and in each case it was associated with a changing paradigm or social construction of the forest. The origins of modern forest science are usually assumed to have been in Central Europe and to have been linked to fears of a wood shortage (Mantel, 1964). The origins of both the science itself and its adoption by the state were thus linked to scarcity, or at least to perceived scarcity, of wood.

In few other areas of life did the Enlightenment project, with its characteristic privileging of rationality and the application of science, leave a clearer landscape expression. In the context of the prevailing obsession with 'progress', unproductive 'wasteland' became a challenge, and reforestation was seen as a means of making it more useful and productive. At another level, specialization and monofunctionality manifested the reductionism that accompanied the rise of rationality. Previously the forest was integral and continuous with farmland. It was used for grazing and collecting fodder, as well as a source of wood for fuel and construction. Now it was a separate category, geared to timber production and enclosed within sharp linear boundaries, which epitomized both the rise of rationality and the dislocation of traditional peasant systems.

The reversal of the long-established trends of deforestation reflected the triumph of the new order. The exclusion of livestock reduced pressure on forests from grazing. 'Scientific' silviculture, including the creation of planted forests, became established. Gradually, the forest began to expand, at least at the national level. Deforestation continued in some areas more favourable for agriculture, but that was more than counterbalanced by reforestation on (agriculturally) marginal land.

Achieving this transition from net deforestation to net reforestation had a cost. Many traditional peasant users of the forest were effectively dispossessed. It is not surprising, therefore, that some of them resisted. Of the three countries examined, the resistance was strongest in France. In Switzerland it was more passive and in Denmark even more subdued. It is perhaps significant that, in the latter case, the government made some provision for the 'dispossessed' at the time of enclosure, through the allocation of some previously common lands to individual dispossessed farmers.

It may be useful to distinguish between 'natural' and 'induced' forest transitions. In the former, market forces unleashed by developments in agriculture and in other sectors lead to the shift from net deforestation to net reforestation. Land is simply released from agriculture, and becomes available for forest expansion through natural regeneration or planting. The case of France, however, suggests that coercion can also be used to accelerate or 'induce' a transition, at the cost of hardship to the dispossessed traditional users of the forest. Presumably, the extent to which agricultural and other conditions approach those required for a 'natural' transition partially determines the degree of coercion required (and of hardship that results).

6. Conclusion

Technological change in agriculture clearly contributed to the transition from net deforestation to net reforestation in the European countries considered in this chapter. It helped to decouple population from agricultural area and encouraged farmers to abandon agricultural land and allow it to return to forest by natural regeneration or planting. It was one of several proximate factors 'driving' the transition, along with the radical change in transport and energy supply and technical change in silviculture. One cannot separate out how much each factor contributed in relative terms, not only because of a dearth of data but, more importantly, because of the synergy between the factors. The factors also operated as both causes and effects of the exodus from rural areas and the emergence of the market economy.

Political and cultural changes played a central role in this process. The state emerged as a legislative and technical agent of environmental management, science was applied to land management, capitalism penetrated even the most remote rural areas and a new social construction of the forest gained acceptance. Previously, woodland and farmland had been largely continuous and multifunctional. Now, they became increasingly separate and specialized, both symbolically and on the ground. The enactment of legislation to protect the existing forest area further weakened the earlier link between farming and forest. Powerful interests used crisis narratives to legitimize their own claims on the forest and its products. State power and the application of science helped achieve a forest transition, but at a cost.

The European countries examined irrefutably demonstrate that deforestation can be halted and reversed. Technical change in agriculture generally favours that outcome, but not always. Improvements in the transport system can open up new areas for logging and accelerate deforestation. But they can also lead to the substitution of fuel wood by fossil fuels and hence alleviate pressures on the forest. Just as the effects of transport changes depend on their nature and circumstance, so do those of agricultural changes. Agricultural change in the European case-studies did not occur in isolation. It was a component of a wider and more deep-seated change amounting to development or modernization. Whether agricultural change could occur in isolation from such changes and, if so, whether it could significantly contribute to reducing deforestation under those circumstances is another question.

Will developing countries that are currently experiencing deforestation experience similar trajectories? In some respects, their situation is comparable. Just as previously occurred in Europe, agriculture in these countries is undergoing technical change and becoming more market-orientated and they are rapidly urbanizing. In some countries, floods and landslides have triggered state intervention in the form of logging bans or other measures, just as they did in France and Switzerland. And, in some countries, farmers are beginning to abandon cropland and allow it to revert to forest, as the younger generations leave the farm and seek a better life in the city. There are also grounds for thinking that the transition might be faster in a modern developing country than in 19th-century Europe. With international concern over deforestation and the influence of an international (as opposed to purely national) civil society, the changes that took decades in Europe might happen more rapidly. On the other hand, the growth of international trade brings its own complications, as agricultural production may gravitate towards the optimal locations at the global scale, and not just at the national level. This could accelerate a forest transition in some countries, or delay or prevent it in others.

Note

1   Much of this chapter is based on work supported by the UK Economic and Social Research Council's programme on global environmental change. This support is gratefully acknowledged, as are the helpful comments of the editors of this volume, Arild Angelsen and David Kaimowitz, and of an anonymous reviewer.

References

Abel, W. (1980) *Agricultural Fluctuations in Europe*. Methuen, London.
Baudrillard, J.J. (1827) *Code Forestier avec Commentaire*, 2 vols. Arthus Bertrand, Paris.
Benoit, S. (1990) La consommation de combustible végétal et l'évolution des systèmes techniques. In: Woronoff, D. (ed.) *Forges et Forêts: Recherches sur la Consommation Proto-industrielle de Bois*. École des Hautes Études en Sciences Sociales, Paris, pp. 87–150.
Blaikie, P. and Brookfield, H. (1987) *Land Degradation and Society*. Methuen, London.
Bonhote, J. and Fruhauf, C. (1990) La métallurgie au bois et les espaces forestiers dans les Pyrénées de l'Aude et de l'Ariège. In: Woronoff, D. (ed.) *Forges et Forêts: Recherches sur la Consommation Proto-industrielle de Bois*. École des Hautes Études en Sciences Sociales, Paris, pp. 11–28.
Bourgenot, L. (1977) Histoire des forêts feuillues en France. *Revue Forestière Française (num. spéc.)*, 7–26.
Bourgenot, L. (1993) Quelques réflexions sur l'histoire des forêts françaises. *Comptes Rendus Académie Agriculture Français* 79, 85–92.
Braudel, F. (1990) *The Identity of France*, Vol. 2: *People and Production*. Fontana, London.
Brosselin, A. (1977) Pour une histoire de la forêt française au XIXe siècle. *Revue d'Histoire Économique et Sociale* 55, 92–111.
Clarenc, L. (1965) Le code de 1827 et les troubles forestiers dans les Pyrénées centrales vers le milieu du XIXe siècle. *Annales du Midi* 77, 293–317.
Clout, H.D. (1983) *The Land of France 1815–1914*. Croom Helm, London.
Corvol, A. (1987) *L'Homme aux Bois: Histoire des Relations de l'Homme et de la Forêt XVIIe–XXe siècle*. Fayard, Paris.
de Planhol, X. with Claval, P. (1988) *An Historical Geography of France*. Cambridge University Press, Cambridge.
Devèze, M. (1979) Le reboisement des montagnes françaises dans la seconde moitié du XIXème siècle. *La Forêt Privée* 126, 27–32.
Fel, A. and Bouet, G. (1983) *Atlas et Géographie du Massif Central*. Flammarion, Paris.
Freeman, J.F. (1994) Forest conservancy in the Alps of Dauphiné 1287–1870. *Forest and Conservation History* 38, 171–180.
Friedmann, K. J. (1984) Fencing, herding and tethering in Denmark, from open field agriculture to enclosure. *Agricultural History* 58, 584–597.

Friis, A. and Glamann, K. (1958) *A History of Prices and Wages in Denmark 1660–1800*, Vol. 1. Longman Green and Institute of Economics and History, London and Copenhagen.

Fritzbøger, B. (1994) *Kulturskoven: Danske Skovbrug fra Oldtid til Nutid*. Gyldendal med Miljøministeriet, Skov og Natur styrelsen, Copenhagen.

Fruhauf, C. (1980) *Forêt et Société: de la Forêt Paysanne à la Forêt Capitaliste en Pays de Sault sous l'Ancien Régime (1670–1791)*. CNRS, Paris.

Gerber, B. (1989) Waldflächenveränderungen und Hochwasserbedrohung im Einzugsgebiet der Emme. *Geographica Bernensia* G33, 1–8.

Grøn, A.H. (1960) Introduction. In: Reventlow, C.D.F. *A Treatise on Forestry*. Society of Forest History, Horsholm, pp. xi–xxxv.

Hauser, A. (1975) Brachland oder Würstung? Zur begrifflichen und historischen Abklärung des Brachlandproblems. *Schweizerische Zeitschrift für Forstwesen* 126, 1–12.

Inspection Fédérale des Forêts (IFF) (1976) *100 Ans de Protection de la Forêt*. IFF, Berne.

Jensen, K.M. (1976) *Opgivene og Tilplantede Landbrugsarealer I Jylland Atlas over Danmark*, Series II, Vol. I. Den danske Kongelkige Geografiske Selskab, Copenhagen.

Jensen, K.M. (1993) Afforestation in Denmark. In: Mather, A. (ed.) *Afforestation: Policies, Planning and Progress*. Belhaven Press, London, pp. 49–59.

Jones, P.M. (1990) Agricultural modernization and the French Revolution. *Journal of Historical Geography* 16, 38–50.

Kjærgaard, T. (1994) *The Danish Revolution 1500–1800: an Eco-historical Interpretation*. Cambridge University Press, Cambridge.

Landolt, E. (1862) *Bericht an den Hohen Schweizerischen Bundesrath über die Untersuchung der Schweiz Hochgebirgswaldungen Vorgenommen in den Jahren 1858, 1859 und 1860*. Weingart, Berne.

Mantel, K. (1964) History of the international science of forestry with special consideration of Central Europe. *International Review of Forestry Research* 1, 1–37.

Marek, D. (1994) Der Weg zum fossilen Enerfgiesystem, Ressourcengeschichte der Kohle am Beispiel der Schweiz 1850–1910. In: Abelshauser, W. (ed.) *Umweltgeschichte; Umweltverträgliches Wirschaften in historischer Perspektive*. Vandenhoek und Ruprecht, Gottingen, pp. 57–75.

Mather, A.S. (1992) The forest transition. *Area* 24, 367–379.

Mather, A.S. and Needle, C.L. (1998) The forest transition: a theoretical basis. *Area* 30, 117–124.

Moriceau, J.M. (1994) Au rendez-vous de la 'Révolution agricole' dans la France du XVIIIe siècle: à propos des régions de grande culture. *Annales: Histoire, Sciences Sociales* 49, 27–63.

Morineau, M. (1970) Was there an agricultural revolution in 18th century France? In: Cameron, R., Mendels, F.P. and Ward, J.P. (eds) *Essays in French Economic History*. Irwin/AEA, Holmwood, pp. 170–182.

Netting, R.M. (1972) Of men and meadows: strategies of alpine land use. *Anthropological Quarterly* 45, 132–144.

Nielsen, S. (1988) Dansk landbrug 1788–1988. *Arv og Eje* 1988, 7–62.

Pfister, C. (1983) Changes in stability and carrying capacity of lowland and highland agro-systems in Switzerland in the historical past. *Mountain Research and Development* 3, 291–297.

Pfister, C. (1990) Food supply in the Swiss canton of Bern, 1850. In: Newman, L.F. (ed.) *Hunger in History: Food Shortage, Poverty and Deprivation*. Basil Blackwell, Oxford, pp. 281–303.

Pfister, C. (1994) Switzerland: the time of icy winters and chilly springs. In: Frenzel, B. (ed.) *Climatic Trends and Anomalies in Europe 1675–1715: High Resolution Spatio-temporal Reconstructions from Direct Meteorological Observations and Proxy Data*. Fischer, Stuttgart, pp. 205–224.

Pfister, C. and Messerli, P. (1990) Switzerland. In: Turner, B.L., II, Clark, W.C., Kates, R.W., Richards, J.F., Matthews, J.T. and Meyer, W.B. (eds) *The Earth as Transformed by Human Action*. Cambridge University Press, Cambridge, pp. 641–652.

Ponchelet, D. (1995) Le débat autour du déboisement dans le département des Basses-Alpes, France (1819–1849). *Revue de Géographie Alpine* 83, 53–66.

Raaschou-Nielsen, A. (1990) Danish agrarian reform and economic theory. *Scandinavian Economic History Review* 38, 44–61.

Rinaudo, Y. (1980) Forêt et espace agricole: exemple du Var au XIXe siècle. *Revue Forestière Française (num. spéc.)*, 136–148.

Sabroe, A.S. (1954) *Forestry in Denmark*. Danish History Society, Copenhagen.

Sahlins, P. (1994) *Forest Rites: the War of the Demoiselles in 19th Century France*. Harvard University Press, Cambridge, Massachusetts.

Schuler, A. (1983) Der Privatwald in der Forstgeschichte. *Schweizerische Zeitschrift für Forstwesen* 134, 687–701.

Schuler, A. (1984) Nachhaltigkeit und Waldfunktionen in der sicht der Schweizer Forstleute des 19 Jahrhunderts. *Schweizerische zeitschrift für Fortwesen* 135, 695–709.

Sclafert, T. (1933) A propos du déboisement des Alpes du Sud. *Annales de Géographie* 42, 266–277, 350–360.

Smout, T.C. (1987) Landowners in Scotland, Ireland and Denmark in the Age of Improvement. *Scandinavian Journal of History* 12, 79–97.

Société Suisse des Forestiers (1874) Société sur la recherche de houille. *Journal Suisse d'Économie Forestière*, Année 8, 146.

Statistischen Bureau (annual) *Statistisches Jahrbuch der Schweiz*. Stämpfli, Berne.

Sutton, K. (1977) Reclamation of wasteland during the 18th and 19th centuries. In: Clout, H.D. (ed.) *Themes in the Historical Geography of France*. Academic Press, London, pp. 247–300.

Tönesson, K. (1981) Tenancy, freehold and enclosure in Scandinavia from the seventeenth to the nineteenth century. *Scandinavian Journal of History* 6, 191–206.

Did a Green Revolution Restore the Forests of the American South?

Thomas K. Rudel

1. Introduction

Recent reports of elevated rates of tropical deforestation in Brazil during the mid-1990s, coupled with the pessimistic report about tropical forests issued by the European Community's Research Centre in 1998, underline the urgency of the search for a policy solution to the problem. In this context, the Borlaug hypothesis, named after its most famous exponent, merits detailed examination. Norman Borlaug and others have asserted that significant increases in the land productivity of agricultural commodities would solve the problem of tropical deforestation by reducing the need to expand the area of cultivated land as demand for crops increases (World Resources Institute, 1986; Rudel with Horowitz, 1993; Southgate, 1998).

The simplicity of Borlaug's argument makes it appealing. It also gains in stature because it draws upon the most coherent body of theory in social science, microeconomics, to make its essential point. The theory also has clear policy implications: to reduce tropical deforestation, governments and international organizations should greatly expand their programmes of research into the land productivity of crops grown in the tropical biome. Despite these attractive features, the theory has not been tested empirically. Under these circumstances, a historical study of changes in crop yields and forest cover in the American South between 1935 and 1975 may provide useful insights about the validity of the Borlaug hypothesis.

Examining forest-cover dynamics in the southern USA, with an eye to the lessons that it might have for forest-cover dynamics in the tropics, may seem like a far-fetched idea, but for two reasons it is not. First, because the Borlaug hypothesis concerns a process of technological innovation and diffusion among thousands of farmers, followed by a period of forest regrowth, any assessment of it must entail a historical study stretching out over several decades. Very few, if any, of the countries in the tropical biome contain the detailed historical records on changes in forest cover that would be necessary to follow changes in the acreage devoted to specific crops over several decades. In contrast, the data on southern US forest cover and its driving forces are complete enough to conduct a fairly conclusive test of the hypothesis.

Secondly, the American South in 1930, at the beginning of the period under study, resembled contemporary developing countries in several crucial respects (Vance, 1932). The south-eastern USA contains red clay soils, much like those commonly found in large parts of the Amazon basin (Sanchez, 1976). Despite the poor soils, a large majority of the regional population, black and white, earned a living from agriculture, usually on small farms devoted to cotton cultivation. Four out of five farmers worked land that they did not own, usually as sharecroppers. They were poor. Farmers in the south-eastern cotton-growing states averaged $143 in income per year from their crops between 1924 and 1929 ($637 in 1989 dollars). Farmers had a commercial orientation, producing cash crops, such as cotton and tobacco, for global markets. In 10–15% of the agricultural districts, farmers had a subsistence orientation, consuming more of their harvests than they sold (Rudel and Fu, 1996: 813). Eleven per cent of the regional population was illiterate in 1930 (Odum, 1936; Johnson *et al.*, 1941). In talking about the South's position in the national economy, analysts anticipated the parlance of contemporary world systems theorists, using terms like 'peripheral' to describe the South's position. Within the USA, the South was a colony of the North (Vance, 1932: 470–481). As a noted regional geographer put it,

> The South is the part of the United States which is most similar to the rest of the world, and the plantation regions are the areas of the South which are most comparable to the new nations that inherited plantation economies. In certain respects the lower Piedmont, the Black Belt, the Loess Plains, and the alluvial Mississippi Valley have more in common with the former colonies of the Caribbean and Central and South America than with the metropolitan United States.
> (Aiken, 1998: 363–364)

Certainly the argument that the experience of the American South between 1935 and 1975 resembles that of contemporary developing countries can be pushed too far. The transportation network of the region – its roads, railways and canals – had been well developed through decades of internal improvements since the Civil War. Unlike many contemporary developing countries, the South had a system of secure property rights in land. The magnitude of industrial job creation in northern metropolitan areas prompted heavy out-migration from the South between 1935 and 1975. Throughout this period, the state supported agricultural production through price supports, subsidized credit and conservation set-aside programmes. None of these factors have historical parallels in developing countries. Nevertheless, the exceptional quality of the data and the existence of some historical parallels between the South and places in the tropical biome argue for using the South as a test case for examining the validity of the Borlaug hypothesis.

To investigate the effect of increases in the land productivity of crops on forest cover, I bring together data on the prevalence of particular crops in counties, trends in the land productivity of those crops and trends in forest cover in those counties. Assuming little change in demand for the different crops, more fields should have reverted to forests in places with agricultural economies organized around the crops that recorded the greatest gains in agricultural productivity between 1935 and 1975.

Three processes of technological change influenced trends in agricultural productivity in Southern agriculture between 1935 and 1975. First, publicly financed land-improvement projects, in particular the drainage of wetlands, changed the land base available to farmers, giving them access to more fertile lands. Secondly, subsidized fertilizer production and, after 1955, the introduction of herbicides in cotton cultivation elevated yields per acre (Aiken, 1998: 109). Thirdly, agribusinesses promoted the mechanization of agriculture, through the introduction of first tractors and later harvesters. The last trend reduced the amount of labour and increased the amount of capital utilized per hectare in Southern agriculture. Not surprisingly, this trend led to a considerable increase in the scale of agricultural operations. In the following analyses, measures of regional soil resources provide a proxy measure for wetlands reclamation and expenditures for fertilizers at the outset of the period measure fertilizer use. I do not have a good measure of the effects of mechanization on land productivity.

As controls in the analysis, I introduce additional data on the human capital of farmers, the size of nearby urban places and government policies. These variables embody plausible alternative explanations for the reforestation of the South during the middle decades of the 20th century. The human capital variable, illiteracy, expresses the idea that farmers with little human capital would face competitive disadvantages brought on by the advent of more scientific agriculture, which would eventually cause them to abandon their lands and allow their fields to revert to forest. In counties with sizeable urban communities, farmers could scale down their agricultural enterprises without completely abandoning them because farmers could more easily secure part-time employment in the non-farm sector. For this reason complete farm abandonment and reforestation should characterize remote rural counties more than counties with sizeable urban communities. The federal government, through several policy initiatives, most prominently the price support–conservation set-aside programme introduced in 1934 and the expansion of national forests during the 1930s, may have played an important role in the reforestation of the South. A multivariate analysis that includes these variables and the productivity variables in a single equation predicting trends in forest cover should tell us something about the relative magnitude of the agricultural productivity effect on forest cover.

2. Data, Variables and Measures

Counties are the units in the analyses reported below. The data on forest cover come from forest inventories conducted by the US Forest Service every 10 years, beginning in the 1930s. The data on crop productivity come from the US Department of Agriculture, and the data on soil resources come from a survey carried out during the 1930s by the Bureau of Agricultural Economics of the Department of Agriculture. The data for all other variables in the analyses come from the *Statistical Atlas of Southern Counties* (Johnson et al., 1941).[1] The data sources and measures are listed in Table 4.1. Several of the measures reported in Table 4.1 require some explanation.

2.1. Changes in forest cover

I calculated the average annual rate of change in forest cover between 1935 and 1975 for approximately 800 counties in the southern USA. The data on forest cover come from successive forest surveys, conducted every 10 years, beginning in the 1930s, by the US Forest Service. The measurement

Table 4.1. Data sources and measures.

Forest cover	Forest survey bulletins published by Southern forest experiment stations, 1930s–1970s. Change in % per annum
Land productivity	US Department of Agriculture. Historical statistics http://usda.mannlib.cornell.edu/data-sets/crops. Increase in % between 1935 and 1965
Price change per unit	US Department of Agriculture. Historical statistics http://usda.mannlib.cornell.edu./data-sets/crops. Change in % between 1935 and 1975
Area planted of major crop	US Department of Agriculture. Historical statistics http://usda.mannlib.cornell.edu/data-sets/crops. Change in % between 1935 and 1970
Illiteracy	Johnson et al. (1941), derived from census data. Proportion of adults in a county in 1930
Fertilizer use	County tables, 1930 census. Expenditures on fertilizer as a proportion of agricultural sales
Size of urban place	1930 census. Population of largest town in a county
National forest	USFS maps, 1975. Proportion of county land in national forests
% Delta	Maps printed in Barnes and Marschner (1933)
% Piedmont	Maps printed in Barnes and Marschner (1933)
Self-sufficient farming	Johnson et al. (1941), typology derived from 1930 census data. Farmers were considered to be self-sufficient if they consumed at home more than half of the agricultural production from their farms

USFS, US Forestry Service.

techniques have evolved over the history of the forest survey from on-ground parallel-line surveys to aerial photos and then to satellite images. The change from on-ground to aerial photo methods does not appear to have biased the measures in a discernible way (Cruikshank and Evans, 1945).

2.2. Land capability (% delta, % piedmont)

At the request of planners in the Department of Agriculture during the depression, C. Barnes and F. Marschner delineated agricultural regions in the USA. They brought together information on the physical geography of agricultural areas, their topography, their soils and their climate, and used this information to construct a map of the agricultural potential of different regions in the USA. In this exercise, the boundaries between regions became, in effect, boundaries between land capability classes. For example, a boundary separating the Mississippi delta from the sandy lands of southern Mississippi is, in effect, a boundary separating a region with high land capability from a region with low land capability.

2.3. Land productivity

This variable measures the gains in yields per acre for the chief commercial crop in a county.[2] Because there are only seven basic commercial crops grown in the South during this period, there are only seven possible values that this variable can assume in a county. Furthermore, the productivity gains reported here are averages for the entire USA, not just the South. While this circumstance creates measurement error for a crop like maize, which was grown extensively outside the South, there is little measurement error for most of the other crops (e.g. cotton, peanuts), because they are grown largely, and sometimes solely, in the South.

In the bivariate and multivariate analyses reported below, I use these variables and the others listed in Table 4.1 to explore the historical relationship between agricultural productivity and forest cover in the South. To avoid problems of simultaneity bias in the analyses, the explanatory variables precede the changes in forest cover or come from the first portion of the four-decade period in which I measure forest cover.

3. Findings

Between 1935 and 1975, forests expanded in extent across much of the South. Overall about 8% of the South's land area reverted to forest during this period (Rudel and Fu, 1996). Dramatic subregional variations marked the regional pattern of forest-cover change. As the map of forest-cover change in Fig. 4.1 suggests, some agricultural regions, such as the piedmont in the Carolinas, virtually disappeared, while others, such as the Mississippi delta, expanded vigorously. To paraphrase the foremost student of Southern land use, John Fraser Hart, the South during this period became 'a splendid laboratory for studying the birth and death of agricultural regions' (Hart, 1991: 276).

The patterns in the data arrayed in Table 4.2 suggest the degree to which these subregional patterns of agricultural expansion and decline correspond to differential rates of change in the land productivity of the crops that dominate in the agricultural economies of the different subregions. The table classifies rural Southern counties by their dominant cash crop and reports data on forest-cover change in those counties. Juxtaposed with the data on forest-cover change are data on changes in the agricultural economy of the dominant crop: increases in yields and in prices and changes in area planted in the USA. A comparison of the first four county types (cotton, tobacco, maize and peanut counties) with the last three county types (sugar cane, rice and orange counties) reveals a pattern that supports the Borlaug hypothesis. Those counties that saw the largest increases in the yields of their dominant crops

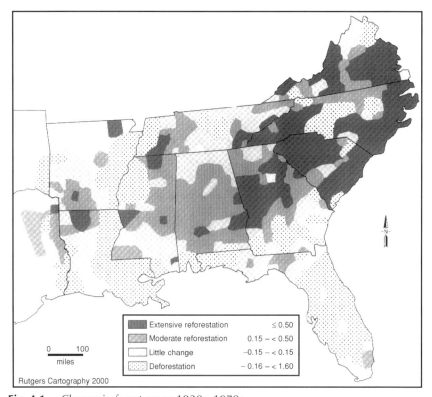

Fig. 4.1. Change in forest cover 1930s–1970s.

Table 4.2. Crops' land productivity, area planted, price and forest-cover trends in counties dominated by those crops, 1935–1975 (based on US Forest Service, Forest Inventories, 1932–1975; US Department of Agriculture, Track Records, Crop Production, 1935, 1965, 1970).

	County forest-cover change, 1935–1975 (in % of land area)	Land productivity change, 1935–1965 (in % of 1935 yield)	Change in area planted in USA, 1935–1970 (% of 1935 acreage)	Price change, 1935–1975 (% of 1935 price)
Cotton	+5.32	+185	−57	+362.6
Tobacco	+14.20	+110	−38	+390.5
Maize	+11.85	+206	−33	+209.0
Peanuts	+20.06	+116	−40	+516.0
Sugar cane	+3.16	+108	+112	+441.6
Rice	−3.56	+96	+123	+379.9
Oranges	−24.16	+88	+60	+149.9

(more than 109% over the 40-year period) showed greater gains in forest cover than those counties that showed lesser increases in yields per acre. The national patterns in acreage planted in the different crops suggest why the patterns of forest-cover change differ across the counties. Those crops with the largest increases in yields saw the largest declines in acreage planted. Interestingly, there is no apparent relationship between trends in the prices of agricultural commodities, productivity increases and reforestation during the 40-year period.

Figure 4.2 reports the results of the multivariate analyses, and it provides more conclusive evidence about the influence of agricultural productivity variables on the pattern of forest-cover change in the South. With the exception of the path from yield increases to changes in crop area to reforestation, Fig. 4.2 presents a simple inventory of causes regressed against the change in forest cover in a county over a 40-year period. The residuals are normally distributed and the levels of multicollinearity are low. Deletion of outliers produces some modest changes in the overall variance explained, but it does not change the relative explanatory strength of the different predictors of the reforestation rate in a county.

The most accurate predictor of reforestation rates, forest cover in a county in 1935, has an artefactual element to it. The highest rates of reforestation between 1935 and 1975 tended to occur in the counties with the lowest levels of forest cover in 1935, presumably because these counties had the most land that could be reforested. Land capability appears to have been an important factor in reforestation, because % piedmont (low land capability) and % delta (high land capability) are strong predictors of the reforestation rate. Human capital variables, loosely expressed here as the proportion of a county's population that is literate, the proportion of farmers engaged in subsistence

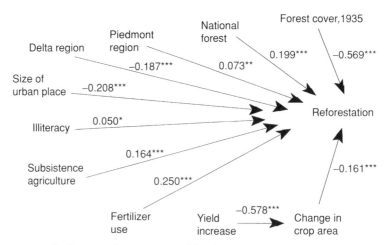

Fig. 4.2. Path diagram: determinants of Southern reforestation, 1935–1975. Numbers are standardized coefficients, $n = 777$, adjusted. $R_2 = 0.594$, P values, $* < 0.05$, $** < 0.01$, $*** < 0.001$.

agriculture and the proportion of the county's population living in its largest urban place, also exercised an important influence on reforestation rates. A government policy of expanding the size of national forests through the purchase of marginal agricultural lands made a direct contribution to the reforestation of the region. Finally, several technological changes associated with the Borlaug hypothesis appear to have played an important role in the reforestation of the South. Where farmers used more fertilizer in 1930, reforestation occurred at higher rates in subsequent decades, presumably because, with the aid of fertilizers, they concentrated production on fewer acres. As depicted in Fig. 4.2, large yield increases in particular crops led to large declines in the amount of land devoted to the cultivation of that crop, and these declines in acreage increased the rate of reforestation in a county. This sequence of events conforms to the logic of the Borlaug hypothesis.

4. Discussion

Does the Southern experience with increasing agricultural productivity and forest-cover change offer lessons for how rising crop yields might curb forest loss in tropical biomes? Certainly, the influence of the American state on forest-cover trends between 1935 and 1975 seems improbable in the current political context of most countries in the tropics. The American state launched more programmes that affected forests than the contemporary neoliberal states of the developing world will ever do. Some of the state's programmes probably had only small effects. Price-support programmes enabled some farmers on marginal lands to remain on their land for a longer period than they would have otherwise. In this sense, the price-support and acreage-control programmes probably slowed the pace of change, rather than reversing its direction (Hart, 1978: 512). The plan for expanding the national forests through the purchase of marginal agricultural lands ensured that some lands, by becoming part of a national forest, reverted to forest. Arguably, these lands would have reverted to forest in any case. The Tennessee Valley Authority's celebrated reforestation programme had little impact on regional land cover (Rudel, 1995).

Other federal programmes, in particular the work of the Army Corps of Engineers, had an important impact on land-cover change. In the 19th and early 20th century, settlers cleared land for farming in the South's upland, but they avoided the low-lying, alluvial land in the Mississippi River delta and along the Gulf coast. The soil was very fertile, but periodic floods and difficulties with drainage prevented agricultural expansion into these areas, and they remained covered with hardwood forests, containing much high-quality wood. In the second half of the 19th century, local groups began to build levees in an efforts to control floods along the Mississippi. Alarmed by the damage wrought by these floods and pressured by local lobbying groups, federal legislators assumed half of the costs of levee construction in 1916, and in 1928, after the particularly disastrous flood of 1927, the federal government assumed the entire cost of levee construction (O'Neill, 1998). A 1944 amendment to the 1928 act extended federal assistance to drainage of lands behind the levees. With these mandates, the Army Corps of Engineers began an ambitious programme of public works in the Mississippi delta and along the Gulf coast during the 1930s, building levees and later draining swamps (Ferguson, 1940; Harrison, 1951; McPhee, 1986; US Army Corps of Engineers, 1989).

With the low-lying lands secured from floods, landowners moved quickly to harvest the valuable timber and plant soybeans in the cleared fields (Sternitzke and Christopher, 1970). The flat, fertile and uniform fields were ideal for the highly mechanized agricultural techniques used in the cultivation of soybeans. The contrasting trends in forest cover in Table 4.3 between cotton counties inside and outside the Mississippi delta testify to the effects of

Table 4.3. Natural land-use areas and forest-cover change in cotton-growing counties, 1935–1975* (based on Barnes and Marschner, 1933; Forest Inventories, US Forest Service, 1935–1965).

	Forest cover, 1935 (% of land area)	Forest cover, 1975 (% of land area)	Forest-cover change (% per annum)
Outside the Mississippi delta	53.99	63.03	+0.245
Inside the Mississippi delta	53.31	38.89	−0.328

*Number of counties: 1935 = 488, 1965 = 509.

government flood-control programmes on forest cover in lowland regions. In the sugar-cane-growing regions of Louisiana, a similar but less pronounced pattern developed, with landowners growing cane in the protected areas behind the levees along the Atchafalaya River (Hart, 1978: 512).

The rapid growth of soybean cultivation in the alluvial lowlands of the South stems in part from the development of new markets for soybean-based animal feeds. A similar set of developments in consumer markets explains the rapid expansion of cultivated acreage in the citrus-producing areas of central Florida between 1935 and 1975 (see Table 4.2). The model in Fig. 4.2 does not incorporate the effects of changes in markets and consumer tastes on forest-cover trends, but clearly they had a significant effect. The increasing returns to human capital in cities also played an important, albeit indirect, role in the pattern of reforestation. The significance of the illiteracy, subsistence farming and urban place variables in the models testify to the rapidly increasing returns to human capital in urban areas, in the form of either new jobs or higher wages (Ruttan, 1984: 151–152). Rural poverty pushed and urban economic growth pulled smallholders off their farms and hastened the return of their fields to forests.

The decline in the agricultural labour force spurred mechanization in Southern agriculture, which, in turn, encouraged land abandonment in areas of low land capability.[3] When farm workers left the land, plantation owners keep the flat, fertile lands of the delta in production by purchasing tractors and harvesters to replace field hands. Farmers who worked the more accentuated terrain of the piedmont did not think that they could use machines to replace labour on these lands (Aiken, 1998: 118–119).[4] Given the more impoverished soils in the piedmont, farmers faced with the problem of labour scarcity in this setting frequently abandoned farming. Other farmers in these regions, faced with declining yields, did not need the spur of labour scarcity to abandon their lands. Because the most capable agricultural lands in the South are concentrated in islands or strips of land surrounded by more extensive areas of less fertile lands, the landscape in the American South began, by the 1970s, to appear like islands of intensive agriculture in a sea of forested and reforested land.

While the findings in Fig. 4.2 provide general support for the Borlaug hypothesis, three issues remain unclear. First, the mid-20th century saw a rapid expansion of cotton cultivation outside the American South, in particular in the western USA. How did the expansion of production in these competing areas affect land abandonment in the South? The Reclamation Act of 1902 authorized the federal government to develop irrigation systems for agriculture in the American West (Lee, 1980). After the Second World War, the state of California supplemented the federal programme with its Central Valley Project. In California, cotton became one of the crops of choice for farmers on these irrigated lands. The yields per acre on these fertile, irrigated fields averaged more than twice the national average for cotton throughout the 1935–1975 period, and California's share of national cotton acreage grew from less than 1% in 1935 to approximately 10% in 1975 (Scheuring, 1983:

117; USDA, 1999). By 1975, cotton had become the most valuable field crop produced in California; only Texas produced more cotton than California.

Several economists have claimed that federal programmes for irrigating the West resulted in the abandonment of 6 to 18 million acres of land in the eastern USA (Howe and Easter, 1971). The growth of cotton cultivation on irrigated lands in California and the decline of rain-fed cotton lands in the South would appear to be a case in point. Historical data on acreage in cotton in the two regions do not, however, support the idea of a simple substitution of western cotton lands for eastern cotton lands. Most of the reforestation on Southern cotton lands begins between 1935 and 1945, a decade in which cotton acreage in California did not increase. The significant increases in California acreage occur during the 1945–1955 and 1969–1975 periods, but they do not coincide with or precipitate significant losses in cotton acreage in the South in a way that is clearly visible. In sum, the increase in cotton cultivation in the American West probably contributed to the abandonment of cotton lands in the South, but the magnitude of this effect on forest-cover dynamics cannot have been particularly large, because it is not apparent in the historical data (Scheuring, 1983: 128; USDA, 1999).

Secondly, questions could be raised about the magnitude of the cause–effect link between yield increases and forest recovery in the South. A comparison of the explanatory power of the different groups of variables in Fig. 4.2 makes it clear that the effects of agricultural productivity on forest recovery were not trivial. The two agricultural productivity variables explain 6% of the total variation in the Fig. 4.2 equation for reforestation, compared with 4.4% for the two human capital variables and 2.8% for the land capability variables. The timing of the reforestation sheds additional light on the influence that increases in agricultural productivity had on reforestation. Virtually all of the reforestation occurred during the first 20 years, 1935 to 1955, of the 40-year period under examination. A historical conjuncture of three watershed events, the Depression, the New Deal and the Second World War, pushed reforestation during this period. Low commodity prices encouraged farmers to abandon marginally productive lands. The recently established Tennessee Valley Authority made low-cost fertilizers widely available, which enabled farmers to concentrate their production on fewer acres. War-induced demands for military service and manufacturing workers spurred the departure of farm workers and increased the use of farm machinery on the flat, fertile soils of the Mississippi delta. The departure of the farm labour force during the war caused farmers in areas of low land capability to allow their lands to remain idle. The poverty and illiteracy of farmers and farm workers posed additional obstacles to the acquisition of credit and the adoption of land-saving technologies, such as fertilizers. In this manner, technological changes interacted with other historical events to produce widespread land abandonment and reforestation in the South. In sum, a conjuncture of events, of which agricultural productivity increase is an important component, contributed to the recovery of the South's forests after 1935.

Could productivity increases alone have produced the widespread conversion of agricultural lands into forest? The answer to this counterfactual question is clearly no, at least in the case of the American South. The related events outlined above, which included increases in land productivity, produced the large-scale conversion of farmlands into forests.

Thirdly, the absence of an obvious relationship between changes in the prices of agricultural commodities during this period and reforestation raises questions, because the Borlaug effect reputably works through changes in prices. Rapid increases in yields per hectare lower the prices of agricultural commodities, which, in turn, encourage farmers to abandon marginal agricultural lands. The political and economic dynamics of agricultural price-support programmes may explain why a causal path from yield increases to price changes and then to reforestation does not exist. Because the federal government intervened to maintain the price of an agricultural commodity when it was in oversupply, productivity gains did not necessarily lead to declines in a commodity's price, but they did lead to an increase in government price-support expenditures. In reaction, government officials may have pushed set-aside programmes more vigorously, in an effort to reduce the government's price-support expenditures. While good historical data to substantiate these claims are scarce, a sequence of events like this one would explain why land productivity changes, but not price changes, associate positively with rates of reforestation.

5. Conclusion: Implications for Patterns of Change in Tropical Forest Cover

In one respect, the American South represents 'the least likely case' (Eckstein, 1975) in which to observe a connection between increases in crop yields, declines in acreage planted and increases in forest cover. The effects of New Deal flood-control programmes, national forest purchases, price supports and acreage controls influenced farmers' decisions about the amount of land to cultivate and, in so doing, these programmes should have obscured the relationship between crop yields and the amount of cultivated land. Despite these dampening effects, crop yield increases did appear to facilitate forest recovery in the South. In the more neoliberal political environments of contemporary developing countries, one should observe a stronger relationship between changes in crop yields and acreage planted.

A second consideration would suggest that the Southern agricultural experience should provide ample evidence of a crop yield–acreage planted connection. The processes of industrialization in American metropolitan areas after 1939 created very large numbers of jobs, which pulled people off farms in a decisive way. When people left the farms in the 1930s, 1940s and 1950s, they usually found full-time employment and did not go back to the farm. The industrialization impulse in most countries in the tropical biome is weaker;

urbanization occurs, but the increase in the number of full-time jobs is smaller relative to the number of migrants than it was in the USA earlier in the century. Under these conditions of 'overurbanization', rural–urban migrants often retain a landholding in rural areas and continue to farm it for subsistence purposes. Because acreage devoted to subsistence cultivation should not be subject to the same crop yield–acreage planted dynamic as acreage producing commodities for the market, an increase in crop yields could produce a muted response in acreage planted, especially during difficult economic periods. People will continue to plant on marginal lands for security reasons, even after calculations of marginal productivity would suggest land abandonment. Because mid-century Americans had a viable economic alternative to agriculture in urban labour markets, they abandoned agriculture on marginal lands more readily when increases in crop yields increased the competitive pressures on marginal farmers. For this reason, we would expect to see a response to crop yield increases in the acreage planted in the American South; it did appear, but only on the marginal lands of the region.

One of the most incontrovertible findings from this investigation of the crop yield–acreage planted relationship involves the way in which the land capability variables mediate the relationship between increases in crop yields and trends in forest cover. The geography of soil fertility influences the elasticity of the acreage-planted variable in response to changes in crop yields. Figure 4.3 portrays this relationship. If fertile soils comprise only a small portion of a region, as in region A in Fig. 4.3, and as much as 67% of the region is cultivated, then an increase in crop yields would, by lowering the price of the agricultural commodity, put the farmers on marginal soils under such competitive pressure that they might decide to allow the land to revert to forest while they seek economic alternatives elsewhere. This land-abandonment response is especially likely if the government has imposed strict acreage restrictions on a particular crop, as, for instance, the American government did on tobacco. Under these circumstances, farmers only cultivate their best

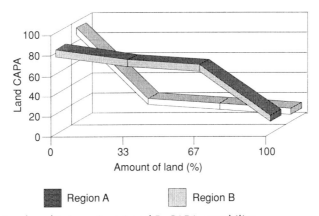

Fig. 4.3. Land quality in regions A and B. CAPA, capability.

lands. This sequence of events approximates what happened in Puerto Rico after 1950, when competitive pressures in global markets for coffee, sugar cane and tobacco, along with other factors, encouraged smallholders to abandon steep hillside farms. Forests increased from 10% to 37% of the land area on the island between 1950 and 1990 (Rudel *et al.*, 2000). In contrast, if fertile lands are distributed as in region B in Fig. 4.3, the increase in crop yields may not produce much of a decline in acreage planted. The large majority of producers are more hard-pressed under the new, more land-productive conditions, but few of them are forced out of the business because more capable lands are more widely distributed, so little land reverts to forest.

These scenarios conform to the familiar geography of forest transitions. As deforestation gives way to reforestation, the cultivated areas retreat to the lands with the most agricultural potential (Mather and Needle, 1998). Answers to questions about the likelihood of this type of transition in tropical biomes probably depend on the geographical distribution of land capability in tropical places. In her work on the Amazon basin, Betty Meggers (1996) drew a sharp distinction between the small area of fertile lands in the *varzea* and the large area of relatively barren lands in the *tierra firme*. Her critics (Whitehead, 1993) have argued for a more variegated understanding of Amazonian soil resources, implying that pockets of fertile land exist in many locales. A resolution of this debate about land capability in the tropics should give us a more precise idea about the conservation potential of increases in the yields per acre of tropical crops. The larger the differences between fertile and infertile areas in their productivity and the more limited the extent of the fertile areas, the more likely it is that increases in yields will produce significant conservation gains.

Notes

1   It would have been useful to include data on the extent of the conservation set-aside programme in each county, but I could not find these data.

2   I classified counties according to the land area planted in different cash crops at the time of the 1930 census. If a county had 45% of its cultivated area in cotton and 30% in maize – a common combination – it was classified as a cotton county. Quite frequently, more acreage was planted in maize than in any other crop, but much of this maize was being grown for subsistence purposes, to feed people or pigs. In those instances where a county reported a considerable amount of acreage (more than 10%) in a cash crop, such as tobacco, cotton or peanuts, the county was defined as a tobacco, cotton, or peanut county. If there was no important cash crop grown in the county and farmers in the county grew a great deal of maize, I defined it as a maize county.

3   Were black and white tenant farmers displaced by machines? Some were and some were not. The widespread adoption of tractors in Southern agriculture begins in 1935, and clearly many tenants evicted after that date were 'tractored out' (Aiken, 1998: 119–132). Black and white tenants begin leaving plantations during the 1920s, well before the introduction of tractors. Others were evicted in 1933/34 by plantation owners, who did not want to share price-support payments with tenants after the passage of the Agricultural Adjustment Act.

4 Later, farmers learned to use machinery on these lands, but only after introducing land improvements, such as terracing (Aiken, 1998: 118).

References

Aiken, C.S. (1998) *The Cotton Plantation South since the Civil War*. Johns Hopkins University Press, Baltimore.
Barnes, C.P. and Marschner, F.J. (1933) *Natural Land Use Areas of the United States*. Bureau of Agricultural Economics, US Department of Agriculture, Washington, DC.
Cruikshank, J.W. and Evans, T.C. (1945) *Approximate Forest Area and Timber Volume by County in the Carolinas and Virginia*. Forest Survey Release No. 19, Appalachian Forest Experiment Station, Asheville, North Carolina.
Eckstein, H. (1975) Case study and theory in political science. In: Greenstein, F. and Polsby, N. (eds) *Handbook of Political Science*, Vol. 7: *Strategies of Inquiry*. Addison-Wesley, Reading, Massachusetts, pp. 79–137.
Ferguson, H.B. (1940) *History of the Improvement of the Lower Mississippi River for Flood Control and Navigation, 1932–39*. Mississippi River Commission, Vicksburg, Mississippi.
Harrison, R.W. (1951) *Levee Districts and Levee Building in Mississippi: a Study of State and Local Efforts to Control Mississippi River Floods*. Mississippi Agricultural Experiment Station, Stoneville, Mississippi.
Hart, J.F. (1978) Cropland concentrations in the South. *Annals of the Association of American Geographers* 68(4), 505–517.
Hart, J.F. (1991) *The Land that Feeds Us*. W.W. Norton, New York.
Howe, C. and Easter, W.K. (1971) *Interbasin Transfers of Water*. Johns Hopkins University Press, Baltimore.
Johnson, C.S., Jones, L., Junker, B., Marks, E. and Valien, P. (1941) *Statistical Atlas of Southern Counties: Listing and Analysis of Socio-economic Indices of 1104 Southern Counties*. University of North Carolina Press, Chapel Hill.
Lee, L.B. (1980) *Reclaiming the American West: a Historiography and Guide*. ABC–Clio, Santa Barbara and Oxford.
McPhee, J. (1986) *The Control of Nature*. Farrar, Straus, and Giroux, New York.
Mather, A. and Needle, C. (1998) The forest transition: a theoretical basis. *Area* 30(2), 117–124.
Meggers, B. (1996) *Amazonia: Man and Culture in a Counterfeit Paradise*, revised edn. Smithsonian Institution Press, Washington, DC.
Odum, H.W. (1936) *Southern Regions of the United States*. University of North Carolina Press, Chapel Hill.
O'Neill, K.M. (1998) State building and the campaign for US flood control, 1924–1936. PhD dissertation, Department of Sociology, University of California-Los Angeles.
Rudel, T.K. (1995) Did TVA make a difference? An organizational dilemma and reforestation in the southern Appalachians. *Society and Natural Resources* 8(6), 493–508.
Rudel, T.K. and Fu, C. (1996) A requiem for the Southern regionalists: reforestation in the South and the uses of regional social science. *Social Science Quarterly* 77(4), 804–820.

Rudel, T.K. with Horowitz, B. (1993) *Tropical Deforestation: Small Farmers and Land Clearing in the Ecuadorian Amazon*. Columbia University Press, New York.

Rudel, T.K., Perez-Lugo, M. and Zichal, H. (2000) From fields to forests: reforestation and development in postwar Puerto Rico. *Professional Geographer* 52(3), 386–397.

Ruttan, V. (1984) The Tennessee Valley Authority in retrospect. In: Hargrove, E.C. and Conkins, P. (eds) *TVA: Fifty Years of Grassroots Bureaucracy*. University of Illinois Press, Urbana, pp. 150–163.

Sanchez, P.A. (1976) *Properties and Management of Soils in the Tropics*. Wiley, New York.

Scheuring, A.F. (1983) *A Guidebook to California Agriculture*. University of California Press, Berkeley, California.

Southgate, D. (1998) *Tropical Forest Conservation*. Oxford University Press, New York.

Sternitzke, H.S. and Christopher, J.F. (1970) Land clearing in the lower Mississippi valley. *Southeastern Geographer* 10(1), 63–66.

US Army Corps of Engineers (1989) *Water Resources Development in Louisiana*. Government Printing Office, Baton Rouge.

US Department of Agriculture (USDA) (1999) *National Agricultural Statistical System*. Archived at http://usda.mannlib.cornell.edu/.

Vance, R. (1932) *The Human Geography of the South*. University of North Carolina Press, Chapel Hill.

Whitehead, N.L. (1993) Ethnic transformation and historical discontinuity in native Amazonia and Guyana, 1500–1900. *L'Homme* 33, 126–128, 285–305.

World Resources Institute (1986) *Tropical Forest Action Plan*, 3 vols. WRI, Washington, DC.

A General Equilibrium Analysis of Technology, Migration and Deforestation in the Brazilian Amazon

5

Andrea Cattaneo

1. Introduction

This chapter seeks to determine how changes in policies and technology affect deforestation in the Brazilian Amazon and to identify strategies to reduce forest clearing. To do this, use is made of a computable general equilibrium (CGE) model, adapted to capture regional economic structures and the environmental processes specific to the tropics.

In the model, economic agents make decisions about production, trade, migration and investment. We assume that relative prices, factor availability, transportation costs and technology influence land use. Biophysical processes change land cover in concert with changes directly ensuing from decisions made by economic agents. We disaggregate agricultural production and other activities by region, sector and size of operation. A sector we call 'deforestation' produces an investment good called 'arable land', which is an input to agricultural production.

The chapter identifies the impact on deforestation of different forms of technological change in Amazon agriculture and compares it with the effects one would expect from: (i) technological change in agriculture outside the Amazon; (ii) a reduction in transportation costs arising from Amazonian infrastructure investments; and (iii) changes in the real exchange rate.

The forces underlying deforestation occur at various geographical scales and are linked to economic processes that range from macroeconomic policies to Amazon-specific conditions, such as technology and tenure regimes. CGE models constitute the best tool for comparing the relative magnitudes of the effects of these forces on deforestation. To fully understand these effects, it is

©CAB *International* 2001. *Agricultural Technologies and Tropical Deforestation*
(eds A. Angelsen and D. Kaimowitz)

important to specify the characteristics of agricultural production in each region, as well as the interplay between regions. We first look at technological change at the Amazon level, and then analyse the impact of interregional and macroeconomic processes. This allows us to demonstrate that, unless deforestation occurs for subsistence needs in an area isolated from markets, multiple and intertwined processes in the non-frontier part of the economy will greatly affect events on the agricultural frontier. To predict how policies will influence deforestation requires a clear understanding of the links between the two, something partial equilibrium analyses generally cannot provide.

At the level of the Amazon, one must analyse how possible technological innovations might affect specific agricultural activities. The short-run and long-run effects differ, as do the potential impacts of various factor-specific productivity changes. In the short run, factors of production are not very mobile and wages are rigid. In the long run, wages are flexible and labour and capital can move between regions. This implies that long-run scenarios that allow technological change in the Amazon to attract economic resources from other regions portray a fuller and, at times, counterintuitive picture of how technological innovation affects deforestation. The livestock sector provides a striking example of this. In the short run, all technological innovations embodied in labour and/or capital appear to both improve smallholder and large-farm incomes and reduce deforestation. Over the long run, innovation in the livestock sector still does the best job of improving incomes, but it also attracts resources from outside the Amazon, which can increase deforestation by up to 8000 km^2 year^{-1}.

The type of technological change alone does not determine whether deforestation increases or decreases. The factor intensities in the activity being improved and in the other activities also matter. In general, our results show that improvements in perennial crops, which already use both labour and capital intensively, reduce deforestation more than livestock improvements, since livestock require little labour per unit of land.

Technological innovation outside the Amazon can strongly affect Amazonian deforestation. If it occurs in a balanced manner across all agricultural sectors, deforestation rates should fall. But, if it changes how intensively producers use each factor, resources will shift around and the 'losing' factor will probably end up on the frontier. Balanced growth is unlikely. Technological innovation usually favours specific sectors and/or factors.

At the interregional scale, our model shows that reducing transportation costs considerably increases deforestation. This scenario is particularly relevant because public investments in roads, railways and waterways are rapidly lowering transportation costs in both the eastern and the western Brazilian Amazon. In the long run, reducing transportation costs by 20% would increase annual deforestation by approximately 8000 km^2. Transport costs affect deforestation a lot, because transportation is a major component of agricultural production costs in the Amazon. Therefore, infrastructure improvements affect the profitability of agriculture a great deal. As Amazonian agriculture becomes more profitable, the price of arable land increases and so does the incentive to deforest.

At the macroeconomic level, exchange-rate fluctuations reverberate through the economy by affecting relative prices. Given enough microeconomic detail, one can follow the effects of a macroeconomic shock throughout the economy, including, in our case, the regional agricultural and logging sectors. Our results indicate that migration between regions within Brazil greatly influences how macroeconomic shocks get transmitted to the agricultural frontier in the Amazon.

The chapter first provides background on the Amazon. Then it explains our modelling strategy and describes the database we used, before it presents the simulation results.

2. Regional Background

Since colonial times, Brazilians have settled new frontiers to obtain access to land and other natural resources. Macroeconomic policies, credit and fiscal subsidies to agriculture and technological change in agriculture have acted as push factors in the migration process. Meanwhile, policies such as road construction, colonization programmes and fiscal incentives to agricultural and livestock projects pulled economic resources towards the region (Binswanger, 1991). Rapid population growth, an economic context in which land is a valuable reserve, unequal income distribution and growing external markets for wood and agricultural goods may be other indirect sources of deforestation (Serrão and Homma, 1993). High transportation costs between the Amazon and the rest of the country, which lead to high agricultural input costs and limit interregional trade, also affect deforestation. Pfaff (1997) confirms this economic intuition by showing that Amazonian locations further from markets south of the Amazon have less deforestation.

In the 1990s, annual deforestation in the Brazilian Amazon ranged between 1,100,000 ha and 1,800,000 ha (with an anomalous peak of 2,900,000 ha in 1994/95). Whether smallholders or large farmers deforest more and whether their primary goal is to plant crops or install pasture remains open to debate. According to Homma *et al.* (1998), smallholders clear at least 600,000 ha each year, implying that they significantly contribute to deforestation. Others say commercial ranching contributes most to deforestation. The fact that the spread of small-scale agriculture may have caused some of the deforestation attributed to pasture expansion further complicates the issue (Mahar, 1988). Rapidly declining crop yields often lead farmers to convert land devoted to annual crops to pasture after a few years.

The 16 million inhabitants of the Brazilian Amazon, 61% of whom are urban, consume mostly local agricultural goods produced on both small and large farms. This implies that decisions regarding policies that affect deforestation rates must take into account their potential impact on regional food security and farmers' livelihoods.

3. Model Characteristics

The model used in this chapter builds on the approach Persson and Munasinghe (1995) applied in a study of Costa Rica. They included logging and squatter sectors and therefore markets for logs and cleared land. We extend their approach to include land degradation as a feedback mechanism into deforestation. The starting-point for the development of this model is a standard CGE model, as described in Dervis *et al.* (1982).[1]

The model centres on the role of land as a factor of production. If land has qualitative characteristics that economic agents perceive as distinct, these characteristics define distinct inputs in the production functions. Based on this type of perception, we divide land into: (i) forested land; (ii) arable land; and (iii) grassland/pasture. We define land transformation as a shift between land types resulting from biophysical processes associated with different land uses. Land conversion describes a change in land type that economic agents bring about intentionally. In the simulations presented below, we allow farmers: (i) to clear forest to obtain arable land; and (ii) to convert arable land into pasture.

The model's biophysical component determines the equilibrium stocks of each land type, given the land uses generated in our simulation scenarios. This represents a first step towards linking biophysical changes to the economic incentives for agents to modify their land use. Biophysical changes, such as soil and pasture degradation, greatly constrain regional development in the Amazon. We assume that they can be modelled as first-order stationary Markov processes that treat land use as exogenous (van Loock *et al.*, 1973; Baker, 1989). The results presented here rely on data collected through farm surveys by researchers from the International Food Policy Research Institute (IFPRI) in Acre and Rondônia.

3.1. Representation of production

Table 5.1 presents the activities the model includes, along with the commodities these activities produce and the factors employed in production.

As noted earlier, we disaggregate agricultural production by region (Amazon, centre-west, north-east, rest of Brazil), activity (annual crops, perennial crops, animal production, forest products and other agriculture) and size of operation (smallholder, large farm enterprise). All factors employed by agriculture are region-specific. We use two-level production functions for sectors that have both activities and individual commodities and assume that the two levels are separable, so that each agricultural activity can produce various agricultural commodities.

Table 5.1. Activities, commodities and factors included in the model.

Activity	Commodities produced	Factors used
Annual crops	Maize, rice, beans, cassava, sugar, soybean, horticultural goods and other annual crops	Arable land, unskilled rural labour, skilled rural labour, agricultural capital
Perennial crops	Coffee, cacao, other perennial crops	Arable land, unskilled rural labour, skilled rural labour, agricultural capital
Animal products	Milk, livestock and poultry	Grassland, unskilled rural labour, skilled rural labour, agricultural capital
Forest products	Non-timber tree products, timber and deforested land for agriculture	Forest land, unskilled rural labour, skilled rural labour, agricultural capital
Other agriculture	Other agriculture	Arable land, unskilled rural labour, skilled rural labour, agricultural capital
Food processing	Food processing	Urban skilled labour, urban unskilled labour, urban capital
Mining and oil	Mining and oil	
Industry	Industry	
Construction	Construction	
Trade and transportation	Trade and transportation	
Services	Services	

The way we specify production activities takes into account the fact that farmers consider certain agricultural commodities substitutes and others complements. Our technological specification captures both price responsiveness, through own-price elasticities, and the technological constraints that limit the possibilities of shifting agricultural output from one commodity to another, through substitution elasticities. We obtained the values for these elasticities from a survey we conducted of IFPRI and Empresa Brasileira de Pesquisa Agropecuaria (EMBRAPA) researchers who are familiar with production processes in Brazilian agriculture. Table 5.2 presents the results.

Except where we have information to the contrary, we assume that producers can easily substitute one commodity with another and we follow the linear programming farm model approach, which assumes that farmers shift production to the most profitable commodity. If, on the other hand, the experts we surveyed believed that farmers consider factors besides prices when making

Table 5.2. Production technology: substitutability between commodities.

Technology	Commodity 1	Commodity 2	Substitutability
Annual-crop production	Maize	Rice, beans	Low
	Maize	Cassava	Low–medium
	Maize	Sugar, soybean, horticulture and other annual crops	Medium–high
	Rice	Beans	Low
	Rice	Cassava	Low–medium
	Rice	Sugar, soybean, horticulture and other annual crops	Medium–high
	Beans	Cassava	Low–medium
	Beans	Sugar, soybean, horticulture and other annual crops	Medium–high
	Cassava	Sugar, soybean, horticulture and other annual crops	Medium
	Sugar	Soybean, horticulture and other annual crops	High
	Horticultural goods	Other annual crops	Medium–high
Perennial-crop production	Coffee	Cacao	High
	Coffee	Other perennial crops	Medium
	Cacao	Other perennial crops	Medium–high
Animal products	Livestock	Milk	Medium
	Poultry	Livestock, milk	Medium–high
Forest products	Deforested land (agric.)	Timber	Low–medium
	Deforested land (agric.)	Non-timber tree products	High
	Non-timber tree products	Timber	High

their decisions about what to produce, then we set the substitution elasticities lower. In this process, we considered non-price factors, such as: (i) relative risk associated with the crops; (ii) subsistence requirements; (iii) crops that require different soil characteristics; (iv) common practice (habit); and (v) whether farmers typically grow the two crops together (intercrop), in which case they have difficulty substituting one for the other.

3.2. Demand for deforested land

The demand for agricultural land determines the price of arable land. If the economic agents act as if they had an infinite time horizon, in equilibrium the return from an asset per unit of time divided by the asset's price must equal the rate of interest. This implies that the land rental rate and producers' discount rate determine the price of the arable land produced by the deforestation sector. If farmers lack secure property rights over their land, one can adjust the discount rate to take into account the risk that they might lose it.

Agricultural productivity influences rental rates. Based on our knowledge of the area, we assume that, over time, arable land degrades and becomes grassland, which farmers can only use for pasture. Since this affects productivity, it also affects the rental rate.

Squatters deforest to supply arable land. They decide how much land to deforest based on the price of arable land, their profit-maximizing behaviour and technology. How they behave depends in part on whether forests are open-access resources or have well-defined property rights that govern their use. In this chapter, we assume that forest is an open-access resource. By assuming that farmers have an infinite planning horizon when they use arable land produced by clearing forest, we implicitly allow squatters to acquire property rights through deforestation.

While a broad consensus exists that the expansion of cropped area and pasture constitutes a major source of deforestation, no similar consensus has emerged about logging. In some contexts, it appears to directly cause deforestation and, in others, to indirectly facilitate farmers' access to forested areas (Uhl and Vieira, 1989; Eden, 1990; Burgess, 1993). In this chapter, we assume that squatters sell arable land to whatever agricultural entity is expanding and that logging does not directly cause deforestation but does facilitate land clearing.

4. Data, Assumptions and Limitations of the Model

We drew the data used in this model from Cattaneo (1998). To construct the social accounting matrix, we originally used the 1995 input–output table for Brazil (IBGE, 1997a) and the national accounts (IBGE, 1997b). We then integrated these sources with the agricultural census data for 1995/96 (IBGE, 1998) to yield a regionalized representation of agricultural activities. We obtained household data from the national accounts and household income and expenditure surveys (IBGE, 1997c, d). We allocated total labour, land and capital value across agricultural activities based on the proportions reflected in the agricultural census. We disaggregated labour into agricultural and non-agricultural labour and further differentiated between skilled and unskilled labour. We allocated part of the gross profits from agriculture to land, based on the return to land being used by the activity (FGV, 1998), and the remainder to capital. All producers maximize profits, subject to their factor endowments and available technology.

We estimated regional marketing margins by calculating the average distance to the closest market and multiplied the ratio of these values relative to the industrial South by the trade and transportation coefficients of each agricultural sector, obtained from transportation cost surveys (SIFRECA, 1998).

We assume deforestation (in hectares) in 1995 equal to average deforestation between 1992 and 1996. The coefficients for the technology used to deforest come from Vosti et al. (1999), and timber production figures from the agricultural census. We based our estimates of the economic rent from timber on a specification proposed by Stone (1998). The elasticities of substitution between production factors for industry came from Najberg et al. (1995). For agriculture, we set the elasticity of substitution between land and capital at 0.4 for smallholders and 0.8 for large farm enterprises. These are judgement-based estimates, which assume that large farmers can substitute more easily between factors. As mentioned previously, we obtained the substitution elasticities for shifting between agricultural commodities from surveys. On the biophysical side, we assumed that arable land sustains annual crop production for 4 years before being transformed into pasture/grassland. Pasture/grassland can sustain livestock for 8 years before degrading the land completely. This implies that, on average, biophysical processes transform 25% of the arable land in annual crops and 12.5% of pastureland each year to other land-use categories.

The data and the model formulation have several limitations. Given the uncertainty surrounding the elasticities, one can only use the simulation results to provide insights into the sign and order of magnitude of the effects and should not interpret them as precise quantitative measures. Although the values we use to assess the impact of technological changes express a reasonable range of possible changes, they are not based on case-studies. Our model is essentially static and the results represent the impact of policy experiments in a timeless world. This chapter considers the two extremes: no factor mobility and perfect mobility. Reality will probably be somewhere in between; therefore these results are meant to give a qualitative representation only.

5. Simulations

Researchers have devoted a great deal of attention to the localized aspects of technological change in agriculture and cattle raising in the Amazon (Serrão and Homma, 1993; Mattos and Uhl, 1994; Almeida and Uhl, 1995; Toniolo and Uhl, 1995). They have shown particular interest in variables such as profitability, credit requirements, sustainability and other factors that determine whether farmers adopt specific technologies. This chapter examines the impacts, at the Amazon basin level, of technological changes that modify the structure of a producing sector as a whole.[2] We assume that technological change is exogenous.

We simulate technological change in annual-crop, perennial-crop and animal production. For each activity, we analyse different types of embodied technological change that increase the productivity of distinct productive factors. We also have a reference run, where we increase total factor productivity (TFP) (disembodied technological change) by up to 70% in 10% increments. To ensure that the technological changes analysed in the factor-specific cases are of the same magnitude as in the TFP case, we make the size of the factor productivity increase inversely proportional to the factor's value share in production. Table 5.3 shows the types of technological change used in the simulations.

In our short-run simulations (1–2 years), we confine agricultural labour and capital to the region where they are currently located. In the long-run simulations (5–8 years), we allow the two factors to migrate between regions. We present results concerning terms of trade for Amazon agriculture, factor rental rates, deforestation rates and value added by smallholders and large farm enterprises. Dividing value added between small and large farms serves as a proxy for regional income distribution. It also suggests which types of technological change each kind of producer is more likely to adopt. Due to space limitations, we present only short-run results for value added. Value-added shares provide a good proxy for income distribution in the short run, because migration is not allowed.

5.1. Improving annual-crop technology in the Brazilian Amazon

In the short run, making annual crops more productive may increase or decrease deforestation, depending on the type of technological change. The TFP case, in which the productivity of each factor increases the same amount, leads to the greatest deforestation, followed closely by capital-intensive technological change (CAP_INT). The reason these two forms of innovation have the strongest push towards deforestation is that arable land appreciates considerably as a consequence of the productivity improvement. To achieve a

Table 5.3. Types of technological change.

Abbreviation	Name	Comments
TFP	Total factor productivity increase	Disembodied technological change. No partial equilibrium effect on factor ratios
LAB_INT	Labour productivity increase	Improves labour productivity: attracts labour
CAP_INT	Capital productivity increase	Improves capital productivity: attracts capital
LABCAP	Labour and capital productivity increase	Improves labour and capital productivity: attracts both factors
DG_LBK	Labour and capital productivity increase with decreased land degradation	Same as above but also reduces the degradation rate by 10% at each step

technological change of a magnitude similar to that simulated in the TFP case in the CAP_INT scenario, capital productivity must improve a great deal, due to the very low capital intensity of annual-crop production in the Amazon.

In the 'labour-intensive' (LAB_INT) and 'labour- and capital-intensive' (LABCAP) cases, the improvement in labour productivity in annual-crop production shifts labour from livestock to annual-crop production. This lowers demand for pastureland, since it is not suited for producing annual crops, and this, in turn, depresses its price. This dampens the rise in arable land prices because, after a few years, arable land degrades into pastureland, which is now worth less. This makes it less attractive to deforest. Once the technological change passes a certain threshold, this effect becomes large enough to significantly reduce deforestation. In the simulation presented here, this threshold is 20% in TFP terms (TFP index = 2) (Fig. 5.1).

In the long run, allowing labour and capital to migrate between regions dramatically changes the result. Technological improvement in annual crop production encourages deforestation, unless farmers widely adopt highly labour- and capital-intensive technologies, and, even if that happens, for smaller technological changes (TFP index 1–3) deforestation still increases. The LAB_INT scenario is particularly interesting, given that it appeared quite

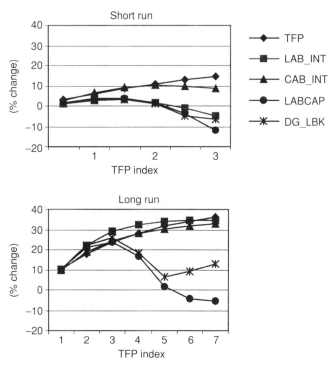

Fig. 5.1. Change in deforestation resulting from technological change in annual-crop production.

promising in the short run. Now annual crops attract labour and capital from outside the Amazon, making arable land the only scarce resource. This drives up the value of arable land. The shift of labour and capital from cattle raising to annual-crop production no longer depresses pasture prices much, since a large portion of the resources that move into annual crops comes from other regions and no longer has to be diverted from other Amazon agricultural activities.

The labour- and capital-intensive scenario (LABCAP) performs well, in deforestation terms, in the higher range of the TFP index. We attribute this in part to the finite amount of rice, manioc and beans that the national market can absorb from the Amazon. When farmers adopt this technology, land availability no longer really constrains production, which increases until the terms of trade seriously deteriorate. The resulting low prices reduce migration into the Amazon. Adjustment outside the Amazon to the growth in annual-crop production also affects the terms of trade for livestock, lowering the return to pastureland and hence the incentive to deforest.

The combination of improving the sustainability of annual-crop production and more labour- and capital-intensive technology (DG_LBK) proves interesting in the long run. Two countervailing processes come into play. Less degradation increases the stock of available arable land and that reduces the demand for deforestation. At the same time, more sustainable agriculture implies that farmers can obtain high revenues from growing annual crops for a longer period of time, increasing the demand for arable land. In the simulation presented here, the first effect is minimal. For TFP indices higher than 4, the second effect clearly dominates.

Given that annual-crop production is labour-intensive, improving labour productivity clearly increases welfare, particularly for smallholders. In fact, it is the only type of technological change in annual-crop production that improves smallholders' condition. This occurs because capital markets are segmented. Smallholders lack access to the capital they would need to adopt more capital-intensive technologies. Therefore, large farm enterprises, which have access to capital, capture most of the gains from new capital-intensive technologies (CAP_INT, LABCAP and DG_LBK). Labour-intensive technologies also considerably improve large farms' value added, since they can hire off-farm labour (Fig. 5.2). But the best option for these enterprises is labour- and capital-intensive innovation (LABCAP and DG_LBK).

5.2. Improving perennial-crop technology in the Brazilian Amazon

With only a few exceptions, increasing perennial-crop productivity reduces deforestation in both the short and long run (Fig. 5.3). In the short run, capital and labour shift from annual crop and livestock production to perennial crops. Perennial crops use labour and capital much more intensively than annual crops. This implies that, when perennial crops draw resources from other agricultural activities, the overall demand for arable land declines. Farmers

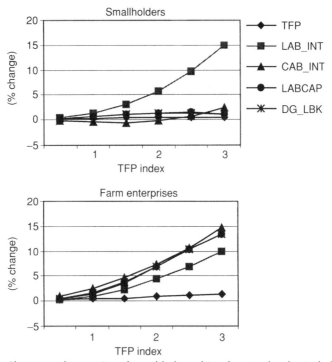

Fig. 5.2. Short-run changes in value added resulting from technological change in annual-crop production.

actually decrease their annual-crop production so much that they decide to convert some of their arable land to pasture, and this depresses pasture prices as well. Deforestation also declines because, unlike annual crops, perennials do not transform arable land to grassland. Thus, the stock of available arable land grows, which reduces the demand for deforestation.

In the short run, improvements in TFP, where factor productivity increases equally across all factors, barely affect deforestation. The effect of the increase in land productivity, which raises the return to arable land, just about offsets the decline in demand for arable land stemming from the factors mentioned above. In contrast, all the technologies that increase labour and/or capital intensity substantially lower deforestation. Given the great differences in the effects of technological changes that increase labour and capital intensity compared with those produced by improvements in TFP, it is important to understand the differences between these two forms of innovation. In the first case, the amount of capital and labour farmers apply to each unit of land increases. An example might be a coffee variety that leads farmers to plant more trees per hectare and use more labour to care for them and harvest the coffee. A typical TFP improvement might be a new marketing strategy that helps farmers get higher prices for their coffee but does not alter the factor intensity of production.

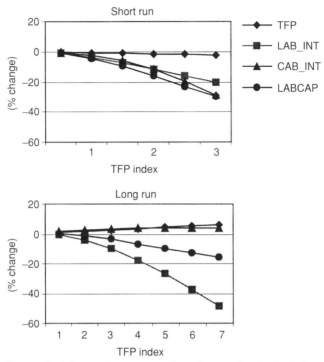

Fig. 5.3. Change in deforestation rates resulting from technological change in perennial-crop production.

In the long run, the results are still encouraging for perennial crops. However, the type of technological change affects the outcomes more. Labour-intensive innovation reduces deforestation even more, because migration allows producers to shift even more from annual-crop to perennial-crop production. The story in regard to technologies that increase both labour and capital intensities changes slightly. Once we allow migration, there is no longer any surplus arable land for farmers to use as pasture. In fact, arable land increases in value. However, deforestation still declines, thanks to the dampening effect of lower returns to pastureland, due to factors shifting towards perennial-crop production. This dampening effect also shows up in the TFP and the capital-intensive scenarios. But it is too small to offset the prospect of higher returns from arable land, so deforestation increases.

In the short run, small farms appear to gain more income than large farm enterprises by shifting their production towards perennials in response to technological improvements in that activity. In part, this occurs because smallholders already produce most of the perennial crops in the Amazon ($620 million compared with $130 million on large farms). However, our results may overstate the potential gains for smallholders, because our framework does not take into account the fact that smallholder capital in

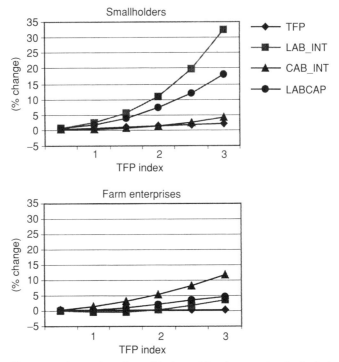

Fig. 5.4. Short-run change in value added resulting from technological change in perennial-crop production.

perennial crops consists mainly of trees, which, in the case of technological change, may have to be replaced for the productivity improvement to occur.

To summarize, labour-intensive change is the best option for smallholders, because of their capital constraints. Conversely, capital-intensive technological change is best for large farmers.

5.3. Improving livestock technology in the Brazilian Amazon

Some researchers claim that pasture improvements in the Amazon will reduce deforestation by allowing production systems to use land more intensively (Mattos and Uhl, 1994; Arima and Uhl, 1997). These authors appear to take a short-term view, but do not take into consideration the long-term effects of a more profitable ranching sector in the Amazon. In the short run, all technological improvements, except an increase in TFP, reduce deforestation (Fig. 5.5). But this does not hold true in the long run.

If we do not allow labour or capital migration, it is straightforward to understand what happens. As the livestock sector becomes more profitable, farmers use some of their arable land as pasture. In fact, with a TFP index equal to 3, farmers demand 70–80% less arable land in all the scenarios except the

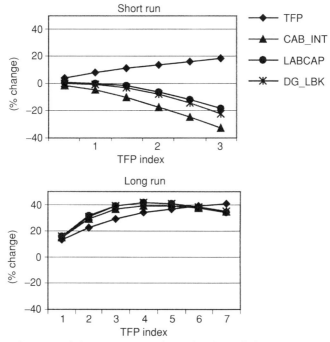

Fig. 5.5. Change in deforestation rates for technological change in animal production.

TFP case. Here too, our results may overstate reality, since we do not consider farmers' food-security constraints and we assume capital is mobile inside both large and small Amazon farms. In reality, the herd embodies capital in the livestock sector and it has a natural growth rate that farmers cannot easily adjust in the short run.

In the long run, the improvements in livestock technology attract resources from outside the Amazon and farmers deforest more to meet the increased demand for pasture. Surprisingly, not only does the return to pastureland increase substantially, but so does the price of arable land. This occurs because annual-crop production degrades the land, which subsequently becomes grassland/pasture. Since owning pasture becomes more attractive, farmers demand more arable land with the expectation that they will use it as pasture in the future. In fact, in all the long-run scenarios, annual crop production increases alongside that of livestock (although at a lower rate). Perennials, which are also produced on arable land but do not cause degradation, do not expand, and may even contract. In all scenarios, improving livestock productivity in any way substantially increases long-run deforestation.

From a farmer's perspective, improved livestock technologies are their highest priority. All farmers in the Amazon would receive extremely high returns from capital-intensive or labour- and capital-intensive technological innovations, compared with those from improvements in annual crops or perennials (Fig. 5.6).

TFP improvements would also provide significant but less pronounced returns. To come back to a familiar theme, improving the productivity of the intensive factor for an activity is bound to make that activity expand more. Since labour scarcity greatly constrains production in the Amazon, livestock, which require little labour and are highly capital-intensive, are a very attractive option, and that is one reason why there are well established in the region. The small wage change for unskilled labour associated with technological change in livestock (+14% for TFP index = 3 in LABCAP for livestock, compared with 47% for the same type of change in annual crops) reflects the highly capital-intensive nature of the former activity.

5.4. Summary of impacts of regional technological change

In summary, technological change in perennial crops offers the best option, in regard to both deforestation and income distribution. However, technological improvements in livestock provide the greatest income gains for both small and large farms. This creates a dilemma, because any technological

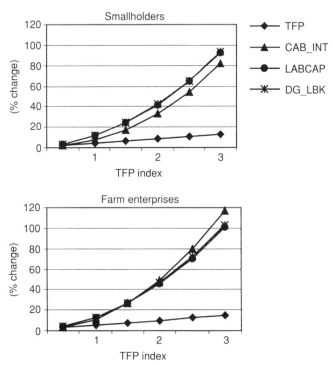

Fig. 5.6. Short-run change in value resulting from technological change in animal production.

improvement in livestock encourages long-run deforestation. Improving annual-crop production, while possible in certain parts of the Amazon, would probably stimulate deforestation, while only increasing returns about as much as would a technological improvement in perennials. Therefore, this alternative does not appear particularly appealing.

When analysing the possible impact of technological innovations, one must bear in mind that short- and long-run effects differ, as do the effects of different types of technological change. TFP scenarios always favour deforestation most, due to the higher returns to land compared with innovations that shift factor intensities towards capital and labour. Innovations that increase the intensities of both labour and capital reduce deforestation in all scenarios except long-run livestock scenarios, in which case they lead to some of the highest deforestation rates observed in our simulations.

5.5. Comparing regional technological change and interregional effects

A diverse set of nationwide phenomena indirectly encouraged deforestation in the Amazon. Here we simulate the effects of three changes outside the Amazon that have immediate policy relevance to the deforestation debate: (i) a technological change in annual production in centre-west, south and south-east Brazil; (ii) a 20% reduction in transportation costs; and (iii) a 30% devaluation of the real exchange rate.

Policy-makers should take an interest in how technological changes outside the Amazon affect deforestation there, because of both past events and what may happen in the future. Some argue that changes in agricultural technology in other regions of Brazil stimulated large-scale migrations to the Amazon frontier in the 1960s and 1970s. Our simulation captures the essence of what has happened with the recent expansion of soybean production, due in part to improved technologies (discussed further in Chapter 11 by Kaimowitz and Smith in this volume). Schneider (1992) observes that, over the last 15 years, livestock producers have sold off their land to soybean producers and moved their livestock operations to frontier areas. Our simulation lends credence to that claim. Soybean farmers use a high-input, capital-intensive production system that can be stylized as improving both labour and capital productivity for annuals production. Our results indicate that combined labour and capital productivity improvements in annuals production outside the Amazon would lead to an increase of up to 10% in the deforestation rate (LBCAP_AN in Fig. 5.7). On the other hand, if the technological innovation had been purely labour productivity improving, deforestation rates would have increased by up to 20% (LABIN_AN in Fig. 5.7) because agricultural capital would have been pushed towards the Amazon leading to further expansion of large from livestock production.

According to our simulation results, 'balanced' technological change outside the Amazon, where all factors become more productive in all agricultural

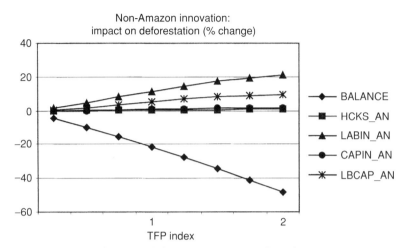

Fig. 5.7. Long-run change in deforestation rates resulting from technological change outside the Amazon.

sectors, reduces deforestation the most (BALANCE in Fig. 5.7). This option slows deforestation most effectively because the technological change involved does not push any factor or activity into the frontier.

The policy relevance of the reduction in transportation costs stems from infrastructure investments, which may considerably modify the incentives that share current land-use patterns in the area. The Brazilian government is currently constructing a road through the Amazon to the Pacific and a port facility in Rondônia to reduce transport costs for soybeans and other goods produced in the region. In all our scenarios, a reduction in transportation costs between the Amazon and the rest of Brazil increases deforestation. A 20% reduction in transportation costs for all agricultural products from the Amazon increases deforestation by 33%.

The real devaluation simulation has special relevance, given Brazil's large devaluation in January 1999. The widespread rumour that states might default on their debt to the central government sent foreign investors fleeing from Brazilian capital markets and forced the government to float the exchange rate. This resulted in a 70% peak nominal devaluation over the following 3 weeks. We simulated the possible effects of a 20–40% real devaluation, on the assumption that, once the currency and inflation stabilize, the real devaluation will probably be around that level. Real exchange-rate fluctuations reverberate through the economy by affecting the relative prices of goods. Prices of export goods rise relative to non-traded outputs produced domestically, such as services and housing, and production correspondingly shifts toward export sectors. General equilibrium frameworks have the advantage of considering all of these processes simultaneously. Our results indicate that devaluation would stimulate logging and this leads to greater deforestation for agriculture. On the other hand, devaluation also affects returns to agriculture in the different regions. What effect this might have on deforestation depends a great deal on whether one assumes that labour can migrate between regions. When we allowed only rural labourers to move between regions, our model predicted that a 30% devaluation would decrease deforestation by 5%. But, when we assumed that even urban labour was willing to migrate to the Amazon in search of rural employment, we obtain a 35% increase in deforestation.

6. Conclusions

The recent changes in exchange rate and transportation costs will probably increase deforestation. However, policy-makers can influence technological change and it is encouraging to note that, if policy-makers carefully choose the technological changes they support, this could reduce deforestation by about as much as we expect the devaluation and infrastructure investments to increase it. Table 5.4 below summarizes our findings.

We base our food-security conclusions on our personal judgement concerning the production structure after technological change occurs. We assumed that, if farmers specialize in activities with small regional (Amazon) markets or with volatile prices, food security decreases. If, on the other hand,

Table 5.4. A qualitative comparison of the impacts of Amazon technological change.

Type of technological change		Deforestation reduction SR	Deforestation reduction LR	Smallholder income (SR)	Large-estate income (SR)	Food security (SR and LR)
Total factor productivity	Annuals	--	----	0	+	++
	Perennials	+	-	+	0	-
	Livestock	--	----	++	++	+
Labour productivity	Annuals	+	----	++	+	+
	Perennials	++	++++	+++	+	--
Capital productivity	Annuals	-	----	0	++	+
	Perennials	++	+	0	++	-
	Livestock	+++	----	++++	++++	+++
Labour and capital productivity	Annuals	+	-	0	++	++
	Perennials	+++	++	++	+	---
	Livestock	++	----	++++	++++	++++
Sustainability + Labour and capital productivity	Annuals	+	--	0	++	++
	Livestock	++	----	++++	++++	++++

SR, short run; LR, long run; + implies a desirable effect; – implies an undesirable effect.

production of Amazon staples rises, regional food security improves. According to this criterion, innovation in livestock production, which increases both annual-crop and livestock production, scores the highest. Technological change in annual crops is also a good food-security option because the production of staples, such as cassava and rice, greatly increases, without adversely affecting livestock. In our classification, we considered perennials risky. Perennial-crop production only declines dramatically when large numbers of farmers adopt much more labour-intensive technologies for the production of annuals. This may decrease perennial-crop production by more than 50% for high levels of technological adoption. Technological innovation in perennial crops leads to specialization in perennial crops and substantial reductions in the production of annual crops and livestock. In the long run, the scenario with technology intensive in both labour and capital reduces annual crops by 20–25% and livestock by 30–40% for high levels of technological adoption.

Table 5.4 points to a significant trade-off between forest-conservation objectives and agricultural growth. Livestock technology improvements provide the greatest returns for all agricultural producers in the Amazon and improve regional food security, but long-run deforestation increases dramatically.

The best alternative would be to pursue improvement in perennial-crop technologies, especially those that are labour-intensive, which could reduce deforestation considerably. Small farmers would gain the most from such technologies. However, food security would suffer and farmers would be more exposed to the risks associated with perennials. Although this option theoretically has potential, non-adoption by large farms (which would have small gains), combined with the risk-averseness of smallholders, would probably limit its effectiveness. None the less, even if adopted only in part, it would still contribute to reducing deforestation.

Improvement in production of annual crops appears to have little potential. In the long run, it would reduce deforestation only if farmers adopted very labour- and capital-intensive technologies and the income effects would be quite small. Before labour and capital intensities got sufficiently high to decrease deforestation, there would almost certainly be, in the early phase of adoption, a period in which forest clearing would rise substantially.

The type of factor intensification alone does not determine whether deforestation rates will increase or decrease. The factor intensities in the activity being improved and in the other activities also matter. Furthermore, the striking difference in deforestation rates between the short run and the long run points to the fact that interregional flows of labour and capital play a crucial role in determining the expansion of the agricultural frontier. Along these same lines, we have seen that processes occurring outside the Amazon can have a strong impact on deforestation. Technological change can reduce deforestation if it occurs in a balanced manner across all agricultural sectors outside the Amazon. However, this is unlikely and, if the innovation is intensive in any specific factor, resources will be moved around and the 'losing' factor will probably end up on the frontier.

Very important interregional transportation links allow for the transmission of economic effects between the Amazon and other regions. The ongoing reduction of transport costs could dramatically increase deforestation. Finally – the last result – a macro shock, in the form of a 40% devaluation, was very sensitive to the migration flows allowed in the model, ranging from a 5% decrease to a 35% increase. Understanding the determinants of capital and labour flows would be a major empirical undertaking, but well worth the effort.

Notes

1   A detailed description of the model is available from the author.
2   At the sectoral level, different levels of technological change can reflect either the magnitude of the changes associated with the new technology on each farm or the number of farmers who adopt the technology. For example, if all producers adopt an innovation that improves total factor productivity (TFP) by 50%, in our framework this is the same as a technology that improves farm-level TFP by 100% but which only 50% of the farmers adopt.

References

Almeida, O.T. and Uhl, C. (1995) Developing a quantitative framework for sustainable resource-use planning in the Brazilian Amazon. *World Development* 23(10), 1745–1764.
Arima, E.Y. and Uhl, C. (1997) Ranching in the Brazilian Amazon in a national context: economics, policy, and practice. *Society and Natural Resources* 10, 433–451.
Baker, W.L. (1989) A review of models of landscape change. *Landscape Ecology* 2(2), 111–133.
Binswanger, H.P. (1991) Brazilian policies that encourage deforestation in the Amazon. *World Development* 19(7), 821–829.
Burgess, J.C. (1993) Timber production, timber trade, and tropical deforestation. *Ambio* 22(2–3), 136–143.
Cattaneo, A. (1998) The interaction between economic incentives, deforestation, and land degradation in Brazil. In: *The Impact of Macroeconomic Policy on Deforestation: a Comparative Study of Indonesia and Brazil*. Progress Report, International Food Policy Research Institute, Washington, DC.
Dervis, K., de Melo, J. and Robinson, S. (1982) *General Equilibrium Models for Development Policy*. Cambridge University Press, Cambridge.
Eden, M.J. (1990) *Ecology and Land Management in Amazonia*. Belhaven Press, London.
FGV (1998) Preços de terra (ARIES on-line database: www.fgv.br/cgi-win/aries.exe). Fundação Getulio Vargas, Rio de Janeiro.
Homma, A.K.O., Walker, R.T., Scatena, F.N., de Conto, A.J., de Amorim Carvalho, R., Palheta Ferreira, C.A. and dos Santos, I.M. (1998) Reducao dos desmatamentos na

Amazonia: politica agricola ou ambiental. In: Homma, A.K.O. (ed.) *Amazonia: Meio Ambiente e Desenvolvimento Agricola*. EMBRAPA, Brasilia.

IBGE (1997a) *Matriz de Insumo-produto Brasil 1995*. Instituto Brasileiro de Geografia e Estatistica, Rio de Janeiro.

IBGE (1997b) *Sistema de Contas Nacionais Brasil 1990–1995/96*. Instituto Brasileiro de Geografia e Estatistica, Rio de Janeiro.

IBGE (1997c) *Pesquisa Nacional por Amostra de Domicílios – 1996*. Instituto Brasileiro de Geografia e Estatistica, Rio de Janeiro.

IBGE (1997d) *Pesquisa de Orçamentos Familiares 1995/1996*. Instituto Brasileiro de Geografia e Estatistica, Rio de Janeiro.

IBGE (1998) *Censo Agropecuário 1995/1996*. Instituto Brasileiro de Geografia e Estatistica, Rio de Janeiro.

Mahar D. (1988) *Government Policies and Deforestation in Brazil's Amazon Region*. Environment Department Working Paper No. 7, World Bank, Washington, DC.

Mattos, M.M. and Uhl, C. (1994) Economic and ecological perspectives on ranching in the eastern Amazon. *World Development* 22(2), 145–158.

Najberg, S., Rigolon, F. and Vieira, S. (1995) *Modelo de Equilibrio Geral Computavel Como Instrumento de Politica Economica: uma Analise de Cambio e Tarifas*. Texto para Discussao n. 30, Banco Nacional de Desenvolvimento Economico e Social, Rio de Janeiro.

Persson, A. and Munasinghe, M. (1995) Natural resource management and economywide policies in Costa Rica: a computable general equilibrium (CGE) modeling approach. *World Bank Economic Review* 9(2), 259–285.

Pfaff, A.S. (1997) *What Drives Deforestation in the Brazilian Amazon? Evidence from Satellite and Socioeconomic Data*. Policy Research Working Paper No. 1772, World Bank, Washington, DC.

Schneider, R. (1992) *Brazil: an Analysis of Environmental Problems in the Amazon*. Report No. 9104 BR, Vol. I, World Bank, Washington, DC.

Serrão, E.A.S. and Homma, A.K.O. (1993) Country profiles: Brazil. In: US National Research Council (eds) *Sustainable Agriculture and the Environment in the Humid Tropics*. National Academy Press, Washington, DC.

SIFRECA, (1998) *Sistema de Informações de Fretes para Cargas Agrícolas: Soja (9/97)*. Escola Superior de Agricultura 'Luiz de Queiroz' (ESALQ), Piracicaba-SP, Brazil.

Stone, S.W. (1998) Evolution of the timber industry along an aging frontier: the case of Paragominas (1990–95). *World Development* 26(3), 433–448.

Toniolo, A. and Uhl, C. (1995) Economic and ecological perspectives on agriculture in the eastern Amazon. *World Development* 23(6), 959–973.

Uhl, C. and Vieira, I.C.G. (1989) Ecological impacts of selective logging in the Brazilian Amazon: a case study from the Paragominas region in the state of Para. *Biotropica* 21, 98–106.

van Loock, H.J., Hafley, W.L. and King, R.A. (1973) Estimation of agriculture–forestry transition matrices from aerial photographs. *Southern Journal of Agricultural Economics* December, 147–153.

Vosti, S., Witcover, J. and Line Carpentier, C. (1999) *Agricultural Intensification by Smallholders in the Western Brazilian Amazon: from Deforestation to Sustainable Land Use*. Draft research report, International Food Policy Research Institute (IFPRI), Washington, DC.

Will Intensifying Pasture Management in Latin America Protect Forests – or Is It the Other Way Round?

Douglas White, Federico Holmann, Sam Fujisaka, Keneth Reategui and Carlos Lascano

1. Introduction[1]

Cattle in tropical Latin America have dual identities. To farmers, they represent status and stable incomes. To environmentalists, they constitute a chewing and belching nemesis that destroys forests and the atmosphere. These two views provoke a spirited debate about whether economic development conflicts with environmental preservation. At the centre of the dispute lies the issue of how advances in livestock and pasture technology influence deforestation rates.

Since markets value forested land modestly in much of tropical Latin America, a private farmer's perspective of raising cattle extensively by converting additional forest for pastures appears perfectly rational. This certainly applies at present to the forest margins of the Amazon. However, in more developed regions with older forest margins in Central and South America, farmers tend to produce livestock more intensively to avoid pasture degradation and the high cost of expanding on to uncultivated land. Thinking about this second type of situation made us realize that we may have our initial research question backwards. Perhaps instead of asking whether pasture intensification increases or decreases deforestation, we should focus on how deforestation influences pasture intensification. From there emerged the unfortunate alternative hypothesis that forest scarcity is a prerequisite for technology intensification.

In a sense, the inspiration for our hypothesis comes from Boserup's early work (1965), which argued that few farmers would intensify their production as long as they could still expand extensively. We reached a similar conclusion, but through a different process. Boserup based her argument on the link between population growth and technological change. We emphasize the more general effects of land and other factor prices on farmers' decisions to adopt intensive or extensive land-use options.

As market access improves and available forest land becomes scarcer, land prices generally rise. Similarly, areas with incipient markets and abundant forests tend to have cheaper land. If land is expensive, farmers will look for ways to increase production that use land more intensively. This led to our second, related, hypothesis that more intensive technologies will only help maintain forest cover if they are a less expensive option than extensive growth.

If our two hypotheses prove to be true, research should focus less on how intensification affects deforestation and more on finding ways to make deforestation and extensive land use less attractive for farmers. In this context, combining technical research designed to increase land productivity with policy research that looks at ways to provide incentives for forest preservation becomes a pressing global need.

This chapter uses data from the Tropileche research and extension consortium to support our two central hypotheses.[2] It provides empirical results from three research sites, in Colombia, Costa Rica and Peru, which allow us to compare the adoption and effects of one particular intensive technology: improved feeding systems for small-scale farmer milk and beef production. The chapter first briefly reviews the literature regarding the link between cattle and deforestation and situates improved pasture technology within the realm of intensive livestock technologies. Section 2 discusses whether and how intensifying pasture management might affect deforestation. Section 3 presents our hypotheses and analytical framework. Section 4 introduces the three study sites and the pasture technology options. Section 5 contains the empirical evidence about technology adoption and the link between pasture technology and forest cover. Section 6 presents policy options and concludes.

2. Livestock, Technology and Deforestation

How much livestock and pasture expansion contributes to the larger phenomenon of tropical deforestation is difficult to determine and varies depending on farm size and region. The following section places the issue of pasture and cattle in the broader context of deforestation and specifies how improved pastures relate to intensive livestock technology more generally.

2.1. Cattle within the deforestation debate

Since the early 1980s, various analysts have used the correlation between pasture expansion and declining forest cover to argue that cattle ranching is the main force behind deforestation (Myers, 1981; Shane, 1986). Although large amounts of primary forest ultimately end up as pasture, many other forces also drive deforestation. Population growth and the exploitation of natural resources, along with perverse government policies and social structures, contribute greatly to forest clearing (Hecht, 1993; Pichón, 1997). While these factors do not necessarily directly drive pasture expansion, often they must be present for it to occur.

The ample literature on the subject points to three main explanations for why pastures replace forests: government policies, features of livestock that appeal to farmers and technological factors. We adapted the following list from Godoy and Brokaw (1994), Kaimowitz (1996) and Faminow and Vosti (1998).

Government policies favouring pasture establishment

- Land tenure polices that require farmers to demonstrate use of the land (often via pastures) to establish and retain property rights (Mahar, 1989; Jones, 1990; Binswanger, 1991; Southgate et al., 1991).
- Government subsidies for livestock credit, input and producer prices and tax breaks for ranching (Mahar, 1989; Binswanger, 1991; Schneider, 1995; Barbier and Burgess, 1996; White et al., 1999).
- Policies that depress timber values and make forest management less profitable (Kishor and Constantino, 1994; de Almeida and Uhl, 1995).
- Reduced violence, which lowers the risk of ranching in isolated areas (Maldidier, 1993).

Favourable markets and attractive features of livestock

- Favourable international (Myers, 1981; Nations and Komer, 1982) and/or national (Schneider, 1995; Faminow, 1996) cattle-product markets.
- Livestock's low labour, purchased input and management requirements, prestige value, ease of transport, biological and financial flexibility and role as an inflation hedge and in risk diversification (Hecht, 1993).

Technological factors

- The lack of other viable income sources because of declining crop yields (Mahar, 1989; Seré and Jarvis, 1992; Hecht, 1993; Thiele, 1993).
- Slow technological change in livestock management, which favours extensive production (Serrão and Toledo, 1992).

- Pasture degradation, leading to abandonment and the further intrusion of pastures into the forest (Toledo *et al.*, 1989; Seré and Jarvis, 1992; Serrão and Toledo, 1992; Schelhas, 1996).

Given tropical Latin America's heterogeneity, one should view the above explanations as broad generalizations. Moreover, since the 1980s, government policies and economic and environmental conditions may have changed.

2.2. Improved pastures for small-scale ranchers

This chapter focuses its analysis on pasture improvements in small-scale dual-purpose (milk and beef) production systems. Small-scale ranchers are very important in tropical Latin America and have a considerable impact on forest margins. In Central America, 40% of the cattle belong to farmers with less than 60 ha (Kaimowitz, 1996). Nearly 46% of all farms in the Peruvian Amazon have cattle and, of these, 95% have fewer than 100 head (Instituto Nacional de Estadística, 1986). In the Brazilian Amazon, small-scale farmers hold only 10% of the land but account for 30% of all deforestation (Fearnside, 1993). Our emphasis on dual-purpose production follows directly from the decision to look mostly at small-scale ranchers, since usually only larger-scale operations tend to specialize exclusively in dairy or beef production (Mattos and Uhl, 1994; Nicholson *et al.*, 1995).

We concentrate on improved pastures because both small- and large-scale producers can adopt them. Many other intensive technologies, such as the use of feed supplements, pasture rotations and artificial insemination, are beyond the reach of small-scale ranchers with limited access to capital and labour and may not address their needs. This is particularly true on the frontier, where ranchers typically have little access to such technologies.

3. Pasture Technology and Deforestation

Researchers have regarded the relation between improved pasture technology and deforestation as a quandary for years. While one school of thought argues that improved pasture technologies increase deforestation, a second school says the opposite; and neither supports its case with much evidence. In the early 1980s, the Centro Internacional de Agricultura Tropical (CIAT) tropical pasture programme came under pressure to expand its research efforts into the forest margins. Yet it faced a dilemma. If the new germplasm and management strategies proved highly productive and sustainable, they might accelerate forest clearing. However, if the programme did nothing, existing ranching practices, which led to rapid degradation and low productivity, might accelerate clearing even more (Spain and Ayarza, 1992).

3.1. Technology decreases the push forces into the forest

Those who argue that intensive pasture technologies reduce deforestation emphasize that inappropriate ranching practices in tropical environments lead to severe productivity declines, thereby forcing ranchers to abandon their existing pastures and clear new forest. They hope that, by developing new low-cost technologies, farmers can maintain their productivity and thereby reduce deforestation.

Unfavourable environmental conditions in many tropical regions make it difficult to maintain the carrying capacity of the pastures using traditional production systems. Declining soil fertility, prolonged dry seasons, soil compaction, insect pests and weeds often rapidly diminish the carrying capacity of pastures, especially on large-scale ranches (Serrão et al., 1979; Nepstad et al., 1991). In Brazil, for example, weeds and soil degradation typically reduce stocking rates from two head per hectare during a pasture's first 4 years to only 0.3 head per hectare 3–6 years later (Serrão and Homma, 1993; Mattos and Uhl, 1994). In the late 1970s, only one-fifth of Brazil's pastures in previously forested regions were degraded or in an advanced stage of decline (Serrão et al., 1979). By 1990, this had risen to at least half of all pastures (Serrão and Homma, 1993). Analysts have also documented substantial pasture degradation in Central America (Kaimowitz, 1996). Those that believe pasture technologies reduce deforestation contend that new low-cost forages and management techniques will reduce pressure on forest cover by making degraded and abandoned land productive again.

When faced with declining crop and (to a lesser extent) livestock production, small-scale pioneer settlers often sell their land to ranchers and migrate deeper into the forest (Jones, 1990; Thiele, 1993; Nicholson et al., 1995; Rudel, 1995). Many pasture specialists assert that low-cost pasture technologies would allow small-scale farmers to earn sufficient income to reduce the need to migrate deeper into the forest. Some researchers also hypothesize that targeting pasture research outside forested areas would reduce pressure on forest cover. Smith et al. (1994: 21) claimed that in South America '[t]he savannah could provide an outlet for the economic objectives of national governments, and for venture capital, while relieving pressure for exploiting the forest margins'.

3.2. Technology increases the pull forces into the forest

The school of thought claiming that improved pasture technology increases deforestation argues that improved pastures lead to higher productivity and therefore more profitable cattle systems. By making cattle ranching more economically attractive, intensive pasture technologies give farmers a greater incentive to convert forest to pasture. This may take the form of existing farmers increasing the portion of their farms they dedicate to pastures, or of outside capital and people flowing into frontier regions to establish new ranches.

3.3. No effect: technology is secondary

A third possibility is that technological change may play a minor role within the overall context of factors that influence the conversion of forests to pastures. This might occur, for example, if the main reason ranchers expand their pasture was to engage in land speculation (Kaimowitz, 1996). Yet, in South America, Faminow and Vosti (1998) have raised doubts about whether land speculation contributes to the spread of cattle ranching. The evidence supporting this widely held belief comes largely from one data set, published by Mahar (1979). Subsequent data and analysis have revealed that the real prices of farmland and pastures in the Amazon have not changed relative to the rest of Brazil. Thus, Faminow and Vosti conclude that large speculative earnings from land ownership have not been consistent and widespread in the Amazon.

Nicholson *et al.* (1995: 719) also raise doubts about the potential of technology to decrease deforestation, stating that 'intensification of cattle systems is unlikely to alter dramatically the deforestation rate in Central America because consumer demand for livestock products is not the principal factor motivating most migration to forest areas'. Rather, they claim that deforestation is the result of pressure from many resource-poor migrants seeking livelihoods at the forest margin.

Pasture technologies may simply be too inaccessible or expensive for many farmers who live near forests to implement. Not every intensive management systems that displays agronomic and financial benefits is widely adopted. For the western Amazon, Faminow *et al.* (1998) argue that high price fluctuations of cattle products have made the activity risky and have inhibited the adoption of new livestock technology.

4. Framework and Hypotheses

Although analysts have recognized for some time that the effects of pasture technology on forest cover were poorly understood, empirical research on the topic did not begin until recently. Several factors contributed to this dearth of research. First, until the early 1990s, much tropical pasture research contained vestiges of a Green-Revolution motivation. Researchers' main goal was to achieve sustainable productivity increases in the face of degrading tropical soils and weed and pest invasions. Secondly, and closely related, few available data linked improved pasture technologies with surrounding forest cover. Few early studies included forest cover with pasture performance data. Thus, to increase the generality of the results given the limited data resources, we have had to follow an alternative approach.

Farmer decisions about how to use their land are central to our approach. More specifically, farmers have a choice between intensive (improved pastures) or extensive (forest clearing) land-use options.[3] The relation between intensive and extensive options leads to the alternative hypothesis: the introduction of intensive technologies will lead to farmers maintaining or expanding forest cover only if adopting such technologies is less expensive than extensive growth. Thus, the financial feasibility of the new technology, its adoption by farmers and farmer incentives to preserve forests all influence the association between improved pasture technology and forest cover.

A variety of studies have examined land-use dynamics. Some focus on the relation between population density and land management intensity (Boserup, 1981; Serrão and Toledo, 1992). Others examine agricultural production changes and market access at the frontier (Henkel, 1971; Maxwell, 1980; Richards, 1997). The following analysis combines both approaches to explain the dynamics of land-use trends in the Tropileche research sites.

In each of our three sites, an array of local and national variables influence land-use decisions. Key biophysical variables include agroecological conditions, such as soil, slope and on-farm forest cover. Farm characteristics, markets and policies (e.g. subsidized credit, technical assistance, protective tariffs and land tenure) constitute important socio-political-economic variables (Table 6.1).

As mentioned earlier, the land price variable captures the effect of two opposing forces: the level of development (i.e. market access) and the amount of forest cover. Areas with low land prices have immature markets and abundant forest. High land prices typically imply more developed markets and scarcer forest cover. Land price also serves as an *ex ante* indicator of whether farmers will adopt improved pasture technologies. When land prices are low, farmers have little incentive to adopt intensive pasture technologies.

Figure 6.1 illustrates the continuum of possible land price, technology adoption and deforestation situations. At one end lie regions with nascent markets, where farmers will not adopt pasture technologies. Since farmers do not adopt the technology, it has no impact on forest cover. At the other end of the spectrum, farmers find adopting intensive pasture technologies attractive and yet the effect on forest cover is small since little forest remains. Nevertheless, the shift to intensive land uses may allow certain areas to revert to forest. Between these two extremes, one encounters situations where farmers are interested in adopting intensive technologies and sufficient forest remains. Here on the continuum, the adoption of new technologies may significantly influence forest cover.

5. The Study Sites and the Improved Pasture Technologies

In 1996 and 1997, the Tropileche research consortium conducted diagnostic surveys of farmers in its three study sites to assess the adoption potential of

Table 6.1. Summary of factors influencing the improved pasture technologies–forest-cover link.

Influencing variable	Costa Rica	Colombia	Peru
Biophysical			
Soil productivity	Good	Poor	Poor
Farm in forest (%)	9	10	32
Pasture degradation			
Erosion	High	Low	Low
Other		Spittlebug	
Socio-political-economic			
Farm characteristics			
Size (ha)	90	150	30–50
Land price (US$)	2400	450	10–200
Off-farm work potential			
Permanent	High	Low	Low
Seasonal	High	Low	Moderate
Market conditions	Good	Fair	Poor
Milk price (US$)	0.28 (cooled)	0.23	0.22
Market demand	High	High	Low
Transport cost (% milk price)	7	10	8
Average distance to processing (km)	60	80	–
Milk processing facility	High value	Industrial grade	None
Public policies			
Milk tariffs (%)	104	30?	0
Credit (% real interest rate)	14	23	34 if possible
Extension service	Good	Limited	Limited
Land tenure problem	No	No	No
Years since initial settlement	200+	40–80	1–50

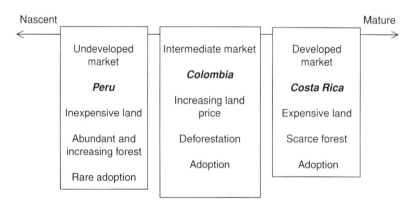

Fig. 6.1. A land-use continuum with respect to markets, land prices, forest cover and technology adoption.

promising new forage germplasm. The farmers interviewed were representative of local dairy producers in terms of farm size, input use, productivity, income and constraints.

5.1. Costa Rica

The study site, Esparza, is located in Costa Rica's Central Pacific region. Settlers arrived there more than 500 years ago, so the site is no longer considered a forest margin. Average farm size is 29 ha, of which 75% is pasture, 11% annual crop systems, 9% forest and 5% perennial crops (Centro Científico Tropical, 1994). Mean herd size is 43 head, with an average stocking rate of 0.9 animal units (AU) per hectare (Fujisaka et al., 1997). Pastures are severely degraded on the steep slopes but less degraded in the lowlands. Land prices are high, averaging $2500 ha^{-1}, in response to a long history of public-funded investment in roads, electrification, health and education and good access to markets.

5.2. Colombia

Florencia Caquetá, in the piedmont of the Colombian Amazon, is a more recently settled forest-margin site. Settlers arrived in the 1900s and cleared most of the forest. The predominant agricultural activity is beef and milk production, 87% of which is dual-purpose. Caquetá accounts for 7% of the national herd, or approximately 1.1 million head. The stocking rate on native pastures is approximately 1.1 AU ha^{-1}. Farms average 150 ha in size. Commercial land prices are moderate, ranging from $400 to $600 ha^{-1}, depending on location and soil quality. The typical farm has approximately 58% of its land in pasture and 10% in forest (Rivas and Holmann, 1999).

5.3. Peru

The most recently settled forest-margin study site, Pucallpa, is in eastern Peru, on the Ucayali River, a major tributary of the Amazon. Households first settled the area in the 1940s after the government constructed a road linking Pucallpa and the capital, Lima. Land use in Pucallpa is more heterogeneous than in the other two sites and the amount of forest that remains on farms is closely related to the number of years since the land was settled. In more recently inhabited areas, 59% of the farmland is still in forest, whereas, in more mature regions, forest cover decreases to 40%. Land area dedicated to pastures generally increases with age of settlement. Recent settlers have about 10% in pasture, whereas the more established farmers have 19%. Farms specifically devoted to cattle ranching have 66% of their land in pasture and 19% in forest (Smith et al., 1999). The stocking rate on traditional pastures is approximately

0.6 AU ha^{-1} (Fujisaka and White, 1998). Land values are relatively low, ranging from US$10 to US$200 ha^{-1}, depending upon the quality of road access.

5.4. Improved pasture technologies: *Brachiaria* and legumes

Our study analyses three technological options promoted by the Tropileche consortium: a new *Brachiaria* grass variety, an association of *Brachiaria* with a legume called *Arachis pintoi* and a cut-and-carry system with *Cratylia*, a leguminous bush that serves as a protein bank during dry months. To establish a hectare of *Brachiaria* only requires $250 and 7 man-days. If one adds *Arachis*, the cash outlay rises to $300. The *Cratylia* option, on the other hand, requires $400 and 19 man-days ha^{-1} year^{-1}. Approximately once every 4 years, farmers may reseed their pastures to maintain them and this costs about 25% as much as the initial establishment.

To varying degrees in all sites, the improved forages increase both stocking rates and milk production. The *Brachiaria* option in Peru increases production by only 0.2 AU and 0.3 kg head^{-1} day^{-1}. However, the use of *Cratylia* in Costa Rica can raise production by 1.6 AU and 2.0 kg head^{-1} day^{-1}. How ranchers manage their systems greatly affects the results. Table 6.2 provides details for each option and site (Holmann and Estrada, 1998; Holmann, 1999; Rivas and Holmann, 1999).

Table 6.2. Costs (labour and capital) and benefits of pasture options.

Country and pasture technology	Labour input (man-days ha^{-1} year^{-1})	Labour costs ($ ha^{-1} year^{-1})	Capital input ($ ha^{-1})	Stocking rate (AU ha^{-1})	Milk production (kg head^{-1} day^{-1})	IRR* (% year^{-1})
Costa Rica						
Native	3	30	–	0.9	4	–
Brachiaria	7	70	270	1.3	5	9.4
Brachiaria/Arachis	7	70	300	1.5	5.8	10.1
Cratylia	19†	190	395	2.5	6.0	12.3
Colombia						
Native	3	39	–	0.9	2.4	–
Brachiaria	7	91	270	1.2	3.0	12
Brachiaria/Arachis	7	91	300	1.5	3.6	19.3
Peru						
Native	3	12	–	0.7	2.5	–
Brachiaria	7	28	50–250	0.9	2.8	0–12
Brachiaria/Arachis	7	28	280	1.3	3.3	9.8

*Lactating herd is 36% in Costa Rica, 40% in Colombia and 20% in Peru.
†Labour input only during the dry season.
IRR, internal rate of return.

6. Empirical Results

Does improved pasture technology decrease pressure on surrounding forest by providing financial incentives to invest in an intensive rather than extensive manner? We address this question by examining the issues of technology adoption, the effect of land prices and the link between technology and forest cover.

6.1. Technology adoption

Each of the three sites presents a different adoption tale. In Costa Rica, the 6-month dry season and the associated low forage output constrains production and influences farmer decisions. To feed dual-purpose cows during those dry months, producers have adopted all three options, the grass, the grass–legume association and the cut-and-carry systems. On average, farmers have improved 15% of their pastures, ranging from 45% on small-scale farms to 5% on large farms (Fujisaka et al., 1997). This potentially counterintuitive situation emanates from the fact that small-scale farms require more intensive land-use strategies. Despite establishment costs, they more readily adopt the new technologies to increase stocking rates.

Over the last 10 years, ranchers in Colombia have had strong incentives to adopt new pasture species resistant to spittlebug (*Aeneolamia* spp. and *Zulia* spp.), as this pest reduces pasture biomass production by 30–35% (Valério and Nakano, 1987; Holmann et al., 2000). The susceptible *Brachiaria decumbens* grass is giving way to the more resistant *Brachiaria humidicola* and *Brachiaria brizantha*, now used on 38% and 25% of the farms, respectively. About 25% of farms have adopted the leguminous *A. pintoi*. Although over 80% of the producers felt satisfied with the performance of *Arachis*, their capital constraints limit further establishment (Rivas and Holmann, 1999).

During the 1970s, technicians in the forest margins of Peru promoted *B. decumbens* to improve pasture performance. Since native grasses, such as *Torourco* species, degrade rather quickly, farmers were attracted by the greater vitality of *B. decumbens* and many adopted it. Between 1982 and 1996, *Brachiaria* use rose from 15.5% to 40% of total pasture cover (Riesco et al., 1986; Fujisaka, 1997; Fujisaka and White, 1998). Nevertheless, continued pasture improvement proved more difficult. Efforts to promote pastures that incorporate leguminous forages faced major challenges, because of the high costs of these systems and farmers' limited access to capital.

6.2. Benefits and costs of pasture investment

The best way to assess whether farmers are likely to adopt improved pastures is by contrasting their financial performance with the alternative of expanding pasture area by purchasing more land or clearing remaining forest. The figures below refer to the financial costs and benefits as perceived by private landholders. While there may be social costs and benefits, they are less likely to affect adoption and the relation between technology and forest cover. Hence, we do not address them in this chapter. For more information regarding social costs and benefits, see Nations (1992) and Toledo (1992).

Although improved pastures require more labour to maintain them, the greatest obstacle to getting small-scale farmers to adopt them is their large initial establishment cost. Intensive pasture systems can be highly profitable, but, when capital is scarce or cannot easily be borrowed, they are not financially feasible. Thus, it is illustrative to compare the establishment costs of each pasture option. We assume the amount of labour required is the same in all three countries, because labour productivity is likely to be similar. To compare the intensive and extensive options meaningfully, we examine production in each, including stocking rates and milk production.

Ranchers realize the benefits from improved pasture investments over time. Thus, to determine how improved pastures perform from a financial perspective, one must sum up and compare the cash flows of both benefits (increased milk and beef production) and costs (establishment and maintenance). One useful measure for examining an investment over time is the internal rate of return or IRR. By definition, the IRR is the interest rate received for an investment, consisting of payments and income that occur at regular periods.

For the three sites, we employed a 12-year time frame to compare the IRR performance of the improved pasture options (see Table 6.2). In Costa Rica, the IRR ranged from 10.1% for the *Brachiaria–Arachis* association to 12.3% for *Cratylia*. For Colombia, the IRR for improved legume-based pasture (*B. decumbens* and *Arachis*) pasture is 19%, whereas the IRR for the *B. decumbens* option is 12%. In Pucallpa, the *B. decumbens* and *Arachis* association has an IRR of 9.8%. Even though the local real interest rate exceeds the IRR of the improved pastures in all sites, farmers who adopt the technologies do so by self-financing the establishment costs.

To compare improved pasture with the extensive alternative also requires examination of the establishment costs. Cost estimates of the improved pasture options include pasture establishment; those for the extensive options also include the cost of fencing. At the Costa Rica site, the total capital investment per hectare for the extensive option is about US$2600. In comparison, ranchers must invest approximately US$400 ha^{-1} to adopt the *Cratylia* option, which permits 89% of stocking capacity and 92% of the milk production that one could obtain by doubling the pasture area.

For Colombia, it costs $830 to establish and stock a hectare of native pasture, while it takes an additional $780 to do the same with improved pasture with *Brachiaria–Arachis*. The legume association can support 83% more cattle and yields 75% more milk than the extensive option. In Pucallpa, an additional hectare of a native pasture costs approximately $150, while establishing an improved forage alternative requires about $250 more. This latter alternative can maintain 93% more stock and yields 56% more milk than the extensive option.

6.3. Land price: an *ex ante* indicator of adoption potential

The Tropileche experiences mentioned above led the researchers involved to raise the issue of what determines whether ranchers will adopt a specific intensive pasture technology. Although there is a long list of potentially relevant factors, land prices appear to largely drive ranchers' decisions about investing in improved forage technologies. As land price rises, farmers find it harder to increase their farm size, since they are typically capital-constrained. Instead, they adopt improved forage alternatives to enlarge their herds for less money. Land costs range from only $150 ha^{-1} in Peru to $2400 ha^{-1} in Costa Rica, with Colombia in the middle with $450 ha^{-1}.

The evidence from the sites suggests that locations with mature markets (a demand for more and higher-quality milk) and greater access (shorter distances and better roads) have higher land prices. In other words, the value of markets and access is capitalized into the price of land. Following the land-use continuum presented in Fig. 6.1, each of our sites represents a different level of market development and associated land price. For example, in the Costa Rica site, farmers supply fluid milk to a central processing facility and capture a high value. In Florencia, Colombia, they produce milk to satisfy industrial demand. To lower transport costs, a satellite processing plant condenses the milk before transport to a metropolitan facility for subsequent processing. In Pucallpa, local groups have sought to attract investment for constructing a milk processing plant but failed. Companies like Nestlé require that the region supply at least some 200,000 l day^{-1} before they would be willing to build a plant, which is beyond the region's current capacity. The poor quality of the surrounding roads and long distances to major markets have also discouraged processing investments.

6.4. Linking improved pasture technology to forest cover

To examine the impact of improved pastures on forest cover, we first present a cross-sectional analysis of the three sites, using a land-use history framework, and then a time-series comparison of one site, Florencia.

Pucallpa lies on the nascent side of the land-use continuum presented in Fig. 6.1, where improved pasture technologies are not a viable option for most farmers. It is much cheaper for them to purchase more land than to intensify their current holdings. Since they never even take the first step of adopting the technology, improved pastures have no or little impact on forest cover. The area finds itself in a trap: demand for dairy products by processors is low because farmers do not supply enough to justify investment in processing, and farmers do not increase their supply because currently no processors demand their products. In light of the frustrated attempts to promote new pasture technologies in the region, Tropileche is redirecting its research efforts to other regions. The western Brazilian Amazon faces a similar situation. Despite 25 years of research and promotion, most small farmers there have not adopted improved pasture technologies and livestock management systems (Faminow et al., 1998).

It is important to note that technology adoption does not always imply intensification, as demonstrated by the use of *Brachiaria* in Pucallpa. Initially, it required intensive investment in capital, but now it easily propagates and grows vigorously. In addition, *Brachiaria* adoption and forest cover have been perversely linked. The region's low stocking rates, combined with political instability during the late 1980s, have led to a supply of pasture biomass that exceeds what the present cattle herd demands (Fujisaka and White, 1998). In this context, *Brachiaria* has sometimes become a weed and flammable fuel, which helps fires to spread into the surrounding forest.

Costa Rica sits on the mature side of the land-use continuum. Farmers can afford to adopt improved pastures and they are financially feasible. Yet here also the technology affects forest cover only marginally. The region was largely deforested decades ago and forest clearing is currently not a major issue. Indeed, the main thrust of government and development agencies efforts at present is to reafforest marginal agricultural land and pasture. While it may be tempting to do so, one cannot attribute these reforestation efforts to the adoption of intensive pasture technologies, since government policies and other factors have had such an active role.[4] Perhaps, as we argue below, to either protect forest cover or reafforest requires government policy initiatives.

Lying between the Peru and Costa Rica sites, we have the intermediate case of Colombia. The way land use has evolved there suggests that improved pasture technology has reduced pressure on forests. A 1986 farm survey found that, on average, farms had 7% of their land in forest and 26% in improved pastures (Ramivez and Seré, 1990). By 1997, the improved pasture area had increased to 58% and the forest area to 10%, although admittedly the change observed fell within the survey's margin of error (Rivas and Holmann, 1999). The improved pasture technologies appear to have increased biomass production so much that they have exceeded the existing cattle herd's capacity to consume it. Thus, farmers have little financial incentive to expand into the surrounding forest.

These Colombian results come with two caveats. First, no one knows whether the land-use outcome will be temporary or permanent. It could be just a matter of time before the natural growth of the herd catches up with the availability of feed supply and ranchers again feel the need to clear additional forest. Hence, the current pasture–forest relation may not represent an equilibrium state, since, as the years go by, the factors constraining farmer land-use decisions are likely to change.

Secondly, we reported the above results in percentages, but absolute farm size has also changed. Over the 11-year span, average farm size grew from 131 to 158 ha. In absolute terms, improved pastures went from 33 to 91 ha, while average forest cover increased from 9 to 16 ha. While most of the growth in farm size appears to have come from consolidation with neighbouring ranches, some encroachment into forests may have occurred. Thus, the aggregate impact on forest cover at the regional level may not be as clear.

7. Conclusions

Our review of the evidence regarding the effect of improved pasture technology on forest cover has led us to an alternative hypothesis. Forest scarcity is a prerequisite for technology intensification. The best way to illustrate this shift in causality is to go back to our land-use continuum. On the side of the continuum with nascent markets and low land values, as in the Peruvian Amazon, continued deforestation and extensive cattle production both appear to be rational private choices. As land-use patterns mature, with less forest and more developed markets, land prices rise. In Costa Rica and, to a lesser extent, Colombia, farmers intensify to avoid pasture degradation and the higher-cost option of expanding on to neighbouring lands. Hence, land price reflects a set of biophysical and socio-political-economic factors, which come together in a simple decision rule that mirrors our second hypothesis. If it is cheaper to intensify production than to cut surrounding forest or purchase more land, then farmers will find improved pasture technologies attractive and adopt them.

It is important to recall that land use in the forest margins is dynamic. Thus, for example, the technology–forest-cover link we identified in Colombia is likely to be ephemeral. Hence, to control deforestation in the long term will probably require policy intervention. In tropical Latin America, where land degradation can spur further deforestation, technological advances that bring degraded land back into production are a critical policy component.

At the same time, one should not forget that policy-makers have other legitimate objectives besides minimizing deforestation and land degradation. They must also be concerned about the welfare of the people living at the forest margin. If properly targeted and coupled with policies that restrict deforestation or make it financially unattractive, technical advances such as improved pastures can achieve these multifaceted objectives of human welfare and environmental sustainability. Below we present some policy options that might meet this dual set of goals (see also Ledec, 1992; von Amsberg, 1994; Nicholson *et al.*, 1995; Kaimowitz, 1996).

7.1. Protected areas

In principle, national parks and reserves can maintain forest cover. Yet issues regarding property rights, governance and land encroachment can be challenging. In the USA, for example, the US cavalry had to patrol and protect the national parks for almost 30 years after the first park was created in Yosemite in 1886 (Hampton, 1971). We do not say this to espouse military involvement but simply to illustrate how much effort it might take to enforce the policy.

7.2. Extractive reserves

These make land with standing forest more valuable. The promotion and development of non-traditional forest products can provide private incentives to use forests (Kishor and Constantino, 1994; Rice *et al.*, 1997).

7.3. Targeted agricultural research

Universities and national and international research centres must continue to develop new agricultural and livestock technologies, but they must also target their research domains better (Loker, 1993). Governments and development agencies can use credit, tax and land-reform policies as incentives to rehabilitate degraded lands for improved pastures, agricultural use or reforestation.

7.4. Conservation payments

In theory, these can allow private landholders to receive monetary compensation for the public services they provide and can come in many forms, including reforestation campaigns and carbon sequestration payments. Managing these interventions is tricky, however, and a lot of work is still needed on market mechanisms, monitoring and accountability (Swisher and Masters, 1992). For example, inappropriate incentive structures may not lead to greater reforestation if the projects involved focus on the number of trees planted, rather than the percentage that survive. To establish a functioning carbon sequestration payments system will be even more challenging.

7.5. Private cattle product certification

In principle, milk processors could require their suppliers to use ranching practices, such as use of silvopastoral agroforestry systems or intensive pasture management, if they had an incentive to do so. In some cases, marketing benefits accruing from producing a 'green' product may be sufficient to cover costs, although the media and the public would probably still need to monitor the claims made by companies.

Although some people may not like the idea, forest-margin regions will continue to have cattle for the foreseeable future, because producers need the incomes and consumers demand the products. In many frontier regions, farmers have no viable use for their land besides cattle. This leads to situations such as we found in Pucallpa, where desperate farmers, with few alternative options, have established pastures without even having cattle in the hope that they might get some in the future. Moreover, consumer demand for animal products will continue to grow rapidly. In developing countries as a whole, the livestock sector expanded so fast between 1982 and 1993 that a recent International Food Policy Research Institute/Food and Agriculture Organization/International Livestock Research Institute study (IFPRI/FAO/ILRI, 1998) has called it *The Next Food Revolution*. Annual growth rates during that period were 7.4% for poultry, 6.1% for pork, 5.3% for all meat and 3.1% for milk. It remains to be seen whether the necessary increases in agricultural and animal production will come from extensive or intensive production systems (May and Segura, 1997).

All of this implies that researchers must move beyond examining how intensification affects deforestation and proactively find ways to improve the feasibility of adopting intensive technologies. Future research should provide alternative land uses so that deforestation and extensive land use will no longer be farmers' most attractive option. Technical research, to increase productivity and prevent land degradation, must go hand in hand with policy analysis and implementation to increase incentives for forest preservation, while addressing farmer objectives. Until then, forest cover will continue to affect the intensification of pastures.

Notes

1   The authors would like to express their appreciation to an anonymous reviewer, David Kaimowitz, Arild Angelsen, Dean Holland, David Yanggen, Douglas Pachico and the workshop participants for their helpful comments.
2   See Holmann (1999) for details on the consortium.
3   A third option involving simultaneous intensification and extensification may also exist. But, for small-scale farmers with very limited capital and labour, this option is unlikely to be feasible, especially in the short term.
4   Government payments to promote secondary forest growth and low earnings from cattle production have led many landowners to let their pastures become forests in northern Costa Rica (Berti Lungo, 1999).

References

Barbier, E.B. and Burgess, J.C. (1996) Economic analysis of deforestation in Mexico. *Environment and Development Economics* 1, 203–239.

Berti Lungo, C.G. (1999) Transformaciones recientes en la industria y la politica forestal Costariquense y sus implicaciones para el desarrollo de los bosques secundarios. MS thesis, CATIE, Turrialba, Costa Rica.

Binswanger, H. (1991) Brazilian policies that encourage deforestation. *World Development* 19, 821–829.

Boserup, E. (1965) *The Conditions for Agricultural Growth: the Economics of Agrarian Change under Population Pressure*. Aldine, Chicago.

Boserup, E. (1981) *Population and Technological Change: a Study of Long Term Trends*. University of Chicago Press, Chicago.

Centro Científico Tropical (1994) *Estudio de Zonificación Agropecuaria en la Región Pacífico Central*. Dirección de Planificación de Uso de la Tierra MAG-MIDEPLAN, San José, Costa Rica.

de Almeida, O.R. and Uhl, C. (1995) Brazil's rural land tax: tool for stimulating productive and sustainable land uses in the eastern Amazon. *Land Use Policy* 12, 105–114.

Faminow, M.D. (1996) Spatial economics of local demand for cattle products in Amazon development. *Agriculture, Ecosystems and Environment* 62, 1–11.

Faminow, M.D. and Vosti, S. (1998) *Livestock–Deforestation Links: Policy Issues in the Western Brazilian Amazon*. International Agriculture Centre, Wageningen, the Netherlands, World Bank, Washington, DC, and FAO, Rome, Italy, pp. 97–115.

Faminow, M.D., Dahl, C, Vosti, S., Witcover, J. and Oliveira, S. (1998) Smallholder risk, cattle, and deforestation in the western Brazilian Amazon. In: *Expert Consultancy on Policies for Animal Production and Natural Resources Management*, FAO Conference, Brasilia.

Fearnside, P.M. (1993) Deforestation in Brazilian Amazonia: the effect of population and land tenure. *Ambio* 22, 537–545.

Fujisaka, S. (1997) Metodología para caracterizar sistemas de uso de tierras: Acre, Rondônia, y Pucallpa en la Amazonía. In: Lascano, C.E. and Holmann, F. (eds) *Conceptos y Metodologías de Investigación en Fincas con Sistemas de Producción Animal de Doble Propósito*. CIAT, Cali, pp. 174–190.

Fujisaka, S. and White, D. (1998) Pasture or permanent crops after slash-and-burn cultivation? Land use choice in three Amazon colonies. *Agroforestry Systems* 42(1), 45–59.

Fujisaka, S., Holmann, F., Escobar, G., Solorzano, N., Badilla, L., Umaña, L. and Lobo, M. (1997) Sistemas de producción del doble propósito el la región Pacífico Central de Costa Rica: uso de la tierra y demanda de alternativas forrajeras. *Pasturas Tropicales* 19, 1–55.

Godoy, R. and Brokaw, N. (1994) Cattle, income and investments among the Tawakha Indians of the rain forest of Honduras: new thoughts on an old link. Unpublished.

Hampton, H.D. (1971) *How the United States Cavalry Saved the National Parks*. Indiana University Press, Bloomington.

Hecht, S.B. (1993) The logic of livestock and deforestation in Amazonia: considering land markets, value of ancillaries, the larger macroeconomic context, and individual economic strategies. *BioScience* 43, 687–695.

Henkel, R. (1971) *The Chapare of Bolivia: a Study of Tropical Agriculture in Transition*. University of Wisconsin, Madison.

Holmann, F. (1999) Análisis ex-ante de nuevas alternativas forrajeras en fincas con ganado de doble propósito en Perú, Costa Rica y Nicaragua. *Pasturas Tropicales* 21(2), 2–17.

Holmann, F. and Estrada, R.D. (1998) *Un Modelo Aplicable a Sistemas de Doble Proposito: Estudio de Caso Sobre Alternativas Agropecuarias en la Region Pacifico Central de Costa Rica*. CIAT, Cali.

Holmann, F., Peck, D. and Lascano, C. (2000) *Economic Damage Caused by Spittlebugs in Colombia: a First Approximation of Impact on Animal Production in* Brachiaria decumbens. Working Paper. CIAT, Cali, Colombia.

IFPRI/FAO/ILRI (1998) *Livestock to 2020: the Next Food Revolution*. Washington, DC.

Instituto Nacional de Estadística (INE) (1986) *Encuesta Nacional de Hogares Rurales (ENAHR): Resultados Definitivos*. INE, Lima, Peru.

Jones, J.R. (1990) *Colonization and Environment: Land Settlement Projects in Central America*. United Nations University Press, Tokyo.

Kaimowitz, D. (1996) *Livestock and Deforestation in Central America in the 1980s and 1990s: a Policy Perspective*. CIFOR, Jakarta.

Kishor, N. and Constantino, L. (1994) Sustainable forestry: can it compete? *Finance and Development*, 36–40.

Ledec, G. (1992) New directions for livestock policy: an environmental perspective. In: Downing, T.E., Pearson, H.A. and Garcia-Downing, C. (eds) *Development or Destruction: the Conversion of Tropical Forest to Pasture in Latin America*. Westview Press, Boulder, pp. 27–65.

Loker, W.M. (1993) The human ecology of cattle raising in the Peruvian Amazon: the view from the farm. *Human Organization* 52(1), 14–24.

Mahar, D. (1979) *Frontier Policy in Brazil: a Study of Amazonia*. Praeger, New York.

Mahar, D. (1989) *Government Policies and Deforestation in Brazil's Amazon Region*. World Bank, Washington, DC.

Maldidier, C. (1993) *Tendencias Actuales de la Frontera Agrícola en Nicaragua*. Nitlapan-UCA, Managua.

Mattos, M.M. and Uhl, C. (1994) Economic and ecological perspectives on ranching in the eastern Amazon. *World Development* 22, 145–158.

Maxwell, S. (1980) Marginalized colonists to the north of Santa Cruz: avenues of escape from the Barbecho crisis. In: Barbira-Scazzocchio, F. (ed.) *Land, People and Planning in Contemporary Amazonia*. Centre of Latin American Studies, Cambridge University, Cambridge, pp. 162–170.

May, P.H. and Segura, O. (1997) The environmental effects of agricultural trade liberalization in Latin America: an interpretation. *Ecological Economics* 22, 5–18.

Myers, N. (1981) The hamburger connection: how Central America's forests become North America's hamburgers. *Ambio* 10(1), 3–8.

Nations, J.D. (1992) Terrestrial impacts in Mexico and Central America. In: Downing, T.E., Pearson, H.A. and Garcia-Downing, C. (eds) *Development or Destruction: the Conversion of Tropical Forest to Pasture in Latin America*. Westview Press, Boulder, pp. 191–203.

Nations, J.D. and Komer, D. (1982) Indians, immigrants and beef exports: deforestation in Central America. *Cultural Survival Quarterly* 6, 8–12.

Nepstad, D.C., Uhl, C. and Serrão, E.A.S. (1991) Recuperation of a degraded Amazonian landscape: forest recovery and agricultural restoration. *Ambio* 20, 248–255.

Nicholson, C.F., Blake, R.W. and Lee, D.R. (1995) Livestock, deforestation, and policy making: intensification of cattle production systems in Central America revisited. *Journal of Dairy Science* 78, 719–734.

Pichón, F. (1997) Colonist land-allocation decisions, land use and deforestation in the Ecuadorian Amazon frontier. *Economic Development and Cultural Change* 45, 707–743.

Ramirez, A. and Seré, C. (1990) Brachiaria decumbens in *Caquetá: Adopción y Uso en Ganaderías del Doble Propósito*. Working Paper No. 67. CIAT, Cali, Colombia.

Rice, R.E., Gullison, R.E. and Reid, J.W. (1997) Can sustainable management save tropical forests? Sustainability proves surprisingly problematic in the quest to reconcile conservation with the production of tropical timber. *Scientific American* April, 44–49.

Richards, M. (1997) *Missing a Moving Target? Colonist Technology Development on the Amazon Frontier*. ODI Research Study, Overseas Development Institute, London.

Riesco, A., de la Torre, M., Reyes, C., Meine, G., Huaman, H. and Garcia, M. (1986) *Análisis Exploratorio de los Sistemas de Fundo de Pequeños Productores en la Amazonia, Región de Pucallpa*. IVITA, Pucallpa, Peru.

Rivas, L. and Holmann, F. (1999) Adopción temprana de *Arachis pintoi* en el trópico húmedo: el caso de los sistemas ganaderos de doble propósito en el Caquetá, Colombia. *Pasturas Tropicales* 21(1), 2–17.

Rudel, T.K. (1995) When do property rights matter? Open access, informal social controls and deforestation in the Ecuadorian Amazon. *Human Organization* 54, 187–194.

Schelhas, J. (1996) Land use choice and change: intensification and diversification in the lowland tropics of Costa Rica. *Human Organization* 55, 298–306.

Schneider, R. (1995) *Government and the Economy on the Amazon Frontier*. World Bank, Washington, DC.

Seré, C. and Jarvis, L.S. (1992) Livestock economy and forest destruction. In: Downing, T.E., Pearson, H.A. and Garcia-Downing, C. (eds) *Development or Destruction: the Conversion of Tropical Forest to Pasture in Latin America*. Westview Press, Boulder, pp. 95–113.

Serrão, E.A.S. and Homma, A.K.O. (1993) Brazil: country profile. In: *Sustainable Agriculture and the Environment in the Humid Tropics*. National Academy Press, Washington, DC, pp. 265–351.

Serrão, E.A.S. and Toledo, J. (1992) Sustaining pasture-based production systems for the humid tropics. In: Downing, T.E., Pearson, H.A. and Garcia-Downing, C. (eds) *Development or Destruction: the Conversion of Tropical Forest to Pasture in Latin America*. Westview Press, Boulder, pp. 257–280.

Serrão, E.A.S., Falesi, I.C., Bastos de Vega, J. and Neto, J.F.T. (1979) Productivity of cultivated pastures on low fertility soils in the Amazon of Brazil. In: Sanchez, P.A. and Tergas, L.E. (eds) *Pasture Production in Acid Soils of the Tropics*. CIAT, Cali, pp. 195–224.

Shane, D.R. (1986) *Hoofprints on the Forest: Cattle Ranching and the Destruction of Latin America's Tropical Forests*. Institute for the Study of Human Issues, Philadelphia.

Smith, J., Winograd, M., Pachico, D. and Gallopin, G. (1994) Dynamics of land use: project TA-02. In: *Tropical Lowlands Program Annual Report 1994*. CIAT, Cali, pp. 1–21.

Smith, J., van de Kop, P., Reategui, K., Lombardi, I., Sabogal, C. and Diaz, A. (1999) Dynamics of secondary forests in slash-and-burn farming: interactions among land use types in the Peruvian Amazon. *Agriculture, Ecosystems and Environment* 76(2–3), 85–98.

Southgate, D., Sierra, R. and Brown, L. (1991) The causes of tropical deforestation in Ecuador: a statistical analysis. *World Development* 19, 1145–1151.

Spain, J.M. and Ayarza, M.A. (1992) Tropical pastures target environments. In: *Pastures for the Tropical Lowlands: CIAT's Contribution*. CIAT, Cali, pp. 1–8.

Swisher, J. and Masters, G. (1992) A mechanism to reconcile equity and efficiency in global climate protection: international carbon emission offsets. *Ambio* 21, 154–159.

Thiele, G. (1993) The dynamics of farm development in the Amazon: the Barbecho crisis model. *Agricultural Systems* 42, 179–197.

Toledo, J., Seré Rabé, C. and Loker, W.M. (1989) Pasture–crop technologies for acid soil savannas and rain forests of tropical Latin America. In: Meyers, R.L. (ed.) *Innovation in Resource Management: Proceedings of the 9th Agriculture Sector Symposium*. World Bank, Washington, DC, pp. 247–274.

Toledo, V.M. (1992) Bio-economic costs. In: Downing, T.E., Pearson, H.A. and Garcia-Downing, C. (eds) *Development or Destruction: the Conversion of Tropical Forest to Pasture in Latin America*. Westview Press, Boulder, pp. 67–93.

von Amsberg, J. (1994) *Economic Parameters of Deforestation*. WPS 1350, Environment, Infrastructure and Agriculture Division, Policy Research Department, World Bank, Washington, DC.

Valério, J.R. and Nakano, O. (1987) Danos causado por adultos da Cigarrinha Zulia entreriana (Berg, 1879) (Homoptera: Cercopidae) na produção de raizes de *Brachiaria decumbens* stapf. *Annals Sociedad Entomología Brasilena* 16(1), 205–211.

White, D., Reategui, K. and Labarta, R. (1999) *El Desarrollo Alternativo: A Qué Precio Viene?* Working Paper. CIAT, Pucallpa, Peru.

Intensified Small-scale Livestock Systems in the Western Brazilian Amazon

Stephen A. Vosti, Chantal Line Carpentier, Julie Witcover and Judson F. Valentim

1. Introduction[1]

> It is increasingly clear that economic alternatives [to traditional pasture systems] need to be provided. One option is to provide credit and technical assistance to make better use of existing pastures... studies undertaken... suggest that well-managed pastures can produce three times more than the average pasture in the Amazon.
>
> (Translated from *Veja*, 7 April 1999, p. 115)

This chapter examines three basic questions regarding the use of more intensive livestock technologies by small-scale farmers in the western Brazilian Amazon. Are farmers likely to adopt them?[2] Would it help protect the forest if they did? What would the effects on the farmers' welfare be? These issues are fundamental, because many people have come to see intensive cattle ranching as a 'win–win' alternative that can simultaneously remove pressure on huge expanses of the Amazon's forests and improve farmers' well-being. Others look at it as a dangerous endeavour, more likely to favour forest destruction than forest conservation, i.e. intensification of this already widespread production system would actually promote the extensive expansion of the agricultural frontier.

In a 'best-case' scenario, intensification increases incomes and reduces deforestation. In a 'worst-case' scenario, farmers do not adopt more intensive systems and their traditional livestock systems deteriorate over time. Incomes decline and deforestation continues or even accelerates, as farmers clear new land to support their herds. In an 'intermediate case', farmers might adopt more intensive systems and thereby increase their incomes, but also clear

©CAB *International* 2001. *Agricultural Technologies and Tropical Deforestation*
(eds A. Angelsen and D. Kaimowitz)

more forest. The latter may occur because the new technology makes it more profitable to plant pasture and generates additional resources to finance expansion. This implies there would be clear trade-offs.

As used in this chapter, the term 'intensify' refers to the adoption of cattle production systems that have higher output per hectare. This can be achieved through the use of various pasture and herd management practices, increased use of purchased inputs and/or improved breeding stock. We focus exclusively on small-scale farmers, because of their large numbers and their importance in cattle management; an estimated 500,000 smallholders live in the forest margins of the Brazilian Amazon and, by 1995, over 40% of the total cattle herd in the state of Acre was held on ranches smaller than 100 ha (IBGE, 1997).

The next section gives background on the Amazon, its development and the policies that have influenced development over the past few decades. Section 3 provides a general overview of smallholder land-use patterns in the western Brazilian Amazon and describes the production systems that generate those patterns. Section 4 describes selected livestock production systems in the western Brazilian Amazon and the capital and labour requirements associated with establishing and managing these systems. It also looks at what these summary statistics can tell us about technology adoption and the links between intensification and deforestation. Section 5 presents a farm-level bioeconomic linear programming (LP) model, which allows us to directly assess the adoptability and impact (if adopted) of more intensive pasture and cattle production systems. Section 6 presents and compares the results of model simulations used to make these assessments, paying special attention to land use (including deforestation), herd dynamics and household income. Conclusions and policy implications appear in section 7.

2. Tapping the Resources of the Amazon

The Amazon basin occupies 7.86 million km^2 in nine countries, covers about 44% of the South American continent and houses the largest tracts of the world's remaining tropical moist rain forests (Valente, 1968). More than 60% of the Amazon forest is located in northern Brazil. This forest covers over 52% of Brazil's entire national territory (IBGE, 1997), an area larger than Western Europe (INPE, 1999).

Since the early 1960s, the Federal Government of Brazil has seen the Amazon region as a depository of huge amounts of natural resources (forests, agricultural land, minerals, etc.) to be used to fuel economic growth. To exploit those resources and integrate the region into the national economy required a substantial workforce. However, the region's low population density (about 0.9 km^{-2} in 1970) made labour scarce. The government also viewed the virtual absence of Brazilian citizens as a threat to national security, particularly given the flourishing illicit drug trade in neighbouring countries (Forum Sobre a Amazônia, 1968; Government of Brazil, 1969, 1981; SUDAM, 1976; Smith *et al.*, 1995; de Santana *et al.*, 1997; IBGE, 1997; Homma, 1998).

Tapping the Amazon's resources and developing the region proved difficult. Huge distances and poor or non-existent infrastructure separated the area from the major markets. This made the region's inputs expensive and its products less valuable. The huge diversity of the Amazon's mosaic of ecosystems saddled planners with the unexpected need for expensive niche-specific projects and programmes. Indigenous people became increasingly vocal about their claims to large tracts of land and the associated resources. Simultaneously, the international community began to pressure the Brazilian government regarding its planned uses of the Amazon, based on its own concerns about greenhouse-gas emissions and biodiversity conservation.

Despite large gaps in knowledge, the Federal Government decided to go ahead with its homogeneous set of policies aimed at developing the Amazon region. To this end, it initiated 'Operation Amazon' in 1966 and set out a broad geopolitical and economic plan for the region (Government of Brazil, 1969; Mahar, 1979; de Santana *et al.*, 1997). To supply the legal framework, financial resources, transportation networks and electric power needed to establish migrants and industry in the Amazon, the government created a plethora of regional development agencies and policy instruments. These included the Amazon Development Agency (SUDAM), the Amazonian Duty-Free Authority (SUFRAMA) and the Amazonian Regional Bank (BASA). Often this support took the form of subsidized credit to agriculture (particularly extensive beef-cattle ranching) and mining projects (Forum Sobre a Amazônia, 1968; Government of Brazil, 1969, 1981; SUDAM, 1976; Smith *et al.*, 1995; IBGE, 1997; de Santana *et al.*, 1997).

In the early 1970s, world economic and oil crises led to a severe economic recession in Brazil. This, combined with agricultural modernization and consequent changes in farm structure, generated large increases in unemployment and landlessness in southern Brazil, as well as social conflicts. The Federal Government saw the opportunity to solve two problems at once. By moving unemployed and landless people to the Amazon and establishing them in settlement projects, it could both reduce social pressures in the south and increase the supply of labour for development activities in the north (SUDAM, 1976; Government of Brazil, 1981; Bunker, 1985). In the efforts to encourage landless people to migrate and colonize, millions of hectares of forested land were turned over to small- and large-scale farmers, despite limited knowledge about whether these areas could support viable agriculture (Valentim, 1989; Wolstein *et al.*, 1998). Incentives to migrate were successful; in the western Brazilian Amazon population grew substantially. The neighbouring State of Acre's 1950 population of about 100,000 jumped to nearly 500,000 by 1996. Rondônia's population went from under 100,000 to over 1.2 million during the same period.

The process of converting forest to agriculture in the western Amazon states of Acre and Rondônia has now been under way for over two decades, and has had major direct and indirect impacts on growth, poverty alleviation and environmental sustainability – a 'critical triangle' of development objectives (Walker and Homma, 1996; Vosti *et al.*, 2001).

Economic growth has been substantial. Rondônia had become the third largest coffee-producing state in Brazil by 1997 and now has some 4 million head of cattle (IBGE, 1997; Soares, 1997). In neighbouring Acre, the area dedicated to agriculture increased from virtually zero in1975 to about 10% of the state's total area by 1999. Acre's cattle herd grew from practically nothing in 1975 to nearly 800,000 head in 1998 (IBGE, 1997). Pasture is the dominant use of cleared land in both states, occupying 1.4 million ha in Acre and about 5.4 million ha in Rondônia (IBGE, 1997).

Progress on poverty alleviation has also been impressive. Between 1970 and 1996, the United Nations Development Programme (UNDP) human development index in Acre rose from 0.37 to 0.75. Over the same period, life expectancy at birth climbed from about 53 years to over 67 and adult literacy shot up from about 47% to over 70% (UNDP, 1998).

The environmental record has been less encouraging. Roughly a quarter of Rondônia's forests have been converted to agriculture over the past 20 years, and about 70% of this is area currently dedicated to low-productivity pastures. Acre has suffered less deforestation (averaging about 0.5% per year over the 1989–1997 period, compared with 1.5% in Rondônia). But declining earnings from traditional extractive activities in Acre may lead to increased forest clearing for agriculture, perhaps even by rubber tappers (Homma, 1998; INPE, 1999).

In summary, forest conversion and subsequent agricultural activities have improved the welfare of many rural families. Nevertheless, questions persist about whether these gains will prove sustainable and replicable. The future role of cattle production in the region is also in doubt and many people are looking for alternative ways to increase growth and reduce poverty that involve less forest conversion (Serrão and Homma, 1993).

The search for alternatives will not be easy. In many ways the 'deck is stacked' in favour of extensive agricultural activities, particularly cattle production. As farmers weigh the relative returns to scarce factors in this generally land-abundant and labour-scarce region, characterized by large distances to major markets and imperfect credit markets, it is not surprising that they have turned to livestock (Vosti *et al.*, 2000). Cattle production systems dominate the landscape, and it is difficult to imagine any production system displacing them. One logical point of departure in the search for alternatives, then, is to ask whether there is any way to modify the current extensive cattle production systems (which consume large amounts of forest) in order to make them both more productive and less destructive to forests. The following sections turn to precisely that question.

3. Smallholder Land Uses and Land-use Systems

According to survey data from smallholders in the western Brazilian Amazon, forest continued to cover about 60% of the land on the average farm in 1994 (Witcover and Vosti, 1996). Pasture dominated the use of cleared land (taking up about 20% of total farm area), followed by fallow (8%), annual crops (6%), perennial tree crops (3%) and intercropped annual/perennial areas (1%). Moreover, the average proportion of cleared land dedicated to pasture and cattle production activities increased by roughly 5% of farm area in the space of 2 years, mirroring state-wide trends (Vosti et al., 2001).

The predominant land-use trajectory (Fig. 7.1) begins with the clearing of the forest and ends in the establishment of pasture (Leña, 1991; Dale et al., 1993; Browder, 1994; Jones et al., 1995; Fujisaka et al., 1996; Scatena et al., 1996; Vosti and Witcover, 1996; Walker and Homma, 1996; Vosti et al., 2001). Newly deforested land (on average, about 4.7 ha every other year) generally goes into annual crop production for about 2 years. After that, three possibilities exist. Farmers can put the land into a fallow rotation lasting about 3 years, after which it can be returned (usually only once) to annual crop production. Or farmers may put the land into perennial tree crops, which, depending on the type of tree crop and its management, can last up to a decade before replanting (some external inputs are required). Or farmers can dedicate the land to pasture, where, depending on herd and pasture management

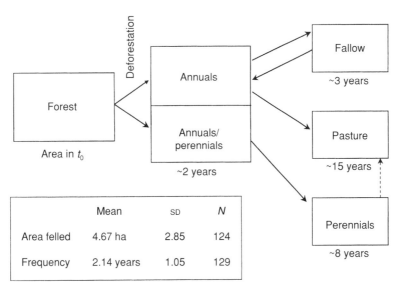

Fig. 7.1. Observed land-use trajectories by small-scale farmers. Number of years noted below each land-use box indicates time continuously in a given land use, and not the time elapsed since t_0 (the year in which deforestation on a given plot of land occurs).

practices, it can remain productive for 15 years or more.[3] This chapter examines how attempts to intensify this 'final' activity of the most common land-use trajectory affect deforestation and farmer livelihoods.

4. Traditional and Intensified Cattle Production Systems

Large farms, which have to a certain degree intensified their production systems, dominate the agricultural landscape in the western Brazilian Amazon. Farms larger than 200 ha accounted for roughly 70% of all planted pastures in Acre in 1995 (IBGE, 1997). Nevertheless, smallholders (less than 200 ha) managed 49% of the state's natural pastures, all low quality and degraded (IBGE, 1998).

Smallholder production systems in the western Amazon tend to have low stocking and calving rates and to generate returns to labour similar to the prevailing rural wage rate (Vosti et al., 2000). In spite of their modest profitability, several features make these systems attractive to many farmers. They are easy to manage and demand little technical expertise. They are inexpensive to establish and maintain, and require few purchased inputs. Cattle can assist farmers in slowing spontaneous forest regeneration, which can be rapid, even on soils depleted by annual crop production. Finally, labour and/or credit constraints limit farmers' ability to expand into more profitable alternatives, such as small-scale coffee production. Often they are left with significant amounts of cleared land that they cannot use for anything but cattle, given the amount of labour and capital available.

Some smallholders are, nevertheless, intensifying their cattle production systems. The remainder of this section defines 'traditional' and 'more intensive' production systems and then examines the capital and labour requirements of establishing and managing each of them.

Tables 7.1 and 7.2 present the technical coefficients for three types of pasture production systems and two types of dairy systems in Acre. Each column in the tables represents a different technological 'package'. The rows show resource requirements and expected production, over a 20-year period. We derived the technical coefficients for all the production systems from focus-group meetings with farmers, agricultural extension agents and researchers, and from field research (EMATER-Acre and Embrapa, 1980; Carpentier et al., 2001).

The first technological package described in Table 7.1 is the traditional pasture system (labelled P1). Farmers with this system use a traditional grass called Brizantão (*Brachiaria brizantha*). They manage the pasture poorly and the pastures display high levels of weed invasion. The more intensive grass-based system (labelled P2) also uses Brizantão, but farmers rotate grazing on and weed these pastures and consequently have fewer weed problems. The third pasture system (labelled P3) is the most intensive and incorporates the use of tropical kudzu, a legume, in addition to Brizantão (see

Chapter 12 by Yanggen and Reardon in this volume). In addition, the pasture is well managed. Ranchers rotate grazing on their pastures adequately and weed invasions are not prevalent.

Table 7.1 shows that P2 and P3 technologies significantly increase the lifespan and carrying capacity of the pasture system, compared with the traditional system. Two factors are chiefly responsible for this. First, P2 and P3 initially use more labour for weeding, green chop and pasture maintenance. Secondly, and perhaps more importantly, they use nearly twice as much

Table 7.1. Small-scale pasture production systems for Acre, by level of technology.

Technical coefficients	Grass		Grass/legume P3
	P1	P2	
1. Inputs			
Seeds (kg ha^{-1})			
Brizantão	15	15	15
Kudzu			1
Labour (man-days ha^{-1} year^{-1})			
Seeding (year 1)	3	3	3
Weeding (year 1)	2	3	3
Weeding and P3 green chop (years 2–4)	2	3	3
Weeding and P3 green chop (years 5–11)	2	3	1.5
Fencing			
Length (km of fence ha^{-1} of pasture)	0.063	0.106	0.106
Oxen time (man-days km^{-1} of fence)	4	4	4
Own chain-saw (man-days km^{-1} of fence)	4.5	1	1
Labour (man-days km^{-1} of fence)	59	56	56
Total costs (R$ km^{-1} of fence)*	302	347	347
2. Production			
Carrying capacity (animal units ha^{-1}, rainy season)			
Year 2–3	1	1	1.5
Year 4	1	1	1.5
Year 5	0.88	0.99	1.5
Year 6	0.79	0.97	1.5
Year 8	0.49	0.9	1.5
Year 9	0.39	0.85	1.5
Year 10	0.29	0.8	1.5
Year 11	0.3	0.85	1.48
Year 15	0	0.65	1.4
Year 20	0	0.15	0.9

*All values are in 1996 Brazilian reais, labelled R$; in 1996, one R$ was roughly equivalent to one US$.

fencing to segment pastures. The same two aspects that contribute to higher yields, however, can be formidable obstacles to adoption. Farmers may lack the labour and expertise required for managing legume-based pastures, as well as the capital to make substantial outlays for fencing.

Table 7.2 presents technical production coefficients for two types of dairy production systems – D1 (traditional, low-input) and D2 (more intensive). The pasture and the dairy packages are 'coupled', i.e. more intensive cattle production can only occur alongside more intensive pasture production, and vice versa. The first block of rows in Table 7.2 shows the herd input requirements for feed supplements, animal health and labour. The second

Table 7.2. Small-scale dairy production systems in Acre, by level of technology.

Technical coefficients	D1	D2
1. Herd inputs		
Feed supplements		
Elephant grass, forage (kg animal^{-1})	0	20
Salt (kg animal^{-1} year^{-1})	110	0
Mineral salt (kg animal^{-1} year^{-1})	0	18.25
Animal health		
Aftosa (foot and mouth disease) (vaccinations animal^{-1} year^{-1})	2	2
Brucellosis (vaccinations female calf^{-1} year^{-1})	0	1
Rabies (vaccinations animal^{-1} year^{-1})	0	1
Carrapaciticida (ml of butox animal^{-1} year^{-1})	5	10
Worm control (ml animal^{-1} year^{-1})	10	25
Antibiotics		
Mata bicheira (cc animal^{-1} year^{-1})	0	0.03
Terramicina (ml year^{-1} to half the herd)	0.06	0.13
Labour for herd management		
Milking (man-days lactating cow^{-1} month^{-1})	0.9	1.5
Other activities (man-days animal unit^{-1} month^{-1})	0.3	0.6
2. Herd dynamics		
Calving rate (% cows giving birth year^{-1})	50	67
Mortality rate (death rates, by age, %)		
< 1 year	10	6
< 2 years	5	3
> 2 years	3	2
Culling/discard rate (% animals discarded year^{-1})		
Cows	0	10
Bulls	6	12
3. Milk production		
Milk production dry season (litres day^{-1})	2.5	4.5
Milk production wet season (litres day^{-1})	3	6
Lactation period (days year^{-1})	180	240

block of rows presents herd demographics. The final block of rows presents milk production coefficients.

As in the case of pasture systems, different production systems involve different levels of investment and changes in management strategies. The traditional dairy system uses low-productivity cattle. Ranchers need little expertise to manage the system, which also makes minimal use of purchased inputs. In contrast, the more intensive dairy system involves an improved breed of cattle, substantial use of purchased inputs and improved animal husbandry techniques.[4] Not only must the rancher purchase animals of higher quality, he or she must also manage the herd more intensively to realize that genetic potential.

The D2 dairy system requires substantially more purchased inputs than the D1 system. Ranchers provide the cattle with mineral salt and elephant grass (green chop) in the dry season, rather than simple salt. The types, number and dosages of vaccinations also increase.

Herd management (culling and discard rates) changes radically in the D2 system. Ranchers using the D1 system do not necessarily discard their cows, although older cows are generally sold, depending on liquidity needs. In contrast, with D2 technology 10% of cows (the oldest and least productive) must be discarded each year to achieve productivity goals.

These changes in the herd genetic composition and management techniques lead to large differences in milk production. Moving from D1 to D2 technology roughly doubles daily milk offtake and increases lactation periods by about one-third.[5]

Tables 7.3, 7.4 and 7.5 summarize the capital and labour requirements for the establishment and maintenance phases of P1 and P2 pasture systems, coupled dairy–pasture systems (D1–P1, D2–P2 and D2–P3) and coupled beef–pasture systems (B1–P1, B2–P2 and B2–P3), respectively.

During the pasture establishment period (Table 7.3), which lasts for about 1 year for all the technologies, switching from P1 to P2 technologies requires substantial (but not proportional) increases in capital and labour. Capital inputs increase by about 60% and labour requirements roughly double. During the maintenance phase, however, no capital is required and, depending on which of the two more intensive technologies the rancher adopts (P2 or P3), labour use can increase or decrease. P2 grass-based pastures require more labour for weeding than do P1 pastures, but P3 legume-based pastures require less. Finally, the capital/labour ratios show that P2 and P3 pastures (but especially P2) are more labour-intensive than traditional pasture technologies.

Adding information on pasture costs to the establishment and operational costs associated with different intensities of dairy production yields Table 7.4.[6] Several results emerge. To establish a D2–P3 system requires about 2.5 times more capital than to establish a traditional dairy/pasture system (D1–P1), primarily due to the costs of acquiring a more productive herd. In addition, the labour required for establishing more intensive systems more than doubles, primarily due to more fence building. Thirdly, due to increased milking

Table 7.3. Capital and labour requirements for establishment and maintenance of pastures, by technology, per hectare.

	Pasture traditional P1	Pasture grass-based P2	Pasture legume/grass-based P3
Establishment period (1 year)			
Capital requirements (R$ ha^{-1} year^{-1})	152	241	252
Labour requirements (man-days ha^{-1} year^{-1})	6.3	11.4	11.4
Labour requirements (R$ ha^{-1} year^{-1})	37.2	64.2	64.2
Maintenance period*	10 years	14 years	19 years
Capital (R$ ha^{-1} year^{-1})	0	0	0
Labour (man-days ha^{-1} year^{-1})	1.1	1.3	0.6
Labour (R$ ha^{-1} year^{-1})	5	8	3.5
Key ratios			
Establishment period			
Capital/labour ratio (R$/R$)	4.1	3.8	3.9
Maintenance period			
Capital/labour ratio (R$/R$)	0	0	0

*Maintenance period is defined as the number of years during which inputs are used to manage pastures. The useful life of pastures can extend a few years beyond the maintenance period.

and herd management costs, it costs nearly three times as much in labour to operate a D2–P3 system than to operate a D1–P1 system. Finally, in the operational phase, the capital and labour costs of the most intensive system (D2–P3) are about seven and two times greater, respectively, than in the traditional system. (These are all dairy-cattle costs. The pastures require no capital during the operational phase.)

The much higher capital and labour requirements of the more intensive systems can limit their adoption, especially in areas with poorly functioning financial and labour markets. But, as we show below, the more intensive systems are much more profitable. So, once established, we expect them to generate sufficient cash to cover all labour and capital costs.

The more intensive D2–P2 systems are more labour-intensive than the D1–P1 systems in the establishment phase (i.e. they have lower capital-to-labour (K/L) ratios), because the labour required to weed the pastures increases substantially. In contrast, the legume-based D2–P3 system is more capital-intensive than the D1–P1 system, this time due to substantial increases in purchased inputs for herd management. In the operational phase, the K/L ratio rises (i.e. the systems become more capital-intensive) as we move from the traditional to more intensive systems.

Table 7.4. Capital and labour requirements for establishment and maintenance of dairy/pasture production systems, on a per-hectare basis.*

	Traditional dairy/pasture system D1–P1	Improved dairy/grass-based pasture system D2–P2	Improved dairy/legume-based pasture system D2–P3
Establishment period (1 year)			
Capital requirements (R$ ha^{-1})	252	479	692
Labour requirements (man-days ha^{-1})	7.6	15.9	19.6
Labour requirements (R$ ha^{-1})	43.2	84.7	102
Maintenance period	10 years	14 years	19 years
Capital (R$ ha^{-1} year^{-1})	1.2	4.6	8.5
Labour (man-days ha^{-1} year^{-1})	3.6	8.2	10
Labour (R$ ha^{-1} yrea)$^{-1}$	21	44.5	48.2
Key ratios			
Establishment period			
Capital/labour ratio (R$/R$)	5.8	5.7	6.8
Maintenance period			
Capital/labour ratio (R$/R$)	0.06	0.10	0.18

*Combined dairy/pasture system requirements are averaged over 20 years, for all systems, to capture declining carrying capacity and the 'zero input' status of P1 and P2 grass systems, which are untouched after years 11 and 15, respectively. P1 pastures become unproductive in year 15, but we continue to use this now idle land to weigh calculations of average input requirements and production.

Finally, Table 7.5 combines information on pasture costs with the establishment and operational costs associated with different levels of intensity of beef-cattle production. Moving from a traditional beef system (B1–P1) to more intensive systems increases the absolute outlays for capital and labour during both the establishment and operational phases of production. Labour costs during the operational phase more than double with a shift from B1–P1 to B2–P2, but the rise is less steep with the adoption of B2–P3, since it has lower pasture management costs than B2–P2. The K/L ratio during the establishment period for beef/pasture systems is basically unchanged by the move from B1–P1 to B2–P2, but increases for the B2–P3 system. Finally, the K/L ratio during the maintenance period increases with the adoption of more intensive systems, due primarily to increased costs of maintaining herd health.

What can these summary tables tell us about technology adoption and the possible links between the intensification of cattle production systems and deforestation? If we keep in mind that small-scale farmers at the forest margins generally operate in labour- and capital-constrained contexts, and if we focus only on how they are likely to allocate their initial available resources and how that might affect deforestation, we can deduce the following.

Table 7.5. Capital and labour requirements for establishment and maintenance of beef/pasture production systems, on a per-hectare basis.*

	Traditional pasture/beef system B1–P1	Improved beef/ grass-based pasture system B2–P2	Improved beef/ legume-based pasture system B2–P3
Establishment period (1 year)			
Capital requirements (R$ ha^{-1})	200	356	464
Labour requirements (man-days ha^{-1})	6.8	13	14.3
Labour requirements (R$ ha^{-1})	39.7	71.5	77.7
Maintenance period	10 years	14 years	19 years
Capital (R$ ha^{-1} year^{-1})	0.6	2.4	4.4
Labour (man-days ha^{-1} year^{-1})	2.8	5.3	4.7
Labour (R$ ha^{-1} year^{-1})	17.5	31.3	23.9
Key ratios			
Establishment period			
Capital/labour ratio (R$/R$)	5.0	5.0	6.0
Maintenance period			
Capital/labour ratio (R$/R$)	0.03	0.08	0.18

*Combined beef/pasture system requirements are averaged over 20 years, for all systems, to capture declining carrying capacity and the 'zero input' status of P1 and P2 grass systems, which are untouched after years 11 and 15, respectively. P1 pastures become unproductive in year 15.

First, traditional beef production systems have the lowest absolute input requirements. In addition, more intensive dairy systems have consistently higher absolute capital and labour requirements than more intensive beef systems – in some cases, substantially higher. Therefore, based on absolute input requirements alone, farmers in severely constrained capital and labour situations should find beef systems in general, and traditional beef systems in particular, most attractive.

Secondly, traditional and more intensive systems have quite similar establishment costs, but the role of capital in maintaining both dairy and beef systems increases markedly as these systems intensify. Based on K/L ratios, the capital constraints to establishing more intensive systems appear relatively similar across all systems, but the more intensive systems impose relatively higher capital constraints faced during the operational phases of production.

Based on absolute input requirements alone, farms adopting dairy production systems of any type should deforest less than those adopting roughly comparable beef production systems. With any given amount of labour and capital the rancher has available, he or she will be able to establish a smaller area with the dairy system than with a beef system that has a comparable level of intensity. Following the same logic, intensifying any livestock production systems that involves grass-based pastures should reduce deforestation, since intensive systems require more capital and labour. However, the most intensive, legume-based pasture management system actually releases labour, which could be used for deforestation.

Dairy systems are slightly more capital-intensive to establish than beef systems with a comparable level of intensity (i.e. they have a higher K/L ratio). However, since the establishment period lasts only a year or so for all systems, the K/L ratio during the operational phase will have a longer (and perhaps greater) influence on deforestation. The latter increases steadily as dairy and beef systems become more intensive, suggesting that, if forest clearing were a relatively capital-intensive activity, intensification of cattle production activities would reduce deforestation by drawing capital away from forest-felling activities.

Nevertheless, this analysis of the links between technology and deforestation, based solely on Tables 7.3 to 7.5, misses several key aspects. First, it fails to address the profitability of the activities, and it is via profits that key farm-level constraints to system adoption and expansion will be overcome. Secondly, it does not specify what the smallholders' objectives are. Thirdly, and perhaps most importantly, the tables present particular activities in isolation of one another and independent of other on- and off-farm activities. The interdependencies among these competing activities may be much more important in determining intensification/deforestation links than the requirements of any specific activity, especially in capital- and labour-constrained environments. To include these elements, we need an approach that looks at the whole farm. The following section takes such an approach.

5. A Farm-level Model

Farmers allocate land, labour and capital based on the expected returns to alternative on- and off-farm activities. Some activities, such as annual cropping, can generate short-term returns. Others, like cattle production, bring returns over the medium term. Still others, including producing timber-trees, offer returns only over the long term. Since poor smallholders prefer short-term returns to long-term returns, timing matters a great deal.

When deciding between activities, farmers also face economic and biophysical constraints. For example, households do not have an unlimited supply of labour to allocate to production and some cropping patterns are simply not feasible on poor soils. The fact that smallholders are often constrained in their access to factors of production implies that different activities compete with each other for household resources. Thus, even if a particular activity like cattle production or agroforestry looks quite promising when examined in isolation, it may turn out to be less profitable than alternative activities. To deal with the timing of returns, the degrees to which biophysical or other constraints limit choices and the extent of on-farm competition among activities for scarce resources requires a long-term, whole-farm view and analytical tools that are based on such a view.

We developed a farm-level bioeconomic LP model to explicitly account for the biophysical and economic factors that determine farmers' land-use decisions and choices of production techniques.[7] The model assumes that farmers maximize the discounted value of their families' consumption streams (directly related to, and hence below referred to as profit stream) over a 25-year time horizon by producing combinations of products for home consumption and sale, subject to an array of constraints. These constraints relate to the technologies available to produce agricultural and forest products, the impact of agricultural activities on soil productivity and the financial benefits associated with different activities, including the potential to sell household labour off-farm and to hire labour for agricultural purposes. Besides producing agricultural products, farmers in our model also have the option of extracting Brazil nuts, an activity that generates a low but constant per-hectare return. The model also includes biophysical constraints, e.g. how soil fertility problems restrict agricultural productivity and soil recovery, and to what extent external inputs can correct these problems.

The model begins from a prespecified set of initial conditions. These include the initial land use on the farm (depicted on the vertical axes of Figures 7.2 and 7.3 at 'year zero'), as well as a number of farm- and household-specific constraints (for example, family size and distance to market) that can influence the allocation of land, labour and cash to alternative land uses.[8] The model also takes into account certain market imperfections, e.g. quotas constrain milk sales and farmers can only acquire 15 man-days of hired labour in any given month. Finally, the model explicitly includes some forestry policies, but excludes others. Small-scale farmers are not allowed to harvest timber products from their forested land. However, the rule that forbids farmers from clearing more than half of their farm for agricultural purposes is not enforced in the model simulations presented here.[9]

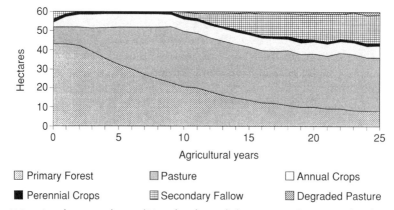

Fig. 7.2. Land uses under traditional-only cattle/pasture production technologies.

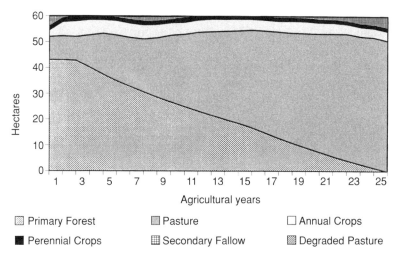

Fig. 7.3. Land uses under the 'free-choice' cattle/pasture production technologies.

6. Results of Model Simulations

We present two sets of simulations to assess how introducing the more intensive pasture and cattle (beef and dairy) production technologies described above may affect deforestation and farm income, as well as whether farmers would find the more intensive systems more economically attractive. First, we constrain our representative small-scale farmer to use only traditional (B1–P1 and/or D1–P1) traditional production technologies. Then we allow the farmer to select whichever pasture and cattle technologies maximize profits. In both simulations, the farmer can choose mixed beef/dairy herds, dairy only or produce no cattle at all if other activities provide higher profits than cattle production.

6.1. Scenario 1: low-intensity technology only

When we restrict our representative small-scale farmer to adopting B1–P1 and/or D1–P1 cattle and pasture production systems, he or she chooses both. Figure 7.2 presents the resulting land uses (including deforestation) over the simulation's 25-year time horizon.[10] Deforestation begins slowly, accelerates from about year 3 to year 15, and then slows substantially (but does not stop). Pasture area expands dramatically between year 3 and about year 10, and remains constant thereafter at roughly 50% of farm area. Area dedicated to annual and perennial crops remains roughly constant over the entire period. Secondary fallow area increases substantially, beginning in about year 8. Finally, beginning in about year 10, small amounts of land are dedicated to rehabilitating degraded (though not necessarily completely unproductive) pastures.

Herd growth (not shown here) under the traditional technology simulation is moderate. Total carrying capacity of pastures reaches a maximum of about 23 animal units in year 9. Milk cows constitute one half of the herd in year 9 and calves and beef cattle account for the other half. The latter become important beginning in year 5 and their number stabilizes after about year 9.

6.2. Scenario 2: free choice between traditional and more intensive technologies

When we allow our representative farmer to choose from a combination of pasture and cattle production technology packages, the model predicts the adoption of D2–P3 and B2–P3 technologies. Figure 7.3 shows how we expect land use to evolve. The amount of forest clearly declines over time, finally disappearing in about year 25. Pasture eventually occupies about 85% of the farm. Annual crops occupy about 8% of the farm throughout the 25-year time horizon. Perennial crops (in this case, manioc, which has a production cycle spanning more than 1 year) consistently take up about 1 ha of land. Secondary fallow fluctuates, becoming significant as forests disappear.[11]

Under the 'free-choice' scenario, herd growth (not shown here) is rapid and sustained. By about year 15, pastures can support roughly twice the number of animal units as the 'traditional technology only' farm. As in the traditional-technology scenario, dairy production using D2–P3 technology begins early on and continues to play an important role throughout. But the scale of milk production is more than double that of the traditional-technology farm. Beef (produced using B2 technology) emerges more slowly than in the traditional-technology case, but still eventually comprises about 25% of the total herd.

Of critical interest to small-scale farmers is the profit stream they can hope to earn in each of these scenarios. The second scenario, which permits farmers to adopt the more intensive technologies, consistently provides higher profit streams than the traditional-technology scenario. The net present values (NPV) of the profit streams for the traditional and 'free-choice' scenarios are R$19,813 and R$50,635, respectively.[12] Savings during the first few years allow for subsequent investments, which boost production (and profits) in later years. To expand P3 pasture areas and purchase high-quality D2 and B2 cattle require large investments (negative savings) in years 5, 9 and 11.

7. Conclusions and Policy Implications

The chapter has addressed three central questions in the context of small-scale agriculture at the forest margins in the western Brazilian Amazon:

1. Do more intensive pasture and cattle production systems exist and, if so, what are their labour and capital requirements?

2. If they exist, will they be adopted, and why/why not?
3. If adopted, what will the impacts be on deforestation and on farm income?

Field research confirms that some types of more intensive, sustainable pasture and cattle production systems exist and smallholders are adopting them. More intensive systems have pastures with higher carrying capacity, which produce more animal products and which can last longer than traditional systems, but require more capital and labour to establish and manage. The K/L ratios during the establishment phases of production are higher for dairy than for roughly comparable beef production systems. The same applies to the operational phases of comparable production systems, except in the case of the most intensive beef and dairy systems, which have similar K/L ratios.

Secondly, many more smallholders are likely to adopt the more intensive pasture and cattle production systems in the future, since the financial returns from the more intensive systems are much higher than those of traditional systems.

Thirdly, more intensive systems will probably increase, rather than decrease, the pressure on the forests that remain on farmers' land. Greater profitability will create a demand for larger milking and beef cattle herds and pasture to support them. The only major constraint on forest conversion at the farm level will be seasonal labour shortages. This, however, only becomes clear once one takes a 'whole-farm' view, which allows comparison of the returns to scarce resources across many possible activities and over time.

There are several caveats, though. Many smallholders in the region may not have enough capital and labour to establish and manage more intensive cattle–pasture systems and so poorly performing capital and labour markets could limit adoption. Credit can help promote adoption, even without high or long-term credit subsidies, since these more intensive systems generally become profitable within a few years of establishment.

Secondly, farmers will have to change their production practices to adopt and effectively use more intensive systems and there is no guarantee that they will have the information and ability they need to make those changes. If they do not establish and manage their intensive systems well, they will get lower returns and cause greater soil and pasture degradation.

Thirdly, in the analysis presented here, it is assumed that the entire technology package was adopted. If only certain components of the packages were adopted, profits and/or environmental sustainability could be undermined.

Fourthly, while the clear trade-off between the greater profitability of the more intensive systems and the higher deforestation associated with them should concern policy-makers, it also provides an entry point for policy action. Policy-makers may now have something to offer to farmers in exchange for reduced forest clearing. More intensive livestock systems will require additional research and extension services for smallholders to properly establish and manage them. Policy-makers can provide smallholders with both. The private sector is actively developing some improved technologies and promoting them to large-scale ranchers, but may not pay much attention to smallholders. Policies that guarantee access to processing facilities for fluid milk may also be needed. Here too, policy-makers can help. In exchange for research, extension services and improved infrastructure, policy-makers could ask farmers to slow deforestation (perhaps by adhering to the 50% rule). Farmers would probably have a financial incentive to agree to such a plan, but problems of monitoring and implementation clearly remain.

Notes

1   This chapter benefited greatly from technical field data provided by Merle Faminow, Támara Gomes, Claudenor Sá and Samuel Oliveira, comments by the editors, an anonymous referee, participants in the CIFOR workshop on agricultural intensification–deforestation links, colleagues in the Alternatives to Slash-and-Burn Agriculture Programme (ASB) and participants at seminars at the International Food Policy Research Institute, the H.A. Wallace Institute, the Empresa Brasileira de Pesquisa Agropecuária and the University of Maryland. Financial support was provided by the Inter-American Development Bank and the World Bank. We dedicate this chapter to the memory of Erennio Giacomazzi, who provided office space and much moral support.
2   Technology adoption issues addressed in this chapter focus primarily on economic viability; for a more comprehensive set of adoption issues in the same socio-economic and agroclimatic setting, see Vosti *et al.* (2000).
3   Earlier reports suggested that cattle production systems and especially the pastures associated with them could not be sustained for more than a few years (Hecht, 1984). More recent evidence on traditional and emerging cattle production systems shows that both are much more sustainable than previously thought (Faminow and Vosti, 1998).
4   The input and output coefficients for traditional and more intensive dairy and beef production systems presented in this section are based on completely specialized production schemes. In reality, mixed herds are quite common among smallholders in the region. These systems are examined in the context of the LP model presented in the next section.
5   We conducted similar analyses of traditional and more intensive beef production systems. These systems are basically calf-purchasing and fattening operations. Space constraints preclude a detailed presentation of these systems here, but more intensive systems increase calf weights by 25%, increase slaughter weight slightly and greatly speed the fattening process. Combined beef–pasture systems are examined at the end of this section.
6   Recall that, by assumption, pasture and cattle production systems (dairy and beef) are 'coupled'. P1 pasture can only support D1 dairy and B1 beef production and P2 pasture is not used in D1 or B1 systems. Field observations support this assumption.
7   For a complete description of the LP model, see Carpentier *et al.* (2001).
8   These initial conditions are based on field data collected in 1994. We used statistical techniques to cluster farm households from the Pedro Peixoto settlement project in Acre into several groups, based on certain characteristics that we felt were exogenous to the farmers' land-use decisions, such as soil type, distance to market and duration of settlement. Several clusters emerged, each of which can be thought to

represent a farm type. We used the average characteristics for the farm type with relatively good access to markets to obtain the initial conditions for our model. The predominant soil types in this cluster of farms had fertility problems and/or mild slope or rockiness partially restricted their agricultural productivity. The model simulations in this chapter take the characteristics of this typical farm as their point of departure.

9 This analysis ignores general equilibrium effects. That may not be justifiable for some products and/or technological changes. For example, to analyse non-timber forest products, which face notoriously thin and seasonal markets, one must take into account the fact that technologies that increase their supply may decrease output prices. In our case, however, which focuses on cattle production, it seems reasonable to ignore general equilibrium effects. Beef is traded internationally and regional supply still does not completely satisfy regional demand, so small-scale farmers can be characterized as price takers in a fairly competitive market (Faminow and Vosti, 1998). Farmers can also increase milk production without significantly depressing prices, since up to 80% of milk processing capacity is idle during at least some part of the year (J.F. Valentim, personal observations).

10 None of the simulations presented in this chapter reach steady-state land uses. Therefore, we cannot assess the potential for any collection of activities (or technologies) to sustain a small-scale farm family over the very long term.

11 Extending this simulation to 35 years shows that the area in secondary fallow continues to increase by approximately 0.20 ha every 2 years and plateaus at 5.5 ha in year 35.

12 We report all values in 1996 Brazilian reais. All the simulations use a constant set of 1993/94 input and product prices for the entire decision time horizon. We used a 9% discount rate to calculate NPV.

References

Brazil (1969) *Amazônia: Instrumentos Para o Desenvolvimento*. Ministério do Interior, Banco da Amazônia, Belem, Pará.

Brazil (1981) *III Plano Nacional de Desenvolvimento – 1980/85*. Presidencia da República, Secretaria de Planejamento, Brasília, DF.

Browder, J.O. (1994) Surviving in Rondônia: the dynamics of colonist farming strategies in Brazil's northwest frontier. *Studies in Comparative International Development* 29(3), 45–69.

Bunker, S.G. (1985) *Underdeveloping the Amazon: Extraction, Unequal Exchange, and the Failure of the Modern State*. University of Illinois Press, Urbana, Illinois.

Carpentier, C.L., Vosti, S.A. and Witcover, J. (2001) *FaleBEM: a Farm Level Bioeconomic Model for the Western Amazonian Forest Margin*. Discussion Paper, Environment and Production Technology Division, International Food Policy Research Institute, Washington, DC.

Dale, V.H., O'Neill, R.V., Pedlowski, M. and Southworth, F. (1993) Causes and effects of land-use change in central Rondônia, Brazil. *Photogrammetry Engineering and Remote Sensing* 59(6), 997–1005.

de Santana, A.C., de Alencar, M.I.R., Mattar, P.N., da Costa, R.M.Q., D'Ávila, J.L. and Souza, R.F. (1997) *Reestruturação Produtiva e Desenvolvimento na Amazônia: Condicionantes e Perspectivas*. BASA, Belém.

EMATER-Acre and Embrapa (1980) *Sistemas de Produção Para: Gado de Corte e Gado de Leite, Microregião Alto Purus.* Boletim No. 221, EMATER, Embrapa, Rio Branco, Acre, Brazil.

Faminow, M.D. and Vosti, S.A. (1998) Livestock–deforestation links: policy issues in the western Brazilian Amazon. In: Nell, A.J. (ed.) *Livestock and the Environment. International Conference, Wageningen, The Netherlands.* World Bank, Washington, DC, Food and Agriculture Organization, Rome.

Forum Sobre a Amazônia (1968) *Rio de Janeiro, RJ. Problemática da Amazônia.* CEB, Rio de Janeiro.

Fujisaka, S., Bell, W., Thomas, N., Hurtado, L. and Crawford, E. (1996) Slash-and-burn agriculture, conversion to pasture, and deforestation in two Brazilian Amazon colonies. *Agriculture, Ecosystems and Environment* 1015, 1–16.

Hecht, S.B. (1984) Cattle ranching in Amazonia: political and ecological considerations. In: Schmink, M. and Wood, C.H. (eds) *Frontier Expansion in Amazonia.* University of Florida Press, Gainesville, Florida.

Homma, A.K.O. (ed.) (1998) *Amazônia: Meio Ambiente e Desenvolvimento Agrícola.* Embrapa Produção de Informação, Brasilia, and Embrapa-CPATU, Belém.

IBGE (Instituto Nacional de Pesquisas Espacias) (1998) *Censo Agropecurária 1995–6,* No. 3. Acre, Rovaima e Amapá. IBGE, Rio de Janeiro, pp. 56–57.

IBGE (Instituto Brasileiro de Geografia e Estatística) (1997) *Anuário Estatistico do Brasil,* Vol.57. IBGE, Rio de Janeiro.

INPE (Instituto Nacional de Pesquisas Espaciais) (1999) Available at http://www.inpe.br/Informacões_Eventos/amz/amz.html. PRODES folder.

Jones, D.W., Dale, V.H., Beauchamp, J.J., Pedlowski, M.A. and O'Neill, R.V. (1995) Farming in Rondônia. *Resource and Energy Economics* 17, 155–188.

Leña, P. (1991) Ritmos e estratêgias de acumulação camponesa em áreas de colonização: um exemplo em Rondônia. *Boletim Museu Paraense Emélio Goeldi, Seria Antropologia* 7(1), 37–70.

Mahar, D. (1979) *Frontier Policy in Brazil: a Study of the Amazonia.* Praeger, New York.

Scatena, F.N., Walker, R.T., Homma, A.K.O., de Conto, A.J., Palheta, Ferreira, C.A., de Amorim Carvalho, R., da Rocha, A.C.P.N., dos Moreira Santos, A.I. and de Oliveira, P.M. (1996) Cropping and fallowing sequences of small farms in the 'terra firme' landscape of the Brazilian Amazon: a case study from Santarem, Pará. *Ecological Economics* 18, 29–40.

Serrão, E.A.S. and Homma, A.K.O. (1993) Country profiles: Brazil. In: *Sustainable Agriculture and the Environment in the Humid Tropics.* National Research Council, National Academy Press, Washington, DC, pp. 265–351.

Smith, N.J.H., Serrão, E.A.S., Alvim, P.T. and Falesi, I.C. (1995) *Amazonia: Resiliency and Dynamism of the People.* United Nations University, New York.

Soares, O. (1997) *Rondônia Produz.* Senado Federal PTB, Brasília, DF.

SUDAM (Superintendência de Desenvolvimento da Amazônia) (1976) *II Plano de Desenvolvimento da Amazônia: Detalhamento do II Plano Nacional de Desenvolvimento (1975–79).* SUDAM, Belém, Pará.

United Nations Development Programme (UNDP) (1998) *Human Development Report.* United Nations Publications, New York.

Valente, M.G. (1968) A Amazônia brasileira e as outras amazônias. In: *Forúm Sôbre a Amazônia, 2, Rio de Janeiro, RJ. Problemática da Amazônia: Conferências.* CEB, Rio de Janeiro, pp. 277–295.

Valentim, J.F. (1989) *Impacto Ambiental da Pecuária no Acre*. Folio 2203, Documento base do Curso de Avaliação do Impacto Ambiental da Pecuária no Acre, 4 a 8 de dezembre 1989. EMBRAPA–UEPAE, Rio Branco.

Veja (1999) 7 April, pp. 108–115.

Vosti, S.A. and Witcover, J. (1996) Slash-and-burn agriculture – household perspectives. *Agriculture, Ecosystems and Environment* 58, 23–38.

Vosti, S.A., Witcover, J., Carpentier, C.L., Magalhães de Oliveira, S.J., dos Santos, J.C., with do Amaral, E., de Araújo Gomes, T.C., Borges de Araújo, H.J., Braz, E.M., Carneiro, J., Castilla, C., Gonçalves Cordeiro, D., Luis Franke, I., Gillison, A., Mansur Mendes, Â., Neves d'Oliveira, M.V., Palm, C., Rodrigues, V., Rossi, L.M., de Sá, C.P., Hermes Vieira, A., Valentim, J.F., Weise, S. and Woomer, P. (2000) Intensifying small-scale agriculture in the western Brazilian Amazon: issues, implications and implementation. In: Lee, D. and Barrett, C. (eds) *Tradeoffs or Synergies? Agricultural Intensification, Economic Development and the Environment*. CAB International, Wallingford, UK.

Vosti, S.A., Witcover, J. and Carpentier, C.L. (2001) *Agricultural Intensification by Smallholders in the Western Brazilian Amazon: From Deforestation to Sustainable Land Use*. Research Report. International Food Policy Research Institute, Washington, DC.

Walker, R. and Homma, A.K.O. (1996) Land use and land cover dynamics in the Brazilian Amazon: an overview. *Ecological Economics* 18, 67–80.

Witcover, J. and Vosti, S.A. (1996) *Alternatives to Slash-and-burn Agriculture (ASB): a Characterization of Brazilian Benchmark Sites of Pedro Peixoto and Theobroma, August/September 1994*. MP-8 Working Paper No. US96-003, International Food Policy Research Institute, Washington, DC.

Wolstein, A.R.P., Lima, E.M., do Amaral, E.F., Braz, E.M., Pinheiro, F.L.N., Franke, I.L., dos Santos, M.H. and Silva, R.F. (1998) *Metodologia Para o Planejamento, Implantação e Monitoramento de Projetos de Assentamentos Sustentáveis na Amazônia*. Documentos, 32, Embrapa–CPAF/AC/INCRA/Funtac, Rio Branco, Acre.

Technological Progress versus Economic Policy as Tools to Control Deforestation: the Atlantic Zone of Costa Rica

Peter Roebeling and Ruerd Ruben

1. Introduction

Analysts often mention technological progress and economic policies as alternatives to promote rural development and reduce deforestation. The effectiveness of these two approaches has rarely been compared. This chapter presents such a comparison, based on a bioeconomic model of three types of farms found in the Atlantic zone of Costa Rica, a tropical lowland region of recent agrarian colonization (Kruseman et al., 1994).

Our modelling framework enables us to assess how farmers may respond to both exogenous technological progress in agriculture and economic policies, and how those responses may affect the competition for land between agriculture and forestry. We designed the model to analyse farm-household reactions to changing production conditions, taking into account the specific objectives of small and medium-sized peasant producers and large livestock haciendas. Production options include growing arable crops for local consumption and export (maize, pineapple, plantain, palm heart and cassava), cattle production (beef and milk) and forestry activities (natural and cultivated trees).

We examine both pure yield-increasing technologies and input-saving technologies. The economic policies we simulate include input price subsidies, increased credit availability and reduced transaction costs. We show that the most appropriate instruments to improve farmers' welfare while controlling deforestation combine capital-saving technological progress, yield increases for arable crops and selective input subsidies. This combination permits farmers to increase their income and invest more resources into input-intensive activities· (non-traditional crops, teak plantations), thereby reducing the pressure on natural forests.

The chapter takes into account forests within farm boundaries and those that lie outside. The former includes both natural forests and teak plantations. In the latter case, we assume that forest cover outside the initial farm boundaries declines when the total farm area for arable cropping and pastures exceeds the existing area used.

Section 2 of this chapter briefly describes agricultural production in Costa Rica's Atlantic zone and the main trends with regard to land use. Section 3 introduces our bioeconomic model. Section 4 presents the results from our farm- and regional-level simulations. Section 5 compares the effects of technological progress and economic policies on farmers' welfare and resource use.

2. Land Use and Deforestation in the Atlantic Zone of Costa Rica

The region we refer to as the Atlantic zone is located in eastern Costa Rica and coincides with the Province of Limón. It encompasses 920,000 ha, of which just over 20% was used for agriculture in 1963 (DGEC, 1966). Between 1963 and 1984, the agricultural area increased by almost 40%, largely at the expense of forest (DGEC, 1966, 1976, 1987). At the end of the 1980s, the main land uses were forest (48%), pasture (39%) and bananas (10%). Non-traditional crops, including plantain, root and tuber crops, palm heart, pineapple and ornamental plants, occupied 3% of the land (Roebeling *et al.*, 2000b). Large haciendas and plantations dominate extensive livestock and banana production. Small and medium-sized farmers mostly produce crops and engage in integrated livestock activities.

Economic policies, technological progress, infrastructure development, demographic factors and various legal and institutional aspects have contributed to widespread deforestation in Costa Rica during recent decades. Over the last 50 years, forest cover fell from 80% to less than 25% (Quesada, 1990; Leonard, 1996). According to the Food and Agriculture Organization (FAO, 1993), farmers cleared roughly 41,000 ha year^{-1} between 1950 and 1977 and 60,000 ha $^{-1}$ year^{-1} during the following decade.

Traditionally, farmers and policy-makers considered forest areas as reserves for agricultural expansion and viewed rural development as practically synonymous with the conversion of land for agriculture. Migration of new settlers to the Atlantic zone caused the regional population to treble between 1950 and 1985. Since part of the deforested land was not suited for long-term crop production, farmers converted a large portion of the land to pasture. Government policies, such as interest subsidies and debt rescheduling for livestock production, further encouraged this trend (Wendlandt and Bawa, 1996).

In the 1960s, the establishment of banana plantations was the main source of forest loss (Veldkamp *et al.*, 1992). In the 1970s, thanks to favourable beef prices and credit policies, pasture expansion became much more important. In the 1980s, declining returns from traditional food crops and incentives to produce non-traditional crops led farmers to shift from the former to the latter, while the pasture area again increased sharply (Kruseman *et al.*, 1994). Even so, at present forests still cover about 35% of the region. This includes natural forests and forest plantations within farm boundaries, as well as forests outside farms (Bulte *et al.*, 1998).

The Atlantic zone supplies nearly half of all roundwood to the national sawmills, which contract independent loggers to obtain most of their supply. This makes it difficult to enforce legal restrictions. Loggers only harvest a few high-value species. The construction of a road infrastructure for logging attracts new settlers to the frontier and encourages deforestation.

National and international agencies have developed various technological options for the Atlantic zone. They have generated yield-improving technologies, such as new varieties and higher-quality seeds. They have also promoted capital-saving technologies, such as selective fertilizer applications and optimal spraying, which improve the efficiency of input use, and labour-saving technologies, such as the mechanization of weeding, harvest and postharvest operations. Given the high labour intensity of peasant production and their limited access to formal credit, these households have a major interest in adopting production technologies that reduce labour demands and economize on capital use.

In the past, the Costa Rican government often used price policies to influence land use (Segura, 1992). Since the advent of structural adjustment policies, this has become less frequent. Nowadays, the government relies more on input delivery schemes, technical assistance, credit policies and public infrastructure investments to influence land-use decisions (SEPSA, 1997).

The following sections examine the potential impact of various technological options and economic policy instruments on household welfare and land use at the farm and regional levels. Any full assessment of these instruments would also have to take into account their budgetary implications, an issue beyond the scope of this chapter.

3. The Modelling Framework

We constructed farm models for three representative types of Atlantic zone producers: small farm households (< 20 ha), medium farm households (20–50 ha) and extensive beef cattle farms or haciendas (> 50 ha) (Roebeling *et al.*, 2000a; Table 8.1). We used the 1984 agricultural census to help identify these farm types, taking into account the dominant land use and farmers' perceived objectives (DGEC, 1987). In 1984, livestock haciendas covered 60% of the total agricultural area in the Atlantic zone and represented 11% of the

Table 8.1. Farm characteristics (from Roebeling et al., 2000a).

Farm type	Farm area (ha)	Labour (days)	Savings (US$)	Cattle (animal units)	Savings coefficient (%)	Number of farms
Small	8.9	491.9	281.8	4.1	25	6480
Medium	39.2	412.7	773.5	48.3	25	1690
Hacienda	170.4	570.0	5524.9	188.1	48	803

farms. Small and medium-sized farms covered 33% of the agricultural area and represented 88% of the farms. Our model takes the area devoted to large-scale banana production to be exogenous.

To determine the aggregate effect of technological progress and economic policies, we have 'scaled up' the results for each farm type to the regional level, using weighted aggregation, with the number of farms belonging to each farm type as our weights. The world market determines the prices of beef and teak (Kaimowitz, 1996; de Vriend, 1998). However, the Atlantic zone supplies a considerable share of the national and even world market for a number of products, including bananas, palm hearts, pineapples and plantains, implying that their prices are not completely exogenous (Schipper et al., 1998; Roebeling et al., 2000b). Agricultural policy models that assume prices to be exogenous tend to overestimate the degree of specialization in crop production (Roebeling et al., 2000b). But the assumption does not affect predictions regarding the choice between crop and livestock production or between beef cattle technologies very much. So, given the focus of the study, it seems reasonable to assume exogenous product prices.

The agricultural labour force in the Atlantic zone provided 1 million days per month of labour in 1995, the most recent year for which we have reliable data. In all the scenarios below, we assume that hired labour and family labour are perfect substitutes. Farmers can hire as much labour as they want for a fixed wage of about US$10 day^{-1}, in part thanks to illegal immigration from Nicaragua. We also assume households can obtain as much off-farm employment as they desire. This is a reasonable assumption, given the significant labour demand from nearby banana plantations.

3.1. Small and medium-sized farm households

Our methodology for modelling the behaviour of the small and medium-sized farm households uses a multiple-goal linear programming optimization procedure to analyse production and an expenditure module with an econometrically derived (non-linear) utility function to analyse consumption (Ruben et al., 1994; Kruseman et al., 1997). Combining a linear programming production framework with a direct expected utility function allows production and consumption decisions to interact in such a way that consumptive preferences determine productive choices, whereas sustainability implications derived from production are part of farm-household objectives. Consequently, iterative procedures are used to optimize the model in a non-separable way, given the existing market imperfections. In this respect, our methodology differs from the traditional household-model approach (Singh et al., 1986). The model assigns weights to each of the households' multiple objectives, which include consumer preferences, as well as farm income and sustainability criteria (Kruseman et al., 1997; Roebeling et al., 2000a).

Peasant households possess land, labour, savings and cattle (Table 8.1). They can use savings and formal and informal credit to finance their labour, input and transaction costs. Our model limits formal credit to 25% of the value of the small and medium-sized farmers' land and 20% of the value of their cattle stock. We limit informal credit to 10% of the value of their crop production. Annual real interest rates for formal and informal credit are 12% and 47%, respectively (Roebeling et al., 2000a).

Farm household options include on- as well as off-farm activities. Off-farm activities refer to external employment possibilities for family labour on banana plantations. On-farm activities include cropping, forestry and beef production systems. We used the LUCTOR expert system to generate technical coefficients for the crop and forestry systems and PASTOR for the cattle systems (see section 3.3).

Households obtain utility by consuming purchased products (Q_j^{buy}) and products from the farm (Q_j^{cons}) and by enjoying leisure. We assume that households seek to maximize the following utility function:

$$Z_1 = UTIL = \sum_j U_j^{max}(1 - e^{-\rho_j(Q_j^{cons} + Q_j^{buy})}) + \sum_j \rho_j(Q_j^{cons} + Q_j^{buy})^{\sigma_j} \quad (1)$$

This utility function has a negative exponential function for basic food crops and an exponential function for other food products, non-food products and leisure. This reflects the fact that households tend to purchase relatively fewer basic foodstuffs as their incomes rise. U_j^{max} denotes the maximum attainable utility with commodity j. ρ_j is the conversion factor of consumption to utility. σ_j is the exponent of consumption commodity j. Our data come from the latest national household income and expenditure survey (DGEC, 1988, 1990).

Households maximize utility subject to net farm income, NFI. NFI is equal to the returns from marketed production (Q_j^{sold}) and off-farm employment (O^{off}), minus the costs related to purchasing labour and inputs (I_i), capital (C_b) and consumption. It is defined as follows:

$$NFI = \{\sum_j p_j Q_j^{sold} + w^{off} O^{off}\} - \{\sum_i p_i I_i + \sum_b p_b C_b + \sum_j p_j(Q_j^{cons} + Q_j^{buy})\} \quad (2)$$

where p refers to the price of commodity j, capital source b and labour and capital input i, and where w is the off-farm wage rate.

In addition to their utility objective, households maximize an income objective. We define that objective, Z_2, as net farm income minus the expected monetary value of nutrient losses (van der Pol, 1993):

$$Z_2 = INCOME = NFI - \sum_n p_n B_n \qquad (3)$$

where B_n represents the change in soil nutrient stock and p_n represents the monetary value the farmer assigns to nutrient n.

Actual and expected market prices determine short- and medium-term production decisions, respectively. Expected prices are based on a weighted average of market prices over the last 3 years. Transaction costs represent the margin between market and farm-gate prices, resulting from transport costs, marketing margins and imperfect market information (Roebeling et al., 2000a).

3.2. Extensive Haciendas

The dynamic linear programming model for haciendas evaluates technical options for beef production, according to a long-term profit objective, subject to resource and liquidity constraints. Its dynamic aspects include a savings and investment module with a 10-year planning horizon and the recognition that livestock production requires several periods to come to fruition. These features allow us to model the evolution of land and cattle stocks and the availability of credit, as well as to analyse fertility, mortality, growth and feed requirements related to buying and marketing strategies in an intertemporal framework (Roebeling et al., 1998).

Ranchers' initial resource endowments constrain their actions, although the availability of resources evolves over time as a result of investments in cattle and land. Ranchers use a fixed proportion of net returns obtained in the previous year to finance these investments, as well as to finance their operating costs. They can also use formal credit, but the most they can borrow is 25% of the value of the land and cattle they owned in the previous year. Hacienda owners pay a real interest rate of 10% per year for their credit (Roebeling et al., 1998).

Ranchers can allocate their capital on or off their farms. We assume the money they invest outside the farm goes into the capital market, with an expected return equal to the opportunity cost of capital. On-farm capital allocation possibilities include beef production and investments in land and cattle. Production options are limited to fattening beef cattle on natural or improved pastures, combined with feed supplements. As with the small and medium-sized farm households, we used PASTOR to generate the technical coefficients. Major options to improve livestock systems include better fertilization and weeding of pastures, adjustment of stocking rates and improved herd management options.

Net returns and the expected long-run salvage value of their land guide hacienda owners' economic decisions as they seek to maximize their total discounted profits over the planning period. By the expected salvage value of land, we mean the price that ranchers expect they will receive when they eventually decide to sell their land. This is important, since many ranchers view land as a hedge against inflation or as a long-run investment opportunity (van Hijfte, 1989; Kaimowitz, 1996). Hacienda owners must choose how much to increase the net returns from beef production on improved fertilized pastures and how much to increase the salvage value of the land by further expansion of the low-cost natural pasture area. In principle, the inclusion of the land salvage value in the hacienda owner's objective function should lead to reduced levels of input use per hectare, as well as lower stocking rates.

The term 'net returns' refers to the present value of the difference between the income from cattle sold (Q_j) and expenditures on inputs (Q_i) (including labour), investments in cattle (I^{cattle}) and land (I^{land}), capital costs (C_b) and tax levies (τ_r) over the hacienda's resource value (R_r). The net return objective (NR) over the ten-year (y) planning period is given by:

$$NR = \sum_{y=1}^{y=10} \frac{1}{(1+i)^y} \{(\sum_j p_j Q_{jy}) - (\sum_i p_i Q_{iy} + \sum_b p_b C_{by} + \sum_j p_j I_{jy}^{cattle} + \sum_s p_{sy} I_{sy}^{land} + \sum_r \tau_r R_{ry})\} \quad (4)$$

where i is the time discount rate and p represents prices related to animal classes j, variable input types i, capital sources b and land units s. Tax levies τ_r are differentiated for the resources (r) land and cattle.

The land salvage value objective, LSV, is given by the expected present value of land assets at the end of the 10-year planning period, as follows:

$$LSV = \sum_{y=1}^{y=10} \sum_{s=1}^{s=3} \frac{1}{(1+i)^{y=10}} (p_{s,y=10} I_{sy}^{land}) \quad (5)$$

As in the model for small and medium-sized farm households, the hacienda model uses expected market prices and takes transaction costs into account. Land prices are differentiated according to fertility characteristics. We assume real land prices grow 12.5% per year. This implies that they grow faster than the discount rate.

3.3. Technical coefficients

As mentioned previously, we used the LUCTOR expert system (Hengsdijk et al., 1998) to determine the input–output coefficients for the on-farm cropping and forestry production activities and we used PASTOR for the pasture, herd and feed supplement systems (Bouman et al., 1998). Cropping systems include cassava, maize, palm heart, pineapple and plantain. Forest production systems include the logging of natural forests and teak plantations. Pasture systems include three fertilized improved grasslands, a grass–legume mixture and a mixture of natural grasses. We combined these land utilization types with the three major land types found in the northern Atlantic zone, each subdivided into areas that can or cannot be mechanized. The information on the existing land-use systems came from interviews with expert farmers in the Atlantic zone. We also created our own set of alternative systems, which met various predefined targets. For crops and forest, the alternative systems had to meet a zero soil nutrient loss restriction. We generated different technology levels by combining levels of fertilizer use, crop protection and substitution between manual weeding and herbicide use. For pastures, we defined seven separate levels of nutrient mining, ranging from 0 to 60 kg ha^{-1} year^{-1}. Weeding, fertilization levels and stocking rate determine pasture technology. We defined four beef-cattle production systems based on target animal growth rates.

Technical coefficients include labour requirements, inputs, yields and sustainability indicators and are expressed on a 'per hectare' basis. Our sustainability indicators were the depletion of nitrogen (N), phosphorus (P) and potassium (K) stocks in the soil and the amount of pesticides and herbicides used.

Technological options for improving arable cropping systems can be divided into pure yield-increasing and input-saving practices. Farmers can improve their yields by using better crop phenotypes that make more efficient use of available water and nutrients (maize, beans) or by producing higher-quality products (pineapple). Capital-saving technologies improve input efficiency by controlling nutrient losses and reducing pesticide use through crop-residue management strategies, erosion control measures and integrated pest management practices. Labour-saving technologies involve better timing of operations and the mechanization of soil preparation, sowing and fertilizer applications. Better fertilization or weeding of pastures, the use of feed supplements, adjustment of stocking rates and improved herd management are some of the options for technological progress in pasture and livestock systems.

4. Model Results

4.1. Base run

Table 8.2 presents base-run results for each farm type. In the small farm type, forest represents more than half of the total farm area and is mostly teak forest. The farmers' main cash crops are pineapple and plantain. Food crops (maize and cassava) as well as beef and milk are important for household consumption. Small farms are the most labour-intensive. The medium-sized type focuses on beef production and the exploitation of natural forests, which take up 50% and 32% of the farm area, respectively. The only cash crop they produce is pineapple. Medium-sized farms have a lower labour intensity than small farms, due to the restricted availability of family labour. Their greater capital resources and better access to credit allow their production systems to be more capital-intensive. The hacienda type specializes in beef production using natural pastures, with an average stocking rate of about 1.6 animal units per hectare. As a result, their cattle-raising activities use little labour and capital.

We aggregated the partial model results for each farm type, weighted by the number of farms of that type, to obtain base-run results at the regional

Table 8.2. Base-run results at the farm and regional level.

	Farm type			Region
	Small	Medium	Hacienda	
Economic return* (US$ 1000 year^{-1})	16.7	126.3	945.2	1,371,515
Production structure (ha)				
Maize	0.0	0.0	0.0	158
Pineapple	1.3	6.4	0.0	19,324
Plantain	2.3	0.0	0.0	14,798
Cassava	0.0	0.0	0.0	161
Pasture	0.3	19.4	190.3	187,218
Forest (natural)	1.3	13.4	0.0	30,917
Forest (teak)	3.8	0.0	0.0	24,344
Agrarian frontier	–	–	–	62,234
Total area†	–	–	–	339,155
Resource use intensity‡				
Labour intensity	50.6	21.4	5.3	41.0
Capital intensity	757.3	864.4	185.3	726.3

*'Economic return' refers to net farm income for small and medium-sized farm types, the value of land and cattle stock for the hacienda farm type and the gross agricultural production value or income at the regional level.
†Total area is calculated as the sum of the agricultural frontier area (Kruseman et al., 1994), plus the crop and pasture areas of our three farm types. This figure does not include other crops grown by other farm types in the Atlantic zone, such as bananas.
‡Labour intensity refers to the number of labour days per hectare. Capital intensity refers to the amount spent (US$) to finance variable inputs and (wage and family) labour per hectare.

level. Table 8.2 shows that pasture covers more than half (55%) of the cultivable area in the Atlantic zone. Forest covers some 35%. The remaining 10% is dedicated to crop cultivation, mainly pineapple and plantain. This simulated land-use pattern reflects actual land use fairly accurately. Regional agricultural income totals about US$1.37 million. On average, agricultural production uses 40 labour days and US$725 worth of input per hectare.

4.2. Technological progress and deforestation

We used our model to simulate various scenarios involving the introduction of both pure yield-increasing and input-saving technologies. In the first case, output increases but input levels do not change. In the second case, fewer inputs are required to produce the same level of output. Our pure yield-increasing simulations postulated a 20% increase in crop, pasture or forestry production. Our input-saving simulations examined situations where there was a 20% decline in labour or capital requirements. Table 8.3 shows the predicted impact of these different types of technological progress on agricultural income, total land use and labour and capital intensity.

Pure yield-increasing technological progress

The 20% yield increase in the production of all crops leads the cash-crop area to expand at the expense of forest and, to a much lesser extent, pasture. Small and medium-sized farmers put more area in cash crops (mainly pineapple for export) and diminish the area devoted to forestry activities. They are able to do this because their higher net margins allow them to obtain more informal credit. They hardly reduce the amount of pastures they have, since livestock provide higher returns than forestry activities. Beef and milk production remain important for household consumption. Improvements in crop yields do not affect the hacienda farm type, which produces only beef. While the crop area rises by more than 8%, the total forest area decreases by almost 5%, due to reduced on-farm forestry production. The forests outside farms remain unaffected. Not surprisingly, production becomes more labour- and capital-intensive as the relative role of crop production increases. Since crop production becomes more profitable, household members find on-farm employment more attractive. Higher yields give farmers greater access to informal credit, which makes it easier for them to hire labour. Regional agricultural income increases by almost 11%.

The 20% increase in the productivity of pastures leads total pasture area to expand. Technological progress in pasture production allows small and medium-sized farmers to produce more beef and milk with the same amount of pasture. Increased beef and milk production enables them to obtain more informal credit and leads to a small increase in cash-crop production, as returns from livestock production are low. The hacienda owners react to the improved returns from pasture by expanding their pasture area through the purchase of additional forested land. The higher profitability of beef production facilitates this and the land salvage value objective makes acquisition of additional land and cattle attractive. The almost 10% increase in pasture area comes largely at the expense of an almost 28% decline in the forest area on the agricultural frontier (i.e. outside existing farms). The income effects of technological progress in pasture production clearly dominate the substitution effects. As a consequence, factor intensity hardly changes. Regional agricultural income increases only slightly (0.3%), since beef production offers low net margins, particularly on the haciendas.

The 20% yield increase in forestry production hardly affects production at all, in part because the net return from forestry activities is low. Small and medium-sized farmers marginally decrease forestry production in favour of

Table 8.3. Technological progress and farmers' response at the regional level.

	Pure yield-increasing (+20%)							Input-saving (−20%)				
	Crops	%*	Pasture	%*	Forestry	%*		Labour	%*	Capital	%*	
Income (US$10⁶ year⁻¹)	1,519	10.8	1,376	0.3	1,372	0.0		1,417	3.3	1,555	13.4	
Production structure (ha)												
Maize	0	−100.0	158	0.0	150	−5.6		187	18.2	2,215	1,299.0	
Pineapple	22,349	15.7	19,510	1.0	19,403	0.4		18,720	−3.1	23,007	19.1	
Plantain	14,798	0.0	14,631	−1.1	14,798	0.0		17,681	19.5	17,390	17.5	
Cassava	180	12.1	157	−2.4	161	0.0		173	7.8	144.2	−10.4	
Pasture	187,143	0.0	204,616	9.3	187,218	0.0		182,622	−2.5	172,316	−8.0	
Forest (natural)	28,754	−7.0	30,990	0.2	30,907	0.0		15,385	−50.2	26,592	−14.0	
Forest (teak)	23,695	−2.7	24,163	−0.7	24,284	−0.2		42,394	74.1	36,221	48.8	
Agricultural frontier	62,234	0.0	44,929	−27.8	62,234	0.0		61,993	−0.4	61,270	−1.5	
Total area†	339,155		339,155		339,155			339,155		339,155		
Resource use intensity‡												
Labour intensity	42.0	2.3	41.0	−0.2	41.1	0.1		38.2	−7.0	47.0	14.4	
Capital intensity	794.6	9.4	729.4	0.4	728.7	0.3		733.3	1.0	747.4	2.9	

*Percentage change as compared with base run.
†See note in Table 8.2.
‡See note in Table 8.2.

cash crops. Instead of shifting more towards forestry activities, the increased returns from forestry allow them to obtain more informal credit, which they use to grow additional crops. Since the haciendas produce only beef, yield improvements in forestry do not affect them. The minimal impact of technological progress in forestry on production also implies that the change in agricultural income and in the demand for labour and capital is virtually insignificant.

Input-saving technological progress

The 20% reduction in labour requirements leads the area in forest and cash crops to expand by 4.6% and 6.7%, respectively, while total pasture area declines by 2.5%. Lower labour requirements in agriculture allow households to increase their off-farm wage earnings and reduce small and medium-sized farmers' hired-labour costs. This, in turn, relaxes their capital constraint and permits them to produce more cash crops and teak. They prefer to produce teak rather than natural forest products or beef, since it is more labour-intensive and the reduction in labour requirements favours labour-intensive activities. As a consequence, pasture and natural forest areas decline and farmers use feed supplements to maintain their beef and milk production. The reduced labour requirements and subsequent lower operating expenditures permit hacienda owners to expand their pasture area by purchasing forested lands on the agricultural frontier. However, the effect is small, since labour costs form a minor portion of their total operating expenditures. The net effect of reduced labour requirements and greater labour-intensive cash-crop production is that production becomes 7.0% less labour-intensive and 1% more capital-intensive. Agricultural income rises by just over 3%.

The 20% reduction in capital requirements leads to similar results, but the responses are stronger, since capital inputs represent a major share of total expenditures. Pasture area declines by 8%, while the cash-crop and forest areas rise by 24.1% and 13.7%, respectively. Whereas labour-saving technological change led farmers to cultivate more labour-intensive cash crops, capital-saving technological change favours the cultivation of cash crops, such as pineapple and plantain, which demand more capital inputs. Higher net margins encourage the use of hired and family labour on small and medium-sized farms. Lower input costs mean that farmers do not have to rely as much on off-farm employment to obtain the funds to finance these costs. The large share of input costs as a portion of hacienda owners' total operating costs means that the new technology saves them money, which they use to expand their natural pasture area at the expense of forests on the agricultural frontier. The net result of new agricultural production technologies that require less capital, combined with an increase in cash-crop production, is that labour and capital intensity rise by 14.4% and 2.9%, respectively. Agricultural income goes up by 13.4%. Four factors explain this strong rise in agricultural incomes: (i) lower input and transaction costs; (ii) higher productivity of owned capital resources; (iii) increased use of informal credit (made possible by the higher net margins); and (iv) increased production resulting from the relaxation of the capital constraint.

To sum up, small and medium-sized peasant households respond to improved crop yields by reducing on-farm forestry production in favour of cash-crop production. In contrast, labour- and capital-saving technological progress leads them to increase cash-crop and forestry production at the expense of beef and milk production. Hacienda owners react to all three types of technological change by converting additional forest in the agricultural frontier to natural grassland for beef production.

4.3. Economic policies and deforestation

After simulating the effects of technological progress, we looked at a number of scenarios involving economic policy. Our economic policy simulations include a 20% input price subsidy, a 20% increase in the availability of formal credit availability and a 20% decline in transaction costs due to infrastructure improvements. Table 8.4 shows the results of these simulations in regard to agricultural income, land use and capital and labour intensity.

The 20% input price subsidy induces a small rise in the area devoted to crops and pasture (3% and 0.5%, respectively) at the expense of forests both outside farms (−1.9%) and within farms (−1.4%). The subsidy favours the production of input-intensive cash crops. Farmers obtain part of the resources they need to expand their cash-crop production by reducing the amount of cultivated forests they maintain and shifting towards less resource-demanding natural forestry and beef production systems. Hacienda owners use the money that the input subsidies allow them to save to expand their pastures, thus pushing out the agricultural frontier. However, since inputs only represent a small portion of their total operating and investment expenditures, the effect on investments in land remains limited. The growth in crop and pasture area comes partly at the expense of a 2% decline in the forest area outside the initial farm boundaries. Reduced input costs stimulate farmers to convert to more capital- and less labour-intensive cash-crop, forestry and beef production systems. Due to labour and capital constraints in crop production and the low incidence of input expenditures in pasture production, agricultural income goes up by less than 1%.

The 20% rise in the availability of formal credit encourages a shift from forestry towards cash crops. The cultivated crop area increases by almost 8%, facilitated by a 2% decrease in total forest area. Relaxing the capital constraint in a context of unchanged relative prices favours plantain and pineapple production on small and medium-sized farms. Farmers obtain the resources for this expansion principally by reducing their teak production, which allows them to devote more labour to cash-crop production. Hacienda owners do not

Table 8.4. Economic policies and farmers' response at the regional level.

	Input price		Credit access		Transaction costs	
	(−20%)	%*	(+20%)	%*	(−20%)	%*
Income (US$10⁶ year⁻¹)	1,380	0.6	1,388	1.2	1,431	4.4
Production structure (ha)						
Maize	0	−100.0	168	6.3	1,088	587.2
Pineapple	21,718	12.4	21,064	9.0	22,380	15.8
Plantain	13,645	−7.8	15,818	6.9	15,627	5.6
Cassava	157	−2.4	161	0.0	144	−10.4
Pasture	188,097	0.5	187,111	−0.1	189,051	1.0
Forest (natural)	32,789	6.1	30,863	−0.2	23,468	−24.1
Forest (teak)	21,695	−10.9	21,736	−10.7	42,467	74.4
Agricultural frontier	61,054	−1.9	62,234	0.0	44,929	−27.8
Total area†	339,155	0.0	339,155	0.0	339,155	0.0
Resource use intensity‡						
Labour intensity	40.1	−2.4	43.9	7.0	44.2	7.7
Capital intensity (US$)	761.4	4.8	803.5	10.6	812.1	11.8

*Percentage change as compared with base run.
†See note in Table 8.2.
‡See note in Table 8.2.

alter their production patterns, since they were not capital-constrained even before the new credit policy went into effect. The shift from forestry and beef production to cash-crop production makes production more capital- and labour-intensive. Agricultural income rises by just over 1%. The fact that small and medium households still have to obtain most of their credit from informal sources, even after the policy change, because they do not have enough collateral to have full access to formal credit markets, partly explains this rather limited growth.

The 20% decline in transaction costs substantially affects the price farmers pay for their inputs, as well as the prices they receive for their outputs. It leads them to expand their crop and pasture area and to invest more in forestry plantations, and to clear more forest on the agricultural frontier. Small and medium-sized farmers increase their cash-crop (especially pineapple) and teak production at the expense of pastures for beef production, since the new policy favours products that use lots of inputs and have high-value outputs. The hacienda owners increase their pasture area at the expense of forest on the agricultural frontier, in response to the higher net margins their beef-fattening production systems provide. The area under cash crops and pastures increases by 14% and 1%, respectively, while the total forest area decreases by 6%. The stronger focus on cash-crop production leads labour intensity to rise by 8% and capital intensity by 12%. Regional agricultural income rises by 4%.

In conclusion, technological progress generally generates larger income effects than the economic policy measures we analysed. Input price subsidies elicit reactions similar to those induced by technological change but lead to far more loss in forest cover. Improved access to formal credit and lower transaction costs greatly affect on-farm and frontier forests, respectively, and should be considered second-best alternatives.

5. Conclusions

Our base-run farm-level scenario indicates that cash crops and forest plantations are the main land uses on small farms. Medium-sized farms focus on beef production, combined with natural forest activities and limited cash-crop production. The haciendas fully specialize in pasture-based beef production and expand their pastures by purchasing additional forested land on the agricultural frontier. Aggregated results at the regional level show that pastures cover more than half of the cultivable area, forests cover about one-third and the remainder is dedicated to cash-crop production.

In recent decades, the Atlantic zone experienced massive deforestation. The expansion of banana plantations, government policies that favoured pasture-based beef production, the immigration of new settlers to agrarian frontier areas and the establishment of road infrastructure contributed to this result. Given the abundance of land during the initial settlement phases, policy-makers paid little attention to technological progress as an alternative strategy for improving welfare while conserving forests.

The Costa Rican government can influence land-use decisions by investing in research, extension and technical assistance services that enable farmers to improve yields or use their resources more efficiently. It can also provide farmers with cheaper inputs, greater access to credit or improved commercial facilities that reduce transaction costs. This chapter compares the likely outcomes of these two strategies, and in particular their implications for deforestation, household and regional welfare and resource use. Ideally, we would like to find an optimal policy mix that allows us to simultaneously increase farmers' incomes and reduce deforestation.

Pure yield increases in crop production lead to low levels of deforestation and substantial welfare growth, due to the shift from forestry to cash-crop production on small and medium-sized farms. However, pure yield increases in pasture production bring about significant deforestation and do little to improve welfare, since hacienda farms tend to use increased returns from pasture-based beef production to purchase additional land for pasture at the expense of frontier forests. Yield increases in forestry production have scant effect on deforestation, welfare and resource use, since net returns per hectare remain low. The commodity orientation of yield-increasing technologies influences the distribution of income among farm types. Investing in attempts to increase arable crop yields is the most effective strategy from a welfare perspective and takes pressure off remaining frontier forest areas.

Labour- and capital-saving technological progress both enhance welfare, promote forestry production and enlarge total forest area. Small and medium-sized farms increase cash-crop and forestry production as a result of the relaxed labour and capital constraints, while the haciendas expand beef cattle production at the expense of agrarian frontier forest areas. Capital-saving technological progress leads to stronger responses than labour-saving technological progress, since labour costs form a smaller share of total operating costs.

Economic policy simulations, which include a 20% input price subsidy, a 20% increase in formal credit availability and a 20% decline in transaction costs, lead to similar levels of deforestation and generate only moderate welfare improvements. Reduced transaction costs and, to a lesser extent, input price subsidies provoke an expansion of cash-crop and pasture area and substantially reduce total forest area. They favour production of high-value and input-intensive products on small and medium-sized farms at the expense of pasture-based beef production. On haciendas, they increase the pasture area at the expense of forests on the agricultural frontier. Improved formal credit availability results in a shift from teak forestry to cash-crop production on small and medium-sized farms.

To summarize, labour- and capital-saving technological progress enhances welfare and at the same time increases total forest cover, because additional resources become available for farmers to invest in forest plantations. It also reduces pressure on the agricultural frontier, although small and medium-sized farmers tend to maintain fewer natural forest areas within their farm boundaries. Pure yield-increasing technological progress involving crop production is an attractive option, because it enhances welfare at a minimum cost to forests. But yield increases in pasture production are detrimental for forest cover in agricultural frontier areas. In regard to economic policies, input and credit policies both present clear trade-offs between welfare growth and deforestation. The strong adjustments in factor intensity arising from the applications of our three economic policy instruments indicate that substitution effects tend to prevail and consequently forest cover is likely to be reduced. This is particularly the case for policies that reduce transaction costs, which lead to a sharp reduction in natural forest cover, both within farms and on the agricultural frontier, and its partial replacement by forest plantations.

To improve farmers' welfare while controlling for deforestation, policy-makers should combine: (i) capital-saving technological progress; (ii) yield increases in arable crops; and (iii) selective input subsidies to safeguard natural forest areas. Combining these instruments permits farmers to increase their income by relaxing the capital and labour constraints. This, in turn, enables them to invest more resources in activities such as non-traditional crops and teak plantations, which are capital- and labour-intensive, thereby reducing the pressure on natural forests.

References

Bouman, B.A.M., Nieuwenhuyse, A. and Hengsdijk, H. (1998) *PASTOR: a Technical Coefficient Generator for Pasture and Livestock Systems in the Humid Tropics, Version 2.0*. Quantitative Approaches in Systems Analysis No. 18, AB-DLO/C.T. de Wit Graduate School for Production Ecology, Wageningen, the Netherlands.

Bulte, E.H., Joenje, M. and Jansen, H.P.G. (1998) *Tropical Forest Functions and the Optimal Forest Stock in Developing Countries: Is There Too Much or Too Little Forest in Costa Rica?* Research Paper REPOSA-WUR, Guapiles/Wageningen.

de Vriend, J. (1998) *Teak: an Exploration of Market Prospects and the Outlook for Costa Rican Plantations Based on Indicative Growth Tables*. Report 134, REPOSA, CATIE/MAG/WAU, Turrialba, Costa Rica.

DGEC (Dirección General de Estadística y Censos) (1966) *Censo Agropecuario 1963*. Ministerio de Economía, Industria y Comercio, San José, Costa Rica.

DGEC (Dirección General de Estadística y Censos) (1976) *Censo Agropecuario 1973*. Ministerio de Economía, Industria y Comercio, San José, Costa Rica.

DGEC (Dirección General de Estadística y Censos) (1987) *Censo Agropecuario 1984*. Ministerio de Economía, Industria y Comercio, San José, Costa Rica.

DGEC (Dirección General de Estadística y Censos) (1988) *Metodología*. Encuesta Nacional de Ingresos y Gastos de los Hogares, Informe No. 2, Ministerio de Economia, Industria y Commercio, San José, Costa Rica.

DGEC (Dirección General de Estadística y Censos) (1990) *Avance de Resultados*. Encuesta Nacional de Ingresos y Gastos de los Hogares, Informe No. 1, Ministerio de Economia, Industria y Commercio, San José, Costa Rica.

FAO (1993) *Management and Conservation of Closed Forests in Tropical America*. FAO Forestry Paper No. 101, Food and Agricultural Organization of the United Nations, Rome.

Hengsdijk, H., Nieuwenhuyse, A. and Bouman, B.A.M. (1998) *LUCTOR: Land Crop Technical Coefficient Generator: a Model to Quantify Cropping Systems in the Northern Atlantic Zone of Costa Rica, Version 2.0*. Quantitative Approaches in Systems Analysis No. 17, AB-DLO/C.T. de Wit Graduate School for Production Ecology, Wageningen, the Netherlands.

Kaimowitz, D. (1996) *Livestock and Deforestation in Central America in the 1980s and 1990s: a Policy Perspective*. Center for International Forestry Research, Jakarta, Indonesia.

Kruseman, G., Ruben, R. and Hengsdijk, H. (1994) *Agrarian Structure and Land Use in the Atlantic Zone of Costa Rica*. DLV Report No. 3, AB-DLO/WUR, Wageningen, the Netherlands.

Kruseman, G., Hengsdijk, H., Ruben, R., Roebeling, P. and Bade, J. (1997) *Farm Household Modelling System for the Analysis of Sustainable Land Use and Food Security: Theoretical and Mathematical Description*. DLV Report No. 7, AB-DLO/WAU, Wageningen, the Netherlands.

Leonard, H.J. (1996) *Recursos Naturales y Desarrollo Economico en America Central: un Perfil Ambiental Regional*. IIED, Washington.

Quesada, M.C. (1990) *Estrategia de Conservacion para el Desarrollo Sostenible de Costa Rica. ECODES*. Ministerio de Recursos Naturales, San José, Costa Rica.

Roebeling, P.C., Ruben, R. and Sáenz, F. (1998) Politicas agrarias para la intensificación sostenible del sector ganadero: una aplicación en la Zona Atlántica de Costa Rica. In: Castro, E. and Ruben, R. (eds) *Políticas Agrarias Para el uso*

Sostenible de la Tierra y la Seguridad Alimentaria en Costa Rica. UNA-CINPE/WAU-DLV, San José, Costa Rica, pp. 156–174.

Roebeling, P.C., Sáenz, F., Castro, E. and Barrantes, G. (2000a) Agrarian policy responsiveness of small farmers in Costa Rica. In: Pelupessy, W. and Ruben, R. (eds) *Agrarian Policies in Central America.* Macmillan, Basingstoke, pp. 76–102.

Roebeling, P.C., Jansen, H.G.P., Schipper, R.A., Sáenz, F., Castro, E., Ruben, R., Hengsdijk, H. and Bouman, B.A.M. (2000b) Farm modelling for policy analysis at the farm and regional level. In: Bouman, B.A.M., Jansen, H.G.P., Schipper, R.A., Hengsdijk, H. and Nieuwenhuyse, A. (eds) *Tools for Land Use Analysis at Different Scales. With Case Studies for Costa Rica.* Kluwer Academic Publishers, Dordrecht (in press).

Ruben, R., Kruseman, G. and Hengsdijk, H. (1994) *Farm Household Modelling for Estimating the Effectiveness of Price Instruments on sustainable Land Use in the Atlantic Zone of Costa Rica.* DLV Report No. 4, AB-DLO/WAU, Wageningen.

Schipper, R.A., Jansen, H.P.G., Bouman, B.A.M., Hengsdijk, H., Nieuwenhuyse, A. and Saenz, F. (1998) Evaluation of development policies using integrated bio-economic land use models: applications to Costa Rica. Paper presented at AAEA Pre-conference on Agricultural Intensification, Economic Development, and the Environment, Salt Lake City, USA, July 1998.

Segura, O. (1992) *Desarrollo Sostenible y Politicas Economicas en America Latina.* DEI, San José.

SEPSA (Secretaría Ejecutiva de Planificación Sectoral Agropecuaria) (1997) *Políticas del Sector Agropecuario (Revisión y Ajuste).* Ministry of Agriculture and Livestock (MAG), San José, Costa Rica.

Singh, I., Squire, L. and Strauss, J. (1986) *Agricultural Household Models: Extensions, Applications and Policy.* Johns Hopkins Press for the World Bank, Baltimore, USA.

van der Pol, F. (1993) Soil mining. an unseen contributor to farm income in southern Mali. *Royal Tropical Institute Bulletin, Amsterdam.* 325, 47.

van Hijfte, P.A. (1989) *La Ganaderia de Carne en el Norte de la Zona Atlantica de Costa Rica.* Field Report No. 31, CATIE/Wageningen/MAG, Turrialba, Costa Rica.

Veldkamp, E., Weitz, A.M., Staristsky, I.G. and Huising, E.J. (1992) Deforestation trends in the Atlantic Zone of Costa Rica: a case study. *Land Degradation and Rehabilitation* 3, 71–84.

Wendlandt, A. and Bawa, K.S. (1996) Tropical forestry: the Costa Rican experience in management of forest resources. *Journal of Sustainable Forestry* 3, 91–155.

9 Land Use, Agricultural Technology and Deforestation among Settlers in the Ecuadorean Amazon

Francisco Pichon, Catherine Marquette, Laura Murphy and Richard Bilsborrow

1. Introduction

The countries of the Amazon basin face the challenge of making their farm sector economically productive and environmentally sustainable. Part of that challenge involves getting small farmers, who are major actors in the region's agricultural development, to clear less forest. As the Introduction to this volume (Angelsen and Kaimowitz, Chapter 1) explains, one influential school of thought considers low agricultural productivity a key factor favouring small-farm forest clearing. According to this view, settlers respond to declines in agricultural productivity by opening up new areas rather than adopting land-saving practices, because they perceive frontier land as abundant. These analysts argue that the limited availability of inputs, such as fertilizer, weak agricultural extension services, policies that discourage adoption of yield-increasing technologies and widespread poverty reinforce this process. From their perspective, increasing the productivity of frontier land would deter settlers from the cycle of continually clearing, so governments should aggressively encourage technologies that have that effect (World Bank, 1992).

Evidence from frontier settlers in the north-eastern Ecuadorean Amazon suggests that the introduction of new, externally generated technologies and production systems that provide more revenue and/or higher yields per hectare is not the only way to reduce forest clearing by small farmers and could be counterproductive. Many settlers in the region have adopted farming systems that minimize forest clearing without introducing high-yielding technologies.

Below, we discuss the land-use patterns and practices used by the settlers of the north-eastern Ecuadorean Amazon and what these imply for the relation between agricultural technology, land use and deforestation. We employ a more inductive and empirical approach to exploring the book's central research question, 'When does technological change in agriculture increase or reduce deforestation?', than some of the other Amazonian studies presented in this book (Cattaneo, Chapter 5; Vosti *et al.*, Chapter 7; Kaimowitz and Smith, Chapter 11). Drawing on our previous research in Ecuador, we demonstrate that frontier farmers sometimes develop land-use patterns and agricultural practices that limit how much land they cultivate and clear (Pichon, 1993, 1996a, b, c, 1997a, b; Marquette, 1995, 1998; Murphy *et al.*, 1997, 1999; Murphy, 1998). They do this in part because they are more concerned with minimizing risk and obtaining stable earnings than maximizing their long-term yields and economic returns. Based on this, we argue that people who wish to reduce forest clearing should pay more attention to agricultural practices currently evolving among Amazon settlers and not focus exclusively on promoting externally generated technologies designed to increase yields.

When we refer to deforestation below, we mean the area (in hectares) or proportion (percentage of total area) of a settler's household plot that no longer remains in primary forest. Based on our knowledge of the area, we assume that primary forest once covered practically the entire north-eastern Ecuadorean Amazon. Settler land-use patterns reflect their agricultural activities. Settlers convert forests to various other land uses and their household plots typically combine multiple land uses (food crops, cash crops, pasture, fallow, forest, etc.) We have classified settler land-use patterns according to the amount of forest clearing they involve (low, medium and high). In keeping with the cultural ecology literature (Netting, 1993), we think of settler land-use patterns as reflecting the particular agricultural technologies households employ. Agricultural technology encompasses the materials (e.g. tools and inputs), practices and decision-making processes settler households use (or do not use) in farming their land.

Section 2 provides background information on the study area and the households studied. Section 3 discusses settlers' land-use patterns, with particular attention to how much of their farms they have kept in primary forest. Coffee production represents a noteworthy feature of farmers' land use and we discuss it in section 4. Finally, we relate our findings to conclusions presented elsewhere in this volume and draw out policy implications.

2. The Study Area and Its Settler Population

Conservationists have designated our study area in the north-eastern Ecuadorean Amazon as one of the world's ten major biodiversity hot spots (Myers, 1988). At the same time, Ecuador currently derives over half of its fiscal revenues and foreign exchange earnings from petroleum extraction in this region (Hicks, 1990; World Bank, 1992). At present, the region has no organized private or public settlement schemes. However, small farmers enter the region spontaneously, settling as close as possible to roads built for the oil industry. As a result, the region's population is growing rapidly, with several districts recording double-digit annual growth rates (INEC, 1992). Small farmers clear most of the forest lost each year. Large-scale plantations and logging are limited.

The information about settler households discussed below comes from a cross-sectional survey of approximately 420 settler households in Napo and Sucumbios in the north-eastern Ecuadorean Amazon conducted in the early 1990s. Pichon (1993, 1997b) provides detailed information on survey design, methodology and sample selection. Murphy and Bilsborrow are currently working on a follow-up survey in the region, which should provide longitudinal information on the 1990 households shortly. In the absence of such longitudinal data, we rely on our analysis of the 1990 survey both to examine cross-sectional land-use patterns and to infer longitudinal patterns across settlers, based on how long the settlers had already been on their plots when the survey was conducted in 1990. Despite the fact that all of our data come from the early 1990s, for convenience we use the present tense throughout the text.

Half of all settlers owned land prior to settlement, but most were either agricultural workers or sharecroppers. The average head of household has some primary education and he and his spouse are in their mid- to late 30s. Once on the frontier, households generally occupy plots of approximately 50 ha. Plot size tends to be rather uniform, since settlers must pay much higher fees to process claims larger than 50 ha. None the less, plot size does vary between settlers and this has important implications for agricultural practices and land use. Households do not extract much timber or non-timber forest products for either household use or sale. Most settlers depend primarily on agriculture, and coffee is their main cash crop. They grow food crops mainly for subsistence. The same applies to their cattle, pigs and chickens.

Median household income is US$680. With an average household size of around seven people, each household typically has three available adult males who do agricultural work, including forest clearing. Half of the settler households use 1–3 months of outside agricultural labour at some point during the year, mainly for planting and harvesting. They may either hire labour or exchange labour with other households. About a third of the households have one member (usually the household head) working off-farm for 1 or 2 months during the year, generally on another nearby settler's farm. The farmers use few yield-increasing and labour-saving technologies, such as fertilizers and chain-saws, although about half occasionally apply herbicides, mainly to coffee. The farmers' main implements are simple hand tools, such as hoes. Few have access to credit or technical assistance.

3. Settler Land-use Patterns

As noted above, settler land use typically combines several land uses. We applied cluster analyses to identify the most frequently occurring land-use combinations among households, based on the percentage of their plot in: (i) forest; (ii) food crops; (iii) perennial crops (mainly coffee); and (iv) pasture and fallow. To get a sense of how much impact the most frequently identified land-use combinations had on forests, we classified the land-use patterns according to the degree of forest clearing they involved. The reader can find more detailed information on settler land use and the cluster analysis in Pichon (1996c) and Marquette (1998). We summarize only the main results here.

Four main land-use patterns or clusters emerge among the settler households, which we describe in Table 9.1. These include a low-cleared-area pattern (50% or more of the plot in primary forest), a medium-cleared-area pattern (20–50% of the plot in forest) and two high-cleared-area patterns, specializing in cattle raising or coffee-growing (< 20% of the plot in forest). The first pattern is the most prevalent, accounting for 61.1% of all households. The medium-cleared-area pattern characterizes 24.1% of the farms. Farms with high-cleared-area patterns specializing in cattle raising and coffee-growing comprise 8.5% and 3.2% of the households, respectively. All four patterns include some pasture and some subsistence food cultivation and in all four patterns coffee occupies the largest portion of cultivated area.

Figure 9.1 shows what percentage of households had plots conforming to each of the four land-use patterns in three groups of households classified according to how long they had been on their plots (recent settlement 0–4 years, longer 5–10 years and longest 10 years or more). In all three groups, most settlers have low-cleared-area land-use patterns. This was the case for 100% of recent settlers, over 50% of settlers who have been there for 5–10 years, and over 60% of those who had been there for over 10 years. Only a minority of households in the longer- and longest-settled groups and none of the more recent settlers had medium- or high-cleared-area land-use patterns. We can read longitudinally across the duration of settlement groups in Fig. 9.1 to infer what individual households might do over time. Most settlers start out and continue with a low-cleared-area pattern, even after they have had their

Table 9.1. Land uses (%) for settler households by land-use pattern.

Land uses	Low-cleared-area pattern (63.1%)	Medium-cleared-area pattern (24.1%)	High-cleared-area cattle pattern (8.5%)	High-cleared-area coffee pattern (3.2%)
Food crops	3	10	4	10
Coffee	13	23	12	69
Pasture	12	39	78	10
Forest	72	28	6	11

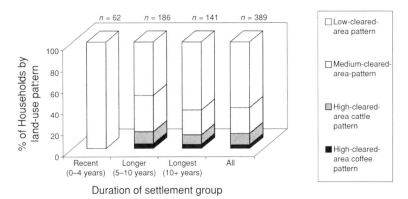

Fig. 9.1. Settler households by land-use pattern and duration of settlement group in the north-eastern Ecuadorean Amazon, 1990s.

plots for more than 10 years. On the other hand, just over a third of the households eventually start clearing larger areas after they have been settled for some time. Interestingly, settlers do not appear to systematically progress from low-, to medium- and high-cleared-area patterns. The proportion of longest-settled households with the low-cleared pattern is actually higher than that of the longer-settled households (64% versus 50%, respectively), while a similar proportion of longer- and longest-settled households have high-cleared-area patterns. (We discuss the limitations of using cross-sectional data to infer longitudinal trends in this context in Marquette (1998).)

Table 9.2 summarizes some key characteristics of the land-use clusters identified above and of the settlers that have each land-use pattern. The majority of households with low-cleared-area land-use patterns have less on-farm labour available than do those in the medium- or high-cleared-area categories. This is due to smaller household size, which implies less available adult male labour, less use of hired labour and more off-farm labour by household members. Households in the low-cleared-area category also tend to have poorer natural resources, less access to roads and credit and lower median annual incomes than do households in the other categories. Correspondingly, households in the medium- and high-cleared-area categories have larger households and fewer labour constraints, greater access to roads and markets, more credit and higher incomes. Among households with high-cleared-area land-use patterns, those that specialize in cattle raising tend to have larger than average plot sizes and household heads from coastal areas, while those that specialize in coffee-growing typically have smaller than average plot sizes.

4. The Role of Coffee in Settler Land Use

Questions arise as to how the characteristics mentioned above might explain why most settlers belong to the low-cleared-area category, even after they

Table 9.2. Links between land use, other household characteristics and deforestation among Ecuadorean Amazon settler households ($n = 389$).

	Land-use pattern (% all households)				
	Low-cleared-area (63%)	Medium-cleared-area (25%)	High-cleared-area cattle (9%)	High-cleared-area coffee (3%)
Land-use characteristics	Smaller-scale mixed agricultural production Coffee most important crop Small-scale cattle raising (< 5 head) Most land-saving	Larger-scale mixed agricultural production Coffee most important crop More cattle (5–10 head) Less land-saving	Primarily cattle raising (15+ head) Least land-saving	Primarily coffee-growing Least land-saving
Temporal characteristics	Most prevalent all durations of settlement All recent households	Emerges only after 4+ years of settlement	Emerges only after 4+ years of settlement	Emerges only after 4+ years of settlement
Labour characteristics	Least household labour Least use of hired labour Most off-farm labour Most restricted on-farm labour	More household labour Most use of hired labour Least off-farm labour Less restricted on-farm labour	More household labour More use of hired labour Less off-farm labour Less restricted on-farm labour	More use of hired labour Less off-farm labour Less restricted on-farm labour

Other characteristics	Smaller household size Poorer-quality natural resource base Further from markets and roads Less use of credit	Larger household size Better-quality natural resource base Closer to markets and roads More use of credit	Larger household size Better-quality natural resource base Closer to markets and roads More use of credit Household head more likely to be from coast Larger plot size	Larger household size Better-quality natural resource base Closer to markets and roads Smaller plot size
Impact on plot-level deforestation	> 50% of plot in forest	20–50% of plot in forest	< 20% of plot in forest	< 20% of plot in forest
Median income per annum (US$)	658	1170	2480	853

have owned their plots for a long time. One might also ask why and how a minority of households adopt medium- and high-cleared-area strategies involving greater forest clearing. We believe that considerable light can be shed in this regard by exploring the role played by coffee in settler production.

In some ways, coffee represents a surprising choice for settlers to include in their production strategies. Labour on the frontier is scarce and coffee generally requires more labour to produce than the other agricultural and livestock activities found in the region. The land clearing, soil preparation, planting, weeding and harvesting required to produce coffee all demand substantial amounts of labour. Moreover, to weed coffee in the north-eastern Ecuadorean Amazon typically requires more labour than in other settings, due to the region's particular soil characteristics (Estrada et al., 1988). Coffee involves a long-term investment that does not offer the immediate returns that cash-short frontier households need. Settlers must wait 4–5 years after planting until their coffee bushes reach full production.

Coffee does, however, offer certain advantages. Since most frontier settlers start off with little capital, activities like growing coffee that require limited initial investment may be their most logical choice. Coffee has a ready market and farmers frequently mention that as an important point in its favour. Coffee has a higher price than other food or fruit-tree crops. It may stand up better to the precariousness of transport in the region, is not too bulky to ship from remote areas and has a better ratio of price to transport costs (Barral, 1987: 103; Estrada et al., 1988: 62–63). Coffee's long lifespan may imply that planting coffee increases the value of the land more than planting other cash crops, such as cocoa (Gonard et al., 1988). Coffee meets settlers' concern for maximizing security in the high-risk environment of the frontier. Settlers feel it provides greater stability and certainty in regard to labour demands, since it requires fairly steady labour inputs throughout the year. Settlers also recognize that, once planted, coffee generates income for many years.

Settlers believe coffee offers long-term security and in their minds that can compensate for the risks associated with short-term variations in its price. Settlers are aware that coffee harvests, prices and associated profits fluctuate, but are reluctant to discuss the short-term economic rationality of their decision to invest in it. Although some farmers make numerical calculations about current yields and prices, they also base their judgements about expected returns and their decisions to invest in coffee on past yields and prices. The unpredictability of market prices does not seem to affect these judgements much. Coffee constitutes a central part of settler production strategies, because it provides regular and secure returns, rather than necessarily having the highest labour productivity or generating the greatest returns.

All this has strong implications for forest cover, since, as noted above, for most settlers coffee-centred production strategies are associated with low-cleared-area land-use patterns that involve less forest clearing. Looking at the characteristics associated with the low-cleared-area pattern, which are presented in Table 9.2, one gets a feeling for why this might be the case. The combination of labour constraints in most settler households and the heavy labour requirements associated with coffee-growing may place a 'brake' on the total area settlers can clear and plant. One study undertaken in the region estimates that a settler household with six persons can handle about 7 ha of coffee, or 14% of a 50-ha plot (Estrada et al., 1988). Our research supports this conclusion and indicates that most settlers have a proportion of coffee on their plots similar to what the Estrada et al. study would suggest (7–13% on average). This implies that these farmers are constrained in regard to how much coffee they can produce and thus how much forest they are likely to clear (Marquette, 1998).

Although the low-cleared-area strategy that prevails among settlers offers some advantages in terms of stable income, it probably offers poorer overall economic returns than the medium- and high-cleared-area strategies. Table 9.2 provides rough estimates for median incomes associated with each strategy. Households with the low-cleared-area pattern (median annual income US$658.00) received much lower incomes than those with the medium-cleared-area pattern (US$1170.00) or the high-cleared-area cattle pattern (US$2483.00). A detailed study of settler income and welfare also confirms that area in pasture, which indicates involvement in cattle raising, is significantly associated with higher income (Murphy et al., 1997; Murphy, 1998).

Contrary to what those who believe that increasing agricultural productivity will reduce forest clearing would expect, higher productivity may actually increase land clearing. Settler households that have the medium- and high-cleared-area cattle pattern tend to have plots with higher quality and hence more productive soils and terrain. In addition, settler households with better-quality land and higher incomes may reinvest their profits in expanding their crop and pasture areas, which may lead to even more forest clearing.

Close to half of all settlers surveyed in the study area indicated that, given the chance, they would increase their involvement in cattle raising. The question is 'Why don't they?' As Table 9.2 and the previous discussion imply, several factors combine to prevent most households from making the shift. Labour constraints, a poorer natural resource base, lower income and lack of access to credit may all lead most settler households to develop the low-cleared-area pattern rather than the medium- or high-cleared alternatives, which involve greater cattle raising or crop production.

5. Discussion

We can think of the three main household land-use patterns identified above (low-, medium- and high-cleared-area patterns) as representing different sets of agricultural technology that have emerged among settlers within the same frontier environment in the north-eastern Ecuadorean Amazon. The interaction of various factors, including the constraints and opportunities created by particular land-use patterns, once adopted, availability of family labour, plot size, market conditions, limited capital and credit availability, soil quality and terrain, shapes these patterns and their impact on forest resources. Our previous analyses suggest, however, that each of these factors may not have equal importance in shaping land-use outcomes (see, for example, Pichon, 1997b). Distance from the nearest road, for example, seems to stimulate additional forest clearing more than labour constraints reduce it.

The previous discussion also implies that frontier farmers' perception of and responses to risk will greatly influence the type of technology and resulting land-use patterns that settler households adopt. Unlike what models where households allocate resources to maximize (expected) profit would predict, settler households may adopt technologies and land-use patterns that prioritize minimizing risk through stable production and income over increasing production and income over time. Thus, many settlers in the north-eastern Ecuadorean Amazon have come to rely on a low-cleared-area pattern that centres on a proved perennial cash crop, coffee, which provides secure and steady, but not necessarily increasing, productivity and income. Settler involvement in coffee growing, which takes at least several years to produce, also suggests that their perceptions of risk do not preclude long-term investment.

Figure 9.2 summarizes the key variables that link settlers' land-use patterns to forest clearing. These are availability of on-farm labour, the land intensity of each product mix and forest area. A key finding of our analysis is that the majority of settlers have adopted the 'low-cleared-area' land-use pattern and thus fall to the left on these three scales (more restricted on-farm labour, product mixes that are more land-saving and larger forest areas). This strongly supports the assertion made in Chapters 1 (Angelsen and Kaimowitz) and 2 (Angelsen *et al.*) in this volume that, when small farmers are constrained

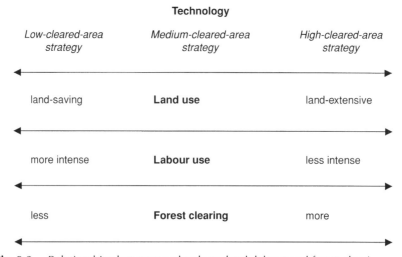

Fig. 9.2. Relationships between technology, land, labour and forest clearing.

by limited labour, capital and credit and labour-substituting technologies, such as tractors, are not widely available, labour-intensive agricultural systems will lead to less forest clearing. Again, however, it is important to bear in mind that other farm characteristics, such as distance to roads, may play stronger, counteracting and possibly overriding roles.

The prevalence of the low-cleared-area strategy among the settlers challenges Boserup's (1992) theory, which asserts that population density and land and labour availability basically drive agricultural intensification. In a Boserupian world, one would not have expected to find crops such as coffee, which require substantial labour inputs, to occur widely under frontier conditions (land abundance, low population density, limited labour availability). However, in frontier settings, such as the north-eastern Ecuadorean Amazon, other factors (natural resource base, household characteristics, market conditions and availability of capital and credit) may be more important in shaping the technologies used and the land-use patterns that develop. These factors may encourage land-use patterns that involve labour-intensive crops, such as coffee, despite abundant land and limited labour. Given the limited labour available to settler households, however, strategies that involve labour-intensive crops like coffee may place a threshold on the total area households clear and farm. Continual land extensification and clearing are not inevitable, even under frontier conditions. Simple Boserupian models cannot fully explain the links between intensification and forest clearing in the Amazon.

Our findings in Ecuador also challenge the view that holds that greater agricultural productivity is necessary to reduce forest clearing among settlers. We found that the minority of settler households that have medium- or high-cleared-area patterns tend to have a more productive natural resource base in terms of soil quality and terrain and that these more productive resources are associated with more, rather than less, clearing. Also, settlers with better land tend to invest their profits from higher productivity into expanding their agricultural areas or land-extensive activities, such as cattle raising. This confirms the possibility raised in the Introduction to this volume (Chapter 1, Angelsen and Kaimowitz) that more productive resources sometimes act as an incentive to forest clearing. Positive-feedback loops connecting profits, access to credit and cattle buying among these households may drive this association in the Ecuadorean context.

6. Conclusions and Policy Implications

The main policy insight we draw from the previous discussion is that the low-cleared-area land-use pattern centring on coffee which many settlers in Ecuador have spontaneously adopted represents an 'endogenous' option for limiting forest clearing. We consider the low-cleared-area pattern observed among many settlers in Ecuador 'endogenous' because it arose largely without outside interventions or technologies linked to agricultural extension or development activities. Everything we know suggests that it resulted from settlers falling back on tried and true practices or observing what other settlers did. Along with others (Boese, 1992; Proano, 1993), we suggest that, rather than introducing or developing new technologies or techniques aimed at increasing productivity to reduce forest clearing, agricultural researchers should concentrate more on improving the endogenous systems that settlers already use to obtain steady and stable returns. These systems may offer important advantages in regard to both settler and forest welfare, since they may already encapsulate successful responses to frontier socio-economic as well as ecological conditions in a way that introduced systems might not.

At the same time, it is important not to over-idealize endogenous land-use patterns among settlers in Ecuador and to seriously evaluate their drawbacks. The low-cleared-area strategy adopted by the majority of settlers places heavy labour burdens on households and subjects them to the vagaries of coffee prices and middlemen. Most importantly, it offers lower incomes than the medium- or high-cleared-area strategies, so any effort to encourage it will inevitably need to make it more profitable. One way to do this may be to develop off-farm employment opportunities and non-agricultural activities that complement the use of low-cleared-area farming strategies, although one would have to seek ways to ensure that settlers did not use the additional resources these activities generate to expand their cattle raising.

Inevitable technological changes on the frontier that fall both in and outside the realm of agriculture may make increased perennial crop growing more or less sustainable or may have conflicting effects on forest and settler welfare. Improved road and market infrastructure, increased use of medicines for improving cattle raising, increased availability and use of herbicides in coffee production that reduce the labour intensity involved in coffee-growing – all these factors could improve household incomes while stimulating increased land clearing. From a market perspective, it is also important to note that the future outlook for coffee in the north-eastern Ecuadorean region and its price prospects will probably remain poor. A major challenge for policy-makers and agricultural researchers is to take into account the more general development trajectory on the frontier, of which agricultural technology is only one facet. In this wider context, they must evaluate what existing endogenous systems have to offer, their capacity to make agriculture more sustainable and improve settler economic welfare and their long-term feasibility in the context of evolving frontier economies and societies.

References

Barral, H. (1987) *Analisis de Los Diversos Tipos de Uso del Suelo en la Amazonia Ecuatoriana (Provincia del Napo)*. Convenio ORSTOM-INCRAE-PRONAREG, Quito, Ecuador.

Boese, E. (1992) *Actividades Agroforestales y Silviculturales en la Region Ammoniac Ecuatoriana. Experiencias y Resultados 1985–1990 en la Region Lumbaqui, Provincia de Sucumbios.* ProFors, Quito, Ecuador.

Boserup, E. (1992) *The Conditions of Agricultural Growth: the Economics of Agrarian Change under Population Pressure.* Earthscan Publications, London.

Estrada, R., Sere, C. and Luzuriaga, H. (1988) *Sistemas de Produccion Agrosilvopastoriles en la Selva Baja de la Provincia de Napo, Ecuador.* USAID–CIAT, Cali, Colombia.

Gonard, P., Leon, J. and Sylva, C. (1988) *Transformaciones Agrarias en el Ecuador.* In: *Geografia Basica del Ecuador: Tomo Geografia Agraria (1).* Centro Ecuatoriano de Investigacion Geografica (CEDIG), Quito, Ecuador.

Hicks, J. (1990) *Ecuador's Amazon Region: Development Issues and Options.* World Bank, Washington, DC.

INEC (Instituto Nacional de Estadistica y Censos) (1992) *V Censo de Poblacion y IV de Vivienda: Resultados Definitivos.* INEC, Quito.

Marquette, C. (1995) Household demographic characteristics, consumption, labor and land use on the northeastern Ecuadorian Amazon Frontier, 1990. Doctoral dissertation, Fordham University, New York.

Marquette, C. (1998) Land-use patterns among small farmer settlers in the north-eastern Ecuadorian Amazon. *Human Ecology* 26(4), 573–598.

Murphy, L. (1998) Making a living in the rainforest: factors affecting economic status of settlers in the Ecuadorian Amazon. Doctoral dissertation, University of North Carolina, Chapel Hill.

Murphy, L., Bilsborrow, R. and Pichon, F. (1997) Poverty and prosperity among migrant settlers in the Amazon rainforest frontier of Ecuador. *Journal of Development Studies* 34(2), 35–65.

Murphy, L., Marquette, C., Pichon, F. and Bilsborrow, R. (1999) Land use, household composition, and economic status of settlers in Ecuador's Amazon: a Review and synthesis of research findings, 1990–1999. Paper presented at Conference on Patterns and Processes of Land Use and Forest Change, 23–26 March, University of Florida.

Myers, N. (1988) Threatened biotas: hotspots in tropical forests. *The Environmentalist* 8(3), 1–20.

Netting, R.M. (1993) *Smallholders, Householders: Farm Families and the Ecology of Intensive, Sustainable Agriculture.* Stanford University Press, Stanford, California.

Pichon, F. (1993) Agricultural settlement, land use, and deforestation in the Ecuadorian Amazon frontier: a micro-level analysis of colonists' land-allocation behaviour. Doctoral dissertation, University of North Carolina, Chapel Hill.

Pichon, F. (1996a) Settler agriculture and the dynamics of resource allocation in frontier environments. *Human Ecology* 24(3), 341–371.

Pichon, F. (1996b) The forest conversion process: a discussion of the sustainability of predominant land uses associated with frontier expansion in the Amazon. *Agriculture and Human Values* 13(1), 32–51.

Pichon, F. (1996c) Land-use strategies in the Amazon frontier: farm-level evidence from Ecuador. *Human Organization* 55(4), 416–424.

Pichon, F. (1997a) Settler households and land-use patterns in the Amazon frontier: farm-level evidence from Ecuador. *World Development* 25(1), 67–91.

Pichon, F. (1997b) Colonist land allocation decisions, land use and deforestation in the Ecuadorian Amazon frontier. *Economic Development and Cultural Change* 45(4), 707–744.

Proano, A. (1993) Problematica ecologica del nor-oriente y estrategias del FEPP. In: *Amazonia: Escenarios y Conflictos*. CEDIME-Abya-yala, Quito, Ecuador.

World Bank (1992) *Brazil: an Analysis of Environmental Problems in the Amazon*. Internal Discussion Paper, Latin American and Caribbean Region, Report No. 9104-BR, World Bank, Washington, DC.

Ecuador Goes Bananas: Incremental Technological Change and Forest Loss

Sven Wunder

1. Introduction[1]

Ecuador is a traditional primary commodity producer and latecomer to economic development. Throughout the 19th century the country relied on cocoa exports. But cocoa declined irreversibly in the 1920s, due to diseases and competition from other suppliers. Two decades later, favourable natural and social conditions helped the country convert bananas into its new lead export and to become the world's largest banana producer in 1954, an expansion that continued until the mid-1960s.

Ecuador has three regions: the coastal lowlands, the highlands and the Amazon lowlands. Only the coast grows bananas for export, where they compete for land with pasture, cocoa, sugar, coffee, rice and other crops and forest. Before humans arrived, forests covered an estimated 90–94% of the country's land area (Cabarle *et al.*, 1989). In 1951, their share was still almost 75%, while crops covered only 4.5%. The coast's entire cultivated area was only 501,021 ha (CEPAL, 1954: 43–48). In this context, the 100,000–150,000 ha of bananas that existed in the early 1960s represented a sizeable portion of the agricultural area. Overall, the expansion of banana production may have augmented the area of coastal agriculture by 20–30%. The area converted to bananas amounted to only 0.5–0.8% of Ecuador's huge forest cover in 1951, but it contributed notably to broader social processes, which eventually reduced the coastal region's forest cover to 33.4% in 1995 (Wunder, 2000). As Larrea (1987: 30) says:

> It is difficult to find a case in the history of the international banana economy where the expansion of the crop produced such ample demographic and migratory effects as in the case of the Ecuadorean coast during 1948–1965. The rapid expansion of production shifted the region's agricultural frontier outwards, until it contained the majority of the area currently under cultivation.
>
> (Author's translation from the Spanish)

The demand for cultivated land and pasture accounts for most deforestation in Ecuador. More than 90% of the deforested areas ends up as pasture, but a large portion of that had already been harvested for timber and used for crops before being converted to grasslands (Wunder, 2000). Forest loss data are unreliable, but it is likely that deforestation in Ecuador rose to between 180,000 ha and 240,000 ha year^{-1} in the mid-1970s. Most forest clearing occurs in the two lowland regions. Estimates of current forest cover range between 11 and 15 million ha, so yearly deforestation rates are between 1.2% and 2.2%.[2]

In assessing how banana production and technologies have affected deforestation, one must distinguish between direct and indirect impacts. During the postwar period, the amount of forested land directly cleared for banana plantations fluctuated heavily and varied from one region to the next. Technological change greatly influenced this process. New varieties and other changes in production and transport technology determined the shifting requirements for, and changing production centres of, banana plantations. Three factors proved vital in setting dynamic comparative advantage: water, soil quality and access to markets (Sylva, 1987: 116–122).

At the same time, banana production indirectly affected deforestation in many complex ways. Bananas were pivotal to the entire economy's growth and transformation. They demanded great amounts of labour and provided the taxes to finance the expansion of railways, roads and credit. They changed the balance of power between political classes and geographical regions and they altered the role of the Ecuadorean state and its institutions (Larrea, 1987; Striffler, 1997).

Against this background, the relevant counterfactual questions – 'how much forest would have been lost without the banana boom?' and 'how much forest would have been lost applying different banana production technologies?' – are very hard to answer. Both questions require speculative judgements on alternative regional and product development options over a period of five decades, and their respective indirect land-use impacts.[3] However, based on sector-wide analyses of banana production (CIDA, 1965; Larrea, 1987), case-studies of banana-led coastal colonization (Brownrigg, 1981; Striffler, 1997) and comparisons with other commodity booms (cocoa and oil) (Wunder, 2000), we conclude that road construction and labour migration encapsulate the banana expansion's main indirect effects on land use. Hence, our discussion of indirect impacts focuses on these two aspects, both of which led to important asymmetries in land-use changes between banana booms and busts.

Sections 2, 3 and 4 analyse three periods in the postwar development of the Ecuadorean banana sector. For each period, an initial subsection describes market and production trends. A subsection on technologies and the regional distribution of banana production follows. Then come a characterization of the indirect impacts and a summary. Section 5 compares the banana technology, production and market characteristics in the three periods, and section 6 the corresponding deforestation impacts. Section 7 discusses the theoretical and policy implications.

2. 'Banana Fever' (1946–1966)

2.1. Markets and production

Several factors facilitated the rapid rise of Ecuadorean banana exports after the Second World War. First and foremost, global demand rose steadily, mainly centred in the US market. Secondly, the country's Central American competitors faced severe problems with 'Panama disease' and other diseases, as well as periodic devastation of their plantations by cyclones. Ecuador's abundant, disease-free, fertile soils, which had sufficient water and were less exposed to tropical storms, gave it a comparative advantage. This helped convince multinationals like United Fruit and Standard Fruit to buy large areas to establish their own banana plantations, as well as providing capital and technical assistance to Ecuadorean banana-growers (Striffler, 1997).

At the time, Ecuador was still suffering from the decline of cocoa. Coastal farmers were diversifying into cattle, sugar and cotton and were searching for ways to reduce production costs (CEPAL, 1954: 52). Underutilized former cocoa plantations, low rural wages and a devalued currency all provided excellent incentives for establishing new lines of production. The government of Galo Plaza (1948–1952) favoured banana producers by expanding the road network and giving them subsidized credit (Sylva, 1987; Acosta, 1997: 92). These advantages outweighed Ecuador's disadvantages, such as its undeveloped port and road infrastructure (CEPAL, 1954: 82) and technological backwardness and its greater distance from the US and European markets, compared with Central America (Larrea, 1987: 47).

The only statistics available prior to 1955 refer to the number of banana racemes exported. From 112,973 in 1920, these rose significantly to 1,181,710 in 1930 and 1,874,595 in 1940. They declined during the war to 693,551 in 1945, but then grew exponentially to 2,686,870 in 1947, 16,755,066 in 1952 and 23,874,310 in 1955 (Riofrío, 1995: 11). From 1945 to 1951, prices rose fourfold and this greatly stimulated production (CEPAL, 1954: 170).

2.2. Technology and regional distribution

'Gros Michel' was the dominant commercial banana variety around the world. Its main advantages were its size and physical robustness. It was simple to plant, maintain, harvest and transport and did not damage easily. This helped it expand widely, both geographically and in terms of the types of farmers that grew it. The requirements for banana production that largely determined their spatial distribution were (Hernández and Witter, 1996; Rios, 1996): (i) fertile, deep, nutrient-rich soils, preferably with loose texture, pH 5.5–7.5; (ii) humid tropical to subtropical temperatures (optimal around 30°C); (iii) abundant, regular availability of water and good drainage; and (iv) access to ports.

Many urban middle-class entrepreneurs invested in land to participate in the boom. The owners of large haciendas, traditionally dedicated to cocoa and cattle ranching, allocated part of their land to bananas. Peasants migrated from highland provinces, cleared forest to gain land rights and planted bananas. Everybody could grow bananas. There were no significant technological or financial barriers to entry (Striffler, 1997: 43). Hence, the impact was much more far-reaching than the cocoa boom, which had been concentrated on haciendas in the Guayas river basin, a fertile lowland area north of Guayaquil.

Two contemporary analyses at the regional and farm level (CEPAL, 1954; CIDA, 1965) give us a detailed vision of the process through which bananas penetrated the rural economy. The first banana plantations were established near navigable rivers – the main transport arteries in the absence of roads. These plantations were often located in or near the old cocoa haciendas in Guayas (see Fig. 10.1). There, bananas constituted one additional element within diversified production systems, which also included sugar, rice, oil crops and cattle. Within this area, one could find both haciendas of over 1000 ha and small to medium-sized lots (CIDA, 1965: 382–392). The area's main advantages for producing bananas were its good soils and accessibility. Its key drawbacks were its deficient rainfall and poor drainage (CEPAL, 1954).

The western Andean foothills, which descend towards the coastal plain, offered the best natural conditions for cultivating bananas. This area offered rich soils and regular abundant rainfall and its hilly topography provided natural drainage. The road network gradually expanded and made new areas of production accessible, especially in the hilly parts of the provinces of Los Ríos and El Oro and, to a lesser extent, in the lower parts of the highland provinces. Migrant farmers colonized and deforested most of these areas, typically claiming a homestead of 50 ha, of which they dedicated up to 30 ha to bananas. Unlike in Guayas province, most of these small- and medium-scale producers established banana monocultures (CEPAL, 1954: 166–169).

Bananas are extremely perishable and cannot withstand more than 5 weeks between harvesting and consumption (López, 1988: 17). Nevertheless, the 'Gros Michel' variety was so robust that, even in places with no direct access to roads, farmers could transport unwashed and unpacked racemes by mule, on shaky trucks and in canoes navigating untamed rivers. When prices were high, the radius of economically feasible cultivation expanded (Sylva, 1987: 118). In the Andean foothills, banana cultivation and deforestation were directly linked. A Comisión Económica para América Latina y Caribe (CEPAL) report from the period noted that with 'the conquest of idle lands in all the hilly zones of the coast, which offered excellent conditions for the new product . . . forests were felled and old gardens destroyed to plant bananas' (CEPAL, 1954: 170, translation from Spanish by the author).

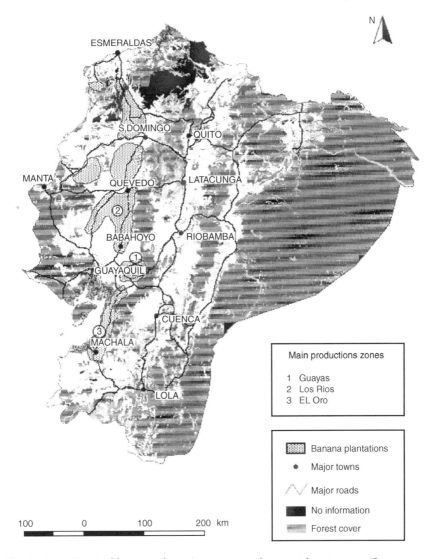

Fig. 10.1. Historical banana plantation zones and current forest cover. (Sources: Wunder, 1999; UPEB, 1990.)

Besides the 'old' (Guayas plains) and 'new' (foothills of Los Ríos and El Oro provinces) production zones, banana cultivation gradually expanded into more marginal production areas with poorer soils, in response to high prices. Already in 1948, the multinational Fruit Trading Corporation had established plantations near the northern port of Esmeraldas (Sylva, 1987: 116). Bananas also expanded into the drier parts of El Oro province. To grow bananas in that area, producers had to both irrigate and drain excess water and the soils were generally less fertile than in Los Ríos and Guayas. The region's only advantage was that it was near the port of Bolívar. In the El Oro lowlands, growers used land particularly extensively. Another report by CEPAL refers to banana cultivation there as 'a bad habit that encroaches on all kinds of soils' (CIDA, 1965: 396).

After banana cultivation depleted the soils, in most cases the growers put the land into pasture and moved their bananas elsewhere, creating a 'semi-migratory production system' (Striffler, 1997: 41), which required access to large areas. On some haciendas in the El Oro lowlands, sharecroppers cleared land for bananas and then abandoned it after several years. Before moving on, the landowners required that they leave the land planted in pasture (CIDA, 1965: 402). One report talks about 'the predatory effect of continued banana cultivation' in reference to El Oro's land-consuming production system, in which farmers grew bananas without fertilizers or drainage infrastructure and constantly shifted the location of their plantations (CIDA, 1965: 414), a practice highly conducive to deforestation. By the end of the period, frequent attacks of Panama disease would lead growers to move out even further, triggering land races with homesteading peasants, who often encroached on the multinationals' banana plantations (Striffler, 1997: 89–136).

2.3. Indirect impacts

Bananas' impact on forests was not restricted to their direct effects. The 'banana fever' epoch also had conspicuous indirect effects. Natural population growth on the coast could not satisfy the mounting demand for wage labour stemming from the rapidly rising production of the highly labour-intensive crop. The growers demanded massive quantities of unskilled labour and paid good salaries, especially the multinationals. Partially in response, over 250,000 people migrated to the coast during the 1950s (Striffler, 1997: 60). The Canadian International Development Agency (CIDA, 1965: 395) mentions that in El Oro 'banana cultivation powerfully influenced the development of the province, increasing the cultivated area and favouring in-migration from the Republic's interior, especially the [highland] provinces of Azuay and Loja' (translated from Spanish by the author).

Since the new production areas were still poorly integrated into the market economy, food crops were largely grown on-farm. Even on one of the largest and most specialized plantations, Tenguel, between the Guayas and El

Oro provinces, workers recall that the plantation produced 'nearly everything, from sugar to cattle, to basic food crops such as yucca and plantains as well as a wide range of fruits' (Striffler, 1997: 34). In all probability, feeding so many workers with locally grown foods sharply exacerbated the demand for land associated with banana cultivation.

By no means all migrants became banana-growers or workers. Many followed other livelihood strategies. For instance, Brownrigg (1981) describes a group of rural–rural migrants who moved from Loja to the El Oro foothills and basically retained the diversified farming systems they practised previously. But these groups' efforts could never have succeeded so much were it not for the growing urban food markets, wage-labour opportunities and other possibilities the banana boom offered. Banana incomes stimulated the transport, construction and service sectors, creating regional development booms in mid-sized coastal towns, such as Naranjal, Machala, Quevedo and Babahoya (Striffler, 1997: 58).

The infrastructure built by the state or banana producers to bring new areas into the plantation economy were key in fomenting other economic activities as well (Striffler, 1997: 59, 239). In several cases in the Guayas and El Oro provinces, the colonization of marginal, hillside areas depended directly on the construction or extension of an existing road or railway designed to promote banana production. Taxes paid by banana producers allowed the state to increase its presence in these newly colonized areas (Striffler, 1997: 56). This helped push the forest frontier forward.

2.4. Summary

Extremely land-extensive technologies (low capital intensity, low yields) characterized the early 'banana fever' period (1945–1966). The rustic nature and technological simplicity of the 'Gros Michel' variety made it possible to grow bananas throughout the coastal lowlands, even in areas far from ports, allowing production to expand widely, both geographically and socially. The growing demand for land led landowners to convert former cocoa plantations and other previously cultivated areas to bananas. But large areas of forest were also converted to banana plantations, especially on the fertile Andean slopes. With their high rainfall, natural drainage and abundant virgin land, these areas provided a perfect setting for a simple banana production system, based on nutrient mining and low investment. Banana production areas frequently shifted, continuously opening up new areas of forest. The technology required a lot of labour, supplied by immigrants from the highlands, attracted by high wages. The banana trade justified an extension of the road and rail networks, which opened up new areas for forest clearing. During this period, production led to substantial deforestation, both directly (land-extensive, shifting banana plantations) and indirectly (immigration, road construction).

3. Stagnation, Variety Shift and Intensification (1967–1985)

3.1. Markets and production

With the spread of banana plantations to marginal soils in the late 1960s, extensive expansion reached its limit. A shift in external conditions changed that. Between 1957 and 1965, Central American producers successfully replaced the 'Gros Michel' by the new, more productive 'Cavendish' variety (López, 1988). Over the next 10 years, mechanization and shifts between 'Cavendish' subvarieties further improved the Central Americans' technology.[4] Central American producers, particularly the multinationals, developed and adopted technology much faster than in Ecuador, where medium-scale domestic growers continued to dominate production. These producers adopted technology more slowly due to financial constraints and their limited know-how. Thus, Ecuador did not shift from 'Gros Michel' to 'Cavendish' until the late 1960s and early 1970s (Larrea, 1987: 57; Ríos, 1996).

The shift from 'Gros Michel' to 'Cavendish' in Central America doubled that region's yields and almost tripled the volume the main producers exported in 6 years (1965–1971). Ecuador's disease- and cyclone-free production environment ceased to give it a major natural comparative advantage, since the new variety made these factors less important (Larrea, 1987: 56–58). During the boom, banana workers had earned continuously higher wages as growers sought aggressively to attract labour (Acosta, 1997: 83). This drove up production costs and eventually proved unsustainable. Banana workers' real wages started to gradually decline, especially after 1969 (Larrea, 1987: 60–61). From 1973 to 1983, the oil boom caused an overvalued exchange rate, which hampered the expansion of agricultural exports in general (Wunder, 1997). The loss of Ecuador's natural comparative advantage, combined with lagging technology and an overvalued exchange rate, kept its banana exports stagnant for a decade. Ecuador came to hold a 'second-class status as a reserve supplier' (Striffler, 1997: 175). Multinationals stopped producing directly and established contract farming arrangements with domestic producers. The crisis, together with the gradual adoption of more land-intensive technologies, sharply reduced the amount of land devoted to banana cultivation in Ecuador, as shown in Fig. 10.2.

A note is in order here regarding Ecuador's banana-area statistics. The National Banana Programme (PNB) annually records the area devoted to bananas for export, while periodic agricultural censuses register the total area with bananas. In theory, the two sources should differ only with respect to the small amount of bananas produced for the domestic market. In practice, the PNB figures include only areas covered by that programme, which must fulfil certain quality standards. Thus, they underestimate the area of bananas produced for export. Census data include banana areas with low planting densities, interplanted with other crops or even abandoned, so they exaggerate the area. For instance, Fig. 10.2 documents the sharp rise in cultivated area

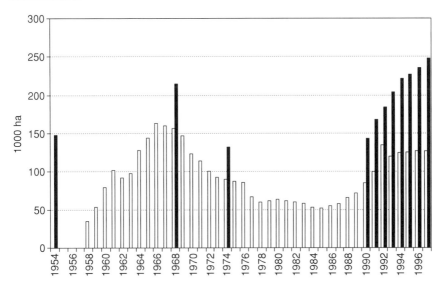

Fig. 10.2. Cultivated area of bananas in Ecuador, 1958–1997 (ha) (from Larrea, 1987; Riofrío, 1995, 1997; MAG, 1998). □, PNB data; ■, 1954 census data.

from the 1950s to 1964, but the PNB data (light-shaded columns) clearly underestimate exports for the late 1950s, since the programme had just begun to sign up producers at the time.[5] At the same time, the 1954 census figure (dark-shaded column) of almost 150,000 ha clearly exaggerates export production, seeing that CEPAL (1954: 167) estimated that the banana export area in 1951 was only 30,530 ha.

Stagnant exports and the adoption of land-saving technologies precipitated a dramatic and continuous fall in the area devoted to producing bananas for export over two decades, from the peak of 163,773 ha in 1966 to 51,796 ha in 1985. Agricultural census figures show a similar trend, although starting from a higher initial level.

3.2. Technology and regional distribution

The 'Cavendish' variety was resistant to Panama disease and could be planted at a higher density, and its lower plant size made it less susceptible to cyclone damage (Sylva, 1987: 118). Figure 10.3 combines the figures on cultivated area with export production data to estimate the trends in physical yields. After the decline in yields that accompanied the extensive expansion of bananas into marginal lands in the 1960s, the gradual introduction of the 'Cavendish' variety brought a pronounced rise in yields, at least up to 1978. As a result, more or less constant overall production levels during this period required less and less land.

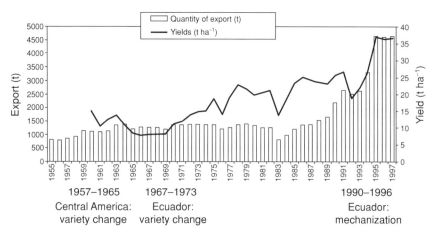

Fig. 10.3. Banana export quantities and per-hectare yields (1955–1997) (from Larrea, 1987; Riofrío, 1995, 1997; MAG, 1998).

This dramatic decline in cultivated area was highly unevenly distributed between producing areas. Table 10.1 shows the evolution of the banana export production areas since 1954 in the five coastal provinces and the lower part of the highland province of Pichincha. In 1954, bananas were still fairly equally distributed among all the coastal provinces. By 1968, this situation had changed. The plantations in the dry and populous province of Manabí receded, while Los Ríos, El Oro and Pichincha increased their participation. Observing Table 10.1, one notes that the regional distribution of banana plantations was markedly unstable and varied sharply from decade to decade. However, by 1983, a clear trend was visible. Banana production had concentrated in three provinces: Los Ríos, Guayas and, in particular, El Oro.[6] Since the 'Cavendish' variety was much more susceptible to transport damage than 'Gros Michel', distance to ports became the deciding factor in where growers located their banana plantations. Excellent access to the small but specialized port of Puerto Bolívar particularly favoured the southern production zone of El Oro. In 1966, 12.1% of Ecuador's banana exports left the country through Puerto Bolívar. Only 4 years later, the share had risen to 50.2%, and in 1978 it peaked at 68% (Larrea, 1987: 238). The country's main port, Guayaquil, which served banana-growers in Guayas and Los Ríos, became a bottleneck. The United Nations Conference on Trade and Development (UNCTAD) figures for 1974 (cited in López, 1988: 20) show that, while Central American ports shipped 10,000–12,000 boxes h^{-1}, Guayaquil only managed to ship 5000 boxes h^{-1}.

3.3. Indirect impacts

Banana production caused more modest indirect deforestation impacts during this period compared with the previous boom. The drop in employment in the

Table 10.1. Geographical distribution of banana plantations, Ecuador, 1954–1983, selected years, six main provinces (ha) (from agricultural census and survey data, cited in Larrea et al. (1987: 125) and Riofrío (1997: 300–301)).

Province	1954	%	1968	%	1974	%	1983*	%	1996*	%
Pichincha	8,270	5.62	39,898	18.59	8,278	6.25	163	0.30	282	0.23
Esmeraldas	36,320	24.66	34,100	15.89	19,235	14.52	1,516	2.75	3,583	2.96
Manabí	27,450	18.64	16,947	7.90	20,532	15.50	249	0.45	50	0.04
El Oro	13,610	9.24	46,851	21.83	25,669	19.38	26,240	47.65	42,515	35.14
Guayas	33,450	22.71	29,201	13.61	25,159	19.00	18,438	33.48	38,396	31.74
Los Ríos	28,170	19.13	47,595	22.18	33,568	25.35	8,464	15.37	36,158	29.89
Total production areas†	147,270	100.00	214,592	100.00	132,441	100.00	55,070	100.00	120,984	100.00
Areas registered for export production*	34,141‡		156,876		90,501		58,317		127,140	

*1983, 1996 and national totals, for export areas.
†1954, 1968, 1974: total production areas.
‡1958 figure.

banana sector ended banana-motivated migration, although population growth continued among families that immigrated during the 'banana fever' period. However, just as the cocoa crisis forced producers to diversify, the banana crisis induced farmers to expand their other crops and their cattle ranching, rather than abandoning the areas released from banana production and allowing the forest there to regenerate. Many laid-off banana workers resorted to colonizing adjacent marginal areas for subsistence agriculture. A former United Fruit worker at Tenguel hacienda recalled that:

> Most of us had just been laid off and had ninety days to leave our houses and the hacienda ... Some talked about going to Guayaquil. No one really had a good idea. Then someone said, 'Why don't we go start working over near the mountains?' Soon after, we went and took a look and decided to struggle for land and form a community.
>
> (Cited in Striffler, 1997: 116)

During the oil-boom period, the government used its abundant foreign exchange to construct many roads into areas with primary forests as part of a deliberate strategy of national integration (Wunder, 1997). But outside the El Oro province, where the expansion of 'Cavendish' production required high-quality roads to the port, these investments had little to do with the banana sector. None the less, just as the after-effects of postwar banana-led immigration continued to cause forest loss even after the banana area contracted, the roads built during the banana fever helped failed banana production areas to survive the crisis by diversifying. As Striffler (1997: 237–238) notes for the La Florida area, forest cover did not return symmetrically after the banana boom:

> To a certain extent ... there was no turning back. Roads were built and lands were cleared. The haciendas retracted but were never again reduced to their 1950s state of near abandonment. Cacao trees and pasture slowly but consistently replaced bananas ... The zone would remain marginal, but never again unpopulated or uncultivated.

3.4. Summary

Ecuadorean banana production stagnated from the mid-1960s to the mid-1980s. Central American growers successfully boosted their region's production by shifting to the more productive 'Cavendish' variety. This dampened world market prices and reduced Ecuador's market share. From 1975 to 1983, the oil boom led to overvalued exchange rates and rising production costs, which made banana exports even less profitable. In a lagged response to the changes that occurred in Central America, Ecuadorean producers gradually shifted to the 'Cavendish' variety. However, the new variety was more fragile, and growers relocated much of their production to areas close to ports, where transport damage of the more fragile variety could be minimized.

The technological package accompanying the 'Cavendish' variety was less labour-intensive and more intensive in financial capital and know-how. This increasingly excluded small producers, who lacked the conditions necessary to adopt the new technologies. Coastal agriculture diversified, and some labour released from the banana sector cleared forest remnants to produce other crops. On the whole, the land-saving 'Cavendish' variety dramatically reduced the direct deforestation impact from bananas, although the relocation of production to areas near ports promoted forest clearing in certain regions. The indirect impacts of banana-motivated road expansion and migration were also weakened (real wages started to decline), but the previous immigrants continued to multiply, which consolidated coastal settlement.

4. Bonanza and Mechanization (1985–present)

4.1. Markets and production

When the Berlin Wall fell in 1989, the banana became a potent symbol of the East German population's desire to gain access to popular Western consumption goods. More generally, the opening of Eastern European markets helped fuel world demand for bananas. World banana prices rose over 40% in the late 1980s (IMF, 1991: 343), although the European Union imposed trade restrictions that harmed Ecuadorean exports. Furthermore, during the economic crisis of the 1980s, Ecuadorean agricultural exports in general experienced some of the fastest growth in Latin America, encouraged by currency devaluations and other macroeconomic policies that favoured agriculture (Southgate and Whitaker, 1996).

As a result of favourable external demand trends and successful internal adjustment, from the mid-1980s and, above all, in the 1990s, Ecuador experienced a new banana bonanza. Production volumes reached unprecedented levels, except in 1992/93 when climatic fluctuations (El Niño) and fungus attacks (*Sigatoka negra*) caused a momentary decline. Up to 1994, this rise was chiefly achieved by expanding the cultivated area. But, from 1995 on, the growth in area levelled off, and production rose solely as a result of growth in land productivity (Figs 10.1 and 10.2). As explained in the following section, Ecuadorean expansion was associated with the gradual mechanization of banana production, which once again lagged in relation to Central America, where similar changes had been under way since the mid-1970s (López, 1988).

4.2. Technology and regional distribution

The new technological package, which gradually diffused among Ecuadorean producers, included greater chemical input use (fertilizers, insecticides, fungicides, herbicides), regular aerial fumigation, on-farm funicular transport of harvested racemes, use of plastic bags and other methods to protect and manipulate flower and fruit development, irrigation systems and subterranean drainage installations. The latter two in particular produced an important rise in yields (V. Espinoza, December 1998, personal communication).[7] The timing of investments needed to implement new technologies reflected both 'push' and 'pull' factors. Mechanization and quality improvements are closely linked and, as banana consumers became increasingly accustomed to purchasing larger fruits with unspotted appearances, this put pressure on Ecuadorean producers, who were slow to modernize.

Table 10.2 shows the revolutionary changes in production technology during the 1990s. From 1990 to 1997, the area under 'mechanized production', involving most of the elements of the technological 'package' described above, rose from 20,343 ha (23.9%) to 90,304 ha (71%). 'Semi-mechanized' areas without mechanized irrigation and drainage installations (MAG, n.d.: 9) doubled in extent, while the non-mechanized plantation area fell from 54,856 ha (64.4%) to 13,817 ha (10.9%).

How did these technological changes affect factor demand? Obviously, mechanization significantly increased the capital intensity of production, in terms of both fixed costs (irrigation, drainage and funicular systems) and inputs (chemicals, plastic bags, etc.). Thus, the ratios of capital to output, land and labour rose. The new production methods also reduced the demand for labour per unit of output, and even per unit of land, by modernizing harvest, transport and maintenance. Even so, banana production remained fairly labour-intensive. The exclusive use of unskilled farm labour increasingly gave way to a more specialized labour force that could handle the new management systems. Total demand for farm labour declined, but there were increasing backward linkages to off-farm activities, such as packaging industries and aerial fumigation services. Some analysts predicted that the new technology would reduce the sector's positive multiplier effects on the national economy (Larrea, 1987: 156), but the most recent estimate (1997) demonstrates that bananas still benefit, directly or indirectly, around 1,250,000 people (MAG, 1998: 3).

Improved infrastructure was vital to the new boom. A recent reorganization of Guayaquil's port facilities allowed it to regain efficiency and importance as a banana port (S. Riofrío, December 1998, personal communication.) Producers greatly improved their postharvest treatment of the fruit (washing, packaging, etc.) and off-farm operations (mechanized port embarkation, refrigerated ship transport). Thanks to these innovations, following the extreme geographical concentration of banana production during the period of stagnation, the radius of production widened once again, making closeness to port less important and favouring the return of bananas to Los Ríos and Guayas provinces (see Table 10.1). Other provinces, such as Esmeraldas, Manabí and Pichincha, have lost ground since their soils and climates do not favour specialized, capital-intensive production (Moreno, 1991).

Table 10.2. The diffusion of technological change in Ecuador in the 1990s (ha) (from National Banana Programme (PNB), published in SICA, 1999).

Technological levels	1990	1991	1992	1993	1994	1995	1996	1997
Mechanized	20,343	40,856	50,793	58,462	58,703	68,059	89,741	90,304
Semi-mechanized	9,989	24,322	38,133	35,824	29,156	26,088	23,524	23,005
Unmechanized	54,856	33,941	45,578	46,703	36,557	31,457	14,145	13,817
Total	85,187	99,118	134,504	140,989	124,416	125,604	127,410	127,126
Percentages								
Mechanized	23.9	41.2	37.8	41.5	47.2	54.2	70.4	71.0
Semi-mechanized	11.7	24.5	28.4	25.4	23.4	20.8	18.5	18.1
Unmechanized	64.4	34.2	33.9	33.1	29.4	25.0	10.9	10.9

Mechanization clearly reduced the ratio of cultivated land to output, especially in the last few years. Thus, the boom continued the ongoing trend towards land-saving technologies, which began with the shift to the 'Cavendish' variety. Technology, soil fertility and unit size were strongly correlated. Mechanization has occurred on the best soils and has favoured medium-sized farms, probably because of their greater ability to mobilize the capital and know-how required for the new methods. The smallest and most capital-constrained farmers, who used to cultivate bananas within diversified farm operations, have increasingly turned to other cash crops, such as cocoa or coffee.

4.3. Indirect impacts

The indirect effects on deforestation linked to bananas during the recent boom were even more restricted than during the previous period. The road network in the prime production zones of the southern coast, where mechanized production was concentrated, was already well established when the boom began (Striffler, 1997: 273). Road construction was less pronounced and, as a result of the intensive but fragile character of mechanized 'Cavendish' production, the specific demands from the banana sector were focused more on the improvement of existing roads than on extending the road network. This new pattern of infrastructure development was less likely to contribute to deforestation.

Mechanization generated a labour surplus in the banana sector, which eliminated the incentives for regional immigration. As in the previous period, this surplus labour typically did not return to the rural highlands, where it originally came from. The cities absorbed part of it. Another group shifted into other crops. Many peasant producers, crowded out of bananas by the new technological and capital requirements, went back to producing cocoa (Larrea, 1987; Striffler, 1997: 273). In this way, the indirect impacts of bananas were largely restricted to long-term trends, which had their origins in the early years of 'banana fever' – notably, continued population growth and settlement among the original migrants to the coast.

4.4. Summary

Ecuadorean banana exports experienced a strong revival after the mid-1980s. Exchange rates became less overvalued, international demand grew and the adoption of mechanized technologies again made Ecuador very competitive. The new technologies are highly intensive in capital, know-how and land, but less labour-intensive. Up to the early 1990s, the steady rise in banana exports involved an expansion in cultivated area, but since then growers have achieved unprecedented levels of production without expanding the total area. The incremental adoption of mechanized technologies by Ecuador's predominantly medium-sized producers made this possible. Improvements in off-farm technologies (packaging, refrigeration) and infrastructure (ports, roads) have again increased the geographical spread of banana production. However, high site-specific on-farm investments in fixed, installed capital (irrigation, drainage and funicular transport systems) have made 'migratory' nutrient-mining technologies unprofitable. The much more intensive and sedentary character of modern banana production has reduced the sector's direct deforestation impact to practically zero. Indirect impacts are now mostly restricted to increasing population, which has its origins in the postwar wave of banana-led migration to the coast.

5. Comparing Production in the Three Periods

Table 10.3 compares the dynamics of the banana sector on the Ecuadorean coast over half a century: the changes in technologies, product and grower characteristics, labour and output markets, the regional distribution of production and the policy environment. First, you have the rapid postwar expansion of simple, labour-intensive and land-extensive production systems into marginal lands. Secondly comes a crisis-cum-adjustment period, during which a shift in banana variety made production more land- and capital-intensive. Thirdly, one observes the recent boom accompanied by mechanization of the plantations, which raised capital and land intensity, but saved labour. Table 10.3 presents how factor intensities (defined in relation to output units) changed during each period. The banana sector went from land-extensive to extremely land-intensive, from migratory to sedentary and from highly labour- to capital-intensive. Technology was initially disembodied, but later embodied into 'packages' during the two latter periods. With increased competition and world market requirements, yields increased and the product and the systems used to produce bananas went from robust and simple to fragile and sophisticated.

Ecuador's growers were much slower to adopt new technologies than their Central American counterparts. In Ecuador, the multinationals withdrew from direct production in the 1960s. Urban investors replaced the smallest farmers, who were pushed out of the banana business, because they lacked the capital and know-how that producing 'Cavendish' bananas required. Medium-scale producers became dominant in Ecuador and technological innovation advanced slowly as a result. Economies of scale may have emerged during this process, but probably more in marketing than in production. Atomized producers generally acted as price takers, but some large trading firms were probably able to influence world prices. *Ceteris paribus*, the growth in Ecuadorean exports lowered prices and thus made farm-level improvements less profitable. Even so, during the last decade, favourable demand trends (e.g. the East European market) and Ecuador's quality advances have sustained the banana boom.

Table 10.3. Changes in the banana sector's production, Ecuador, 1945–1999.

	Period		
	Banana fever	Stagnation and variety shift	Boom and mechanization
Years	1945–1966	1967–1984	1985–1999
Main technological change	'Gros Michel' extends to marginal lands	Adoption of high-yield 'Cavendish' variety	Drainage, irrigation, chemical inputs, etc.
Factor intensity	Level / Trend	Trend	Trend / Level
L/Y (labour intensity)	High / 0	–	– – / Medium
K/Y (capital intensity)	Low / 0	+	++ / High
(Installed K)/Y	Nil / 0	+	+++ / High
H/Y (land intensity)	High / +	–	– – / Low
Production type	Extensive	Semi-intensive	Intensive
	Shifting plantations	Shifting plantations	Sedentary plantations
Product type	Low yield, robust	High yield, fragile	High yield, fragile
Technology	Disembodied	Embodied	Embodied
Off-farm technology	Rudimentary transport	Improved port handling systems	Improved packaging Refrigeration in ships
Main producers	All types of farmers	Medium-sized farms Urban investors	Medium-sized farms Urban investors
Producers' adoption of new technologies	Negligible	Lagged, gradual	Lagged, gradual

Ecuador Goes Bananas

Export markets	Rising demand	Saturation	Rising demand
Main factors of comparative advantage	1. Rainfall, drainage 2. Soils, transport distance	1. Transport distance 2. Soils, rainfall, drainage	1. Soils 2. Transport, rainfall
Regional concentration: leading provinces*	Low: Los Ríos, Esmer., Pich., El Oro, Guayas	High: El Oro, Guayas, Los Ríos	Medium: El Oro, Guayas, Los Ríos
Favoured production zones	1. Hilly frontier 2. Old cocoa farms	Areas near ports and roads	Prime agricultural areas
Labour-market constraints and population	Labour shortages High wages Seasonal migration Low population density	Demand saturation Falling real wages Seasonal migration Medium population density	Demand saturation Wage differentiation Seasonal migration Medium population density
Main policies affecting the banana economy	Credit subsidy (+) Road building (++) Exchange rate (++)	Credit subsidy (+) Road building (+) Exchange rate (−−)	Credit subsidy (0) Road building (0) Exchange rate (+)
Direct deforestation impact of bananas	Frontier expansion +++	Bust/reduced area −−−	Boom/intensification +/0
Indirect impacts	Roads ++ In-migration ++ Pop. growth +	Roads + In-migration 0 Pop. growth ++	Roads 0 In-migration 0 Pop. growth ++

*By the end of the respective period.
Esmer., Esmeraldas; Pich., Pichincha.

Innovations in transport technology, the unique requirements of each new variety and the geographical distribution of diseases combined to bring about frequent shifts in banana production between regions. Initially, rain-fed production and natural drainage favoured the clearing of hilly frontier areas. Nowadays, irrigation and drainage systems have pushed production towards the more accessible prime agricultural areas with fertile soils. The 'banana fever' spread production equally over large parts of the coast, but disease problems in Esmeraldas province and other producing regions and the demanding transport requirements of the 'Cavendish' variety subsequently concentrated banana production on the southern coast. In 1983, almost half of all production came from El Oro province (see Table 10.2). Improvements in transport technologies and packaging methods facilitated a more even distribution in the 1990s, but the three provinces with the best soil and humidity conditions, El Oro, Guayas and Los Ríos, continued to produce most of the bananas. Even though at any given moment banana plantations only occupied a relatively small area, one must keep in mind that historically fruit production frequently changed location and thus affected land use in much larger areas.

Given the initial very high labour intensity of banana production, labour shortages on the coast severely constrained the expansion of exports in the 1950s. Growers continuously offered high wages to attract both seasonal and permanent workers. Together with the moderately labour-saving technological changes and natural population growth among settlers, this gradually saturated labour demand in the second period. Real wages declined and labour demand in the banana sector became more differentiated. The inflationary pressures from the oil boom and an overvalued exchange rate kept production costs high. However, the economic crisis from the 1980s onwards again turned policies in favour of agro-export interests.

6. Comparing the Impact on Deforestation in the Three Periods

The last two rows in Table 10.3 summarize the direct and indirect deforestation impacts associated with the banana sector in the three periods. The direct impact – new, previously forested areas converted for banana production – varied greatly. The banana area initially expanded sharply, then contracted, then grew moderately and now seems to have halted. Two factors magnified the direct impact beyond what one might expect from the cultivated-area figures – 150,000–250,000 ha, at its peak. The first was the migration of banana production from one location to another during the initial boom. Growers typically mined and degraded the soils and then abandoned the location and moved on. The second involved the repeated relocation of plantations, more related to sudden structural shifts in the requirements of different banana varieties and technological packages. Together, these two factors explain the historically 'semi-migratory' character of banana production, which critically aggravated its deforestation impact.

Typically, banana production provided the economic justification for the initial clearing of forest. Once the plantations moved on, however, these areas rarely reverted back to forests. Farmers used most abandoned banana areas for other crops or pastures. This created an asymmetry in land-use conversion. The dynamic character, or instability, of the technologies used thus ended up promoting deforestation. Large areas were initially cleared for bananas in the Ecuadorean provinces of Esmeraldas, Manabí and Pichincha, which were later abandoned. Hernández and Witter (1996) report a similar process in Central America.[8]

How much deforestation does banana production directly cause today? In 1997, bananas occupied an area of between 127,126 ha (PNB figures) and 248,350 ha (census figures) and that area shows little or no sign of expanding. Total crop and pasture area in Ecuador in 1997 was 1,878,500 ha and 5,008,000 ha, respectively (SICA, 1999), implying that bananas occupy 7–13% of the area in crops and 2–4% of the total agricultural area. Nobody can predict whether a banana disease or a new variety will cause renewed shifts in the spatial distribution of banana cultivation, but this seems less likely now. The high fixed investments in irrigation, funicular and draining systems make capital-intensive banana production much less mobile than in the past.

Banana production's indirect impacts on deforestation are more difficult to analyse over such a long period, since they necessarily involve difficult judgements about what might have happened without bananas. Clearly, the crop's high labour intensity induced a mass migration to the coast and helped sustain the long-run population growth that established Ecuador as the most densely populated country in South America. Over the long run, population growth is not fully exogenous, but rather responds positively to the income opportunities that trade and development provide. Food demand from the growing population of banana workers and the various local multiplier effects it involved created a demand for land that took an additional toll of forest resources. In addition to these demographic factors, road construction associated with banana production contributed to forest clearing beyond what was needed for bananas alone. However, except for population growth, other indirect deforestation impacts have dampened over time.

To assess the true impact of bananas on land demand, one should compare the land-use intensities of different agricultural products. Table 10.4 presents a tentative attempt in that direction. We used 1997 production (column 2) and harvested area (column 3) figures from the agricultural census to calculate the yields (column 4) of Ecuador's ten most important crops. In terms of harvested biomass, only sugar cane surpasses bananas. We put together farm-gate prices from Guayas province, a banana production area, and prices from other provinces (column 5) to calculate gross income per hectare (column 6).[9] At US$3236 ha^{-1}, bananas generate by far the highest gross income per unit of

Table 10.4. Comparative yields and intensities of land use for main crops, Ecuador, 1997 (from own calculation from SICA, 1999).

Products	Production (metric tonnes)	Harvested area ('000 ha)	Yields (t ha^{-1})	Farm-gate prices (sucres kg^{-1})	Gross income per land unit (US$ ha^{-1})*	Ha to produce US$1000 income	Ranking
Banana[†]	5,750,262	248.35	23.15	559	3236.02	0.309	10
Sugar cane	2,527,215	24.47	103.31	61	1575.87	0.635	8
Rice	992,971	320.20	3.1	939	727.91	1.374	5
African palm	1,357,616	91.05	14.91	374[§]	1394.43	0.717	7
Plantain[†]	894,091	73.88	12.1	314	950.09	1.053	6
Hard maize[‡]	546,448	278.80	1.96	638	312.70	3.198	2
Cotton	23,703	18.23	1.3	1,904	618.95	1.616	4
Potatoes[†]	601,838	66.27	9.08	809[§]	1836.89	0.544	9
Soybeans[‡]	6,750	5.00	1.35	886[||]	299.10	3.343	1
Cocoa[‡]	89,862	345.62	0.26	5,272	342.77	2.917	3
Total	12,790,756	1,471.87					

*1997 exchange rate US$1 = 3999 sucres; prices Guayas province, unless indicated otherwise.
[†]Fresh fruit/vegetable.
[‡]In dried form.
[§]Farm-gate prices Pichincha province.
[||]Farm-gate prices Los Ríos province.

land, followed by potatoes (US$1837), sugar cane (US$1576) and African palm (US$1394).

The inverse measure – how many hectares an activity requires to produce a gross income of US$1000 (column 7) – and the corresponding ranking from most to least extensive land use (column 8) make interpretation more straightforward. To generate US$1000, a farmer needs only 0.3 ha of bananas, but 3.3 ha of soybeans, 7.6 ha of wheat and 10.7 ha of coffee. In other words, if farmers decided to transfer US$1000 of gross income from coffee to bananas, they could earn the same amount from 0.3 ha of bananas as they had been earning from 10.7 ha of coffee, leaving 10.4 ha that they could put to other uses, including forest. Although this argument is oversimplified, it does have some validity. If one were to include cattle ranching in the calculations, which accounts for 5 million ha, the differences in land intensity would be even more dramatic. This type of calculation is particularly relevant when farmers are capital- and/or labour-constrained so that forested areas serve as a sort of 'reserve' for future occupation. The figures in Table 10.4 give one a feeling for how important what crop a region specializes in is for explaining the variations in forest loss in different regions. In regard to bananas, they show that, with current technologies, a shift from any of the other crops analysed to bananas would significantly intensify land use, which would tend to reduce deforestation pressures.

Even if banana production currently has almost no direct impact on deforestation, its long-term indirect impacts have been important. Economic historians in Ecuador generally agree that bananas had a much larger impact on the development of the coastal region than cocoa (Benalcázar, 1989; Abril-Ojeda, 1991; Acosta 1997). Ecuadorean banana production remained in the hands of small- to medium-scale national producers (80% of the banana area was in units of less than 30 ha) and technologies remained highly labour-intensive for much longer than in Central America. As a result of the historical sequence of technological change, labour absorption was followed by labour release, land absorption by land release and low capital requirements by high fixed investments. This implies that the labour influx to the coast and subsequent population growth were higher than they would have been without bananas and this additional population eventually cleared more forest on the coast. On the other hand, the rural families that moved to the coast no longer cleared forest in their regions of origin, nor did they move to the Amazon.

Between 1950 and 1962, coastal population grew an impressive 4.11% per year and it continued to rise by 3.48% yearly between 1962 and 1974. The share of the national population living in coastal provinces increased from 40.5% in 1950 to 47.5% in 1962 (Acosta, 1997: 245). Of course, not all lowland colonization was tied to bananas. For instance, the settlement of the Santo Domingo area reflected increasing trade integration with the nearby highlands and the capital Quito (Casagrande *et al.*, 1964; Wood, 1972). None the less, even coastal areas not dominated by bananas benefited from the associated improvement of the road network and the growth of agricultural markets.

Without wishing to take the analogy too far, it may be relevant here to apply an approach originally developed by Rudel with Horowitz (1993) for Ecuador's Amazon region and to distinguish between forest clearing in large compact forests and the subsequent clearing of forest fragments. The initial banana boom led to agricultural frontier expansion, providing the overriding economic rationale for forest clearing in previously inaccessible areas. The subsequent crisis and diversification periods are more likely to have involved the clearing of forest remnants. In the latter case, incremental factors, such as population growth and domestic market integration, had greater influence. Road building and migrant settlement appear to 'bridge' boom-and-bust periods and to provide asymmetries for land demand and forest conversion. Their occurrence during boom periods has lasting repercussions on forest clearing even during busts.

7. Conclusions

What policy lessons can we derive from the half-century of banana expansion in the coastal region? For the period as a whole, bananas had a catalytic role in promoting coastal deforestation. At first, this was mostly through direct banana frontier expansion. Later, the gradual settlement effects proved to be of key importance. Modest credit subsidies, the large-scale construction and improvement of roads and ports and a devalued exchange rate were probably the most important policies that contributed to the expansion of banana production, though they varied in importance during the different periods. How one evaluates this process depends greatly on the relevant policy objectives. Ecuadorean policy-makers clearly considered deforestation, sustained coastal settlement and integration with the highland economy to be positive contributions to economic development.

Short-run, 'predatory' use of marginal soils for banana production might be seen as an inappropriate land use, but it can equally be seen as an individually rational strategy in a capital-scarce, land-abundant economy. One may conjecture that, had cheap external credits and significant R&D investments been available for banana producers throughout the postwar period, farmers would have adopted new technologies faster, thus accelerating intensification. This probably would have reduced plantation mobility, labour attraction and settlement, and hence coastal deforestation. However, it might also have increased the scale of banana production, since capital constraints greatly impeded further expansion of the crop. On aggregate, the employment and income opportunities bananas provided, combined with their comparatively intensive use of land and labour, would probably lead most observers to conclude that – historically, but even more so today – bananas have played a positive role.

In regard to the theoretical framework and working hypotheses set out prior to the elaboration of this book, the Ecuadorean banana experience provides important lessons. It shows that, in the medium run, the use of labour-intensive technologies may actually increase deforestation if it encourages in-migration and population growth. In a standard economic theory, comparative-static story, adopting labour-intensive technologies with a given factor endowment should reduce deforestation. But, on the Ecuadorean coast, labour pull and demographic adjustment were endogenously determined by changes in the productive sphere, which created a rural proletariat. The long-run impact of greater settler food demand and other multiplier effects actually stimulated deforestation.

Technology intensive in fixed, installed capital (such as mechanized 'Cavendish' production) may reduce deforestation, by making production more stationary. Migratory production systems can have particularly strong deforestation effects, because of asymmetries that keep forests from returning to abandoned production areas. The gradual and unequal diffusion of new banana technologies among farmers confirms the importance of capital constraints, although the adoption of innovations may have been equally constrained by the differential access to know-how, in an increasingly complex production system. These changes tended to crowd out small producers, who were then forced into other products. However, even small producers were market-orientated and clearly responded to pull incentives. Subsistence-orientated, 'full-belly' behaviour played no role (cf. Angelsen *et al.*, Chapter 2, this volume). Banana producers became increasingly integrated into the market economy through improvements in infrastructure, which reinforced deforestation. The initial, simple technologies gave a natural comparative advantage (soils, water) to hilly frontier areas, meaning that conversion of forests was particularly strong in these zones. Here, homesteading rules (land rights as a reward for clearing) provided a strong complementary motivation for deforestation.

More generally, the Ecuadorean case suggests six points that may be relevant in other settings:

1. One needs to distinguish between the direct and indirect deforestation impacts of technological change. In the long run, the latter may be larger than the former.
2. Boom-and-bust export-product cycles lead to asymmetries in forest clearing, whereby forests cleared in the boom do not return in the bust.
3. Technological changes in other supplier regions that compete for the same markets may influence global prices, redistribute market shares and affect land demand and forest conversion pressures.
4. Technologies intensive in fixed, installed capital can make agriculture more stationary, which tends to reduce forest conversion.
5. Off-farm technologies, especially in the transport sector, may greatly affect the regional patterns of land use.

6. Shifts from one agricultural product to another can have a strong impact on deforestation.

Notes

1 I thank the editors and an anonymous referee for useful comments. Funding from the Danish International Development Assistance (Danida) and help from my research assistant, Mr Breno Piectracci is greatly appreciated.

2 Wunder (2000) discusses the various estimates of Ecuadorean forest cover and deforestation in detail.

3 For instance, one may conjecture that, in the absence of a banana boom, highland surplus labour would have caused more deforestation both in their region of origin (the highlands) and in regions that provided alternatives for colonization (the Amazon). But this depends on what other sectors might have been developed in the absence of the banana boom.

4 'Giant Cavendish' increasingly replaced the 'Robusta' ('Valery') variety. The former allows higher planting densities, with larger fruits and less farm labour input per unit of output, but also demands better soils and its higher curvature requires greater packaging efforts. These pros and cons meant that 'Robusta' was not fully replaced, but rather was combined with 'Giant' (López, 1988: 98-100).

5 As late as 1991, the Ministry of Agriculture (MAG, 1994: 2) estimated that about 30,000 ha of export plantations were not registered in the PNB, amounting to an underestimation of about 15%.

6 Bromley (1981: 20) claims that 'Cavendish' had a 'low tolerance to wet, cloudy conditions', which would be an extra benefit in the drier El Oro province. However, other sources do not confirm this. Both varieties seem equally demanding in regard to water management.

7 I am indebted to Victor Espinoza, Guayaquil, for his patient on-site explanations on shifting banana production and marketing methods, during a visit to his plantation between La Troncal and El Triunfo (Guayas province) in December 1998.

8 For example, Panama disease problems led United Fruit to shift its plantations from the Atlantic to the Pacific coast before the Second World War. But in the 1980s, under the name of United Brands, it returned to the Pacific coast (Hernández and Witter, 1996: 172–173).

9 In addition to being a banana area, Guayas province has a diversified agriculture, which allows for substitution between crops. This is important for the interpretation of results. Some crops, however, are exclusively highland crops (e.g. potatoes), so no direct land substitution could occur.

References

Abril-Ojeda, G. (1991) Export booms and development in Ecuador. In: Blomström, M. and Meller, P. (eds) *Diverging Paths: Comparing a Century of Scandinavian and Latin American Economic Development*. Johns Hopkins University Press, Baltimore, pp. 157–179.

Acosta, A. (1997) *Breve Historia Económica del Ecuador*. Corporación Editora Nacional, Quito.
Benalcázar, R.R. (1989) *Análisis del Desarrollo Económico del Ecuador*. Ediciones Banco Central del Ecuador, Quito.
Bromley, R. (1981) The colonisation of humid tropical areas in Ecuador. *Singapore Journal of Tropical Geography* 2(1), 15–26.
Brownrigg, L.A. (1981) Economic and ecological strategies of Lojano migrants to El Oro. In: Whitten, N.E. (ed.) *Cultural Transformation and Ethnicity in Modern Ecuador*. Urbana, Chicago, pp. 303–306.
Cabarle, B.J., Crespi, M., Dodson, C.H., Luzuriaga, C., Rose, D. and Shores, J.N. (1989) *An Assessment of Biological Diversity and Tropical Forests for Ecuador*. USAID, Quito.
Casagrande, J.B., Thompson, S.I. and Young, P.D. (1964) Colonization as a research frontier: the Ecuadorian case. In: Manners, R.A. (ed.) *Process and Pattern in Culture: Essays in Honor of Julian H. Steward*. Aldine, Chicago.
CEPAL (1954) *El Desarrollo Económico del Ecuador*. Naciones Unidas, Mexico.
CIDA (1965) *Tenencia de la Tierra y Desarrollo Socio-económico del Sector Agrícola – Ecuador*, Unión Panamericana and OEA, Washington.
Hernández, C.E. and Witter, S.G. (1996) Evaluating and managing the environmental impact of banana production in Costa Rica: a systems approach. *Ambio* 25(3).
IMF (1991) *International Financial Statistics, Yearbook 1991*. International Monetary Fund, Washington, DC.
Larrea, M.C. (1987) Chapters 1, 2, 3, 6 and 7. In: Larrea, C.M., Espinosa, M. and Sylva, C.P. (eds) *El Banano en el Ecuador*. Bibliotèca de Ciencias Sociales 16, Corporación Editora Nacional, Quito, pp. 9–110, 237–286.
López, J.R. (1988) *La Economía del Banano en Centroamérica*. Colección Universitaria, San José.
MAG (n.d.) *Situación y Pronóstico del Cultivo de Banano en el Ecuador 1980–1991*. PRSA and Ministerio de Agricultura y Ganadería, Quito.
MAG (1994) *Situación y Perspectiva del Cultivo de Banano en el Ecuador 1990–1993*. Proyecto PRSA, Ministerio de Agricultura y Ganadería, Quito.
MAG (1998) *Ecuador Producción Bananera*. Ministerio de Agricultura y Ganadería, Guayaquil.
Moreno, M.C. (1991) *El Fin del Boom Bananero?* Vistazo, Quito.
Riofrío, S. (1995) *Banano en Cifras . . . y Otras Novedades 1995*. Acción Gráfica, Guayaquil.
Riofrío, S. (1997) *J. Banano Ecuatoriano, Perspectives*. Producciones Agropecuarios, Guayaquil.
Ríos, P.R. (1996) Estudio de la estructura y marco regulador del banano. Unpublished consultancy report to the Ministry of Agriculture, Quito.
Rudel, T. with Horowitz, B. (1993) *Tropical Deforestation: Small Farmers and Land Clearing in the Ecuadorian Amazon*. Columbia University Press, New York.
SICA (1999) Proyecto SICA-BIRF/MAG-Ecuador, www.sica.gov.ec.
Southgate, D. and Whitaker, M. (1996) *Economic Progress and the Environment: One Developing Country's Policy Crisis*. Oxford University Press, New York.
Striffler, S. (1997) In the shadows of state and capital: the United Fruit Company and the politics of agricultural restructuring in Ecuador, 1900–1995. PhD thesis, Graduate Faculty of Political and Social Science, New York School for Social Research, New York.

Sylva, C.P. (1987) Los productores de banano. In: Larrea, C.M., Espinosa, M. and Sylva, C.P. (eds) *El Banano en el Ecuador*. Bibliotèca de Ciencias Sociales, Corporación Editora Nacional, Quito, pp. 111–186.
UPEB (1990) *Informe UPEB*. Unión de Países Exportadores de Banana, Panama.
Wood, H.A. (1972) Spontaneous agricultural colonization in Ecuador. *Annals of the Association of American Geographers* 62(4), 599–617.
Wunder, S. (1997) *From Dutch Disease to Deforestation – a Macroeconomic Link? A Case Study from Ecuador*. CDR Working Papers No. 6, Centre for Development Research, Copenhagen.
Wunder, S. (2000) *The Economics of Deforestation: the Example of Ecuador*. St Antony's Series, Macmillan and St Martin's Press, Oxford, London and New York.

Soybean Technology and the Loss of Natural Vegetation in Brazil and Bolivia

11

David Kaimowitz and Joyotee Smith

> Ten years ago, you couldn't find Mimosa on a map of Brazil. Back then, the town consisted of little more than a Shell truck stop on an asphalt highway, a backwater in the midst of 500 million acres of untamed scrub trees and grassland. That was before soybean farmers conquered the Cerrado. Today, this frontier boom town in the state of Bahia boasts a population of 15,000, a farm cooperative, two soybean processors, a phosphate fertilizer plant, three machinery dealers, half a dozen chemical dealers, a branch of the Bank of Brazil, a $49-a-night motel and a brand-new country club for the families of the nouveau riche. Dozens of young soybean tycoons traded their fathers' small stakes in Southern Brazil for 30 or 50 times more land in the north. Some quit comfortable $70,000-a-year white-collar jobs in Sao Paulo; others are descendants of Japanese immigrants, subsidized by Asian money . . . Gold may have drawn settlers to California's Wild West, but Mimosa owes its prosperity to soybeans and agricultural technology.
>
> (Marcia Zarley Taylor, Farming the last frontier, *Farm Journal Today*, 16 November 1998)

1. Introduction

Thirty-five years ago, South American farmers grew virtually no soybeans. Now, Brazilian farmers plant almost 13 million ha of soybeans and Brazil ranks as the world's second largest exporter (Waino, 1998). Bolivian farmers cultivate an additional 470,000 ha (Pacheco, 1998).

Soybean expansion in southern Brazil contributed to deforestation by stimulating migration to agricultural frontier regions in the Amazon and the

Cerrado. Since producing soybeans requires much less labour than producing coffee or food crops, when soybeans replaced those crops many small farmers and rural labourers lost their jobs and moved to the frontier. Elsewhere, in the Brazilian Cerrado and in Bolivia, farmers cleared large areas of Cerrado vegetation (natural savannah and open woodlands) and semi-deciduous forest to plant soybeans.

Technology was the key in all this. In a sense, soybeans themselves were a new technology, since, up to the 1970s, Brazilian and Bolivian farmers knew little about how to produce them. The development of new varieties adapted to the tropics and the use of soil amendments permitted farmers to grow soybeans in the low latitudes and poor acid soils of the Brazilian Cerrado. More generally, new varieties, inoculants, pest control agents, postharvest technologies and cultural practices made growing soybeans more profitable in both Bolivia and Brazil and stimulated their expansion.

Favourable policies and market conditions reinforced the new technologies' effect. Together, they helped soybean production attain a level that justified establishing the associated services and infrastructure competitive soybean production requires. High international prices and government subsidies encouraged the spread of soybeans in Brazil. Export promotion policies, favourable exchange rates and preferential access to the Andean market stimulated Bolivia's production. In both countries, road construction, government land grants and rising domestic demand for soybeans accelerated the crop's advance. This in turn increased the political power of the soybean lobby and enabled farmers and processors to obtain further government support.

This chapter examines the relation between soybean technology and the loss of natural vegetation in south Brazil, central-west Brazil (the Cerrado) and Santa Cruz, Bolivia. We first present our theoretical framework. Then, for each case, we show how technology and other factors interacted to stimulate soybean expansion, look at the general equilibrium effects this generated in labour and product markets, assess the impact on forest and savannah and briefly comment on the resulting costs and benefits.

2. The Theoretical Framework as it Applies to our Case

Technological change makes agricultural activities more profitable and that leads to their expansion. In southern Brazil, improved soybean technologies mostly led to soybeans replacing other crops. In the Cerrado, they replaced mostly Cerrado vegetation, while in Santa Cruz, Bolivia, it was mostly semi-deciduous forest.

Potentially, general equilibrium effects in either the product or labour markets can dampen the expansionary effects of technological change. In the product market, rising soybean production can push down international prices, thus discouraging further expansion. This effect was significant in Brazil, due to the huge production increases involved. Since the early 1970s,

Brazil has ceased to act like a 'small country' in the world soybean market (Frechette, 1997). Bolivia finds itself in a similar circumstance in regard to the Andean market, where its soybean exports have privileged access.

In regard to labour markets, the technology used to produce soybeans is highly capital-intensive and requires little labour. This means that rapid growth is unlikely to provoke labour shortages that push up wages and curtail subsequent growth. In situations, such as in southern Brazil, where soybeans replaced more labour-intensive crops, the advance of soybean production actually displaced labour. That labour then became available to migrate to the agricultural frontier. In other contexts, such as in the Brazilian Cerrado and the Santa Cruz expansion zone, where farmers have removed natural vegetation to plant soybeans, the demand for labour rises, but only slightly.

The profits resulting from technological change can also provide the capital required to expand agricultural production. Many farmers in southern Brazil used the profits obtained from soybeans to move to frontier regions and clear additional forest.

Three unique features of our theoretical framework compared with other chapters in this book are the roles we attribute to: economies of scale, the interaction between technology and other policies and the impact of technology on the political economy. To produce soybeans competitively, you need a large and modern processing, transportation, storage, financial, technological and marketing system. This implies that major economies of scale exist at the sector level. Technological progress can make it easier to profitably reach levels of production that justify installing ancillary services and infrastructure. Since one piece of agricultural machinery can cultivate a large area, mechanized soybean production also exhibits economies of scale at the farm level.

Technological advances and government policies interact in a non-linear fashion. For example, credit subsidies in the Brazilian Cerrado induced farmers to adopt agricultural machinery and soil amendment technologies that made growing soybeans more profitable than extensive cattle ranching. Once this process had begun, the economies of scale in soybean production accelerated it.

Figure 11.1 illustrates this process. The isoquant CR1 represents land and capital combinations for the Cerrado's traditional land-use system: extensive cattle ranching on natural pastures, which maintains most of the natural vegetation. The three SB isoquants represent the new soybean technology. In this case, farmers totally remove the natural vegetation. The numbers attached to each isoquant refer to how much revenue is generated. Hence, SB1 gives the same gross revenue as CR1. The SB isoquants show increasing returns to scale resulting from the use of agricultural machinery, and SB technologies enable farmers to get higher returns from their land, compared with CR1, by using more capital.

With cheap land, shown by a flat factor price ratio (FP), farmers produce at point X on CR1. As long as the capital/land price ratio remains high, farmers

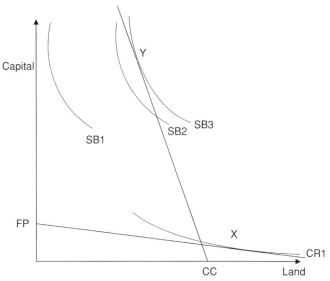

Fig. 11.1. Technology choice and land use under soybean and extensive cattle ranching (Brazilian Cerrado): impact of factor price ratios and economies of scale.

will not adopt SB technologies. But subsidized credit can tilt the factor price ratio to CC and persuade farmers to grow soybeans. Thus, policy can stimulate farmers to adopt a capital-intensive technology in a land-abundant area.

Even though soybeans increase the returns to land, their potential for 'land saving' is diluted because of economies of scale. Thanks to increasing returns to scale, the new factor price ratio of CC resulting from subsidized credit allows farmers to move to point Y on the SB3 isoquant, rather than to some point on SB1 or SB2. Thus, even though subsidized credit makes capital cheap compared with land, rather than using more capital and less land, farmers are inclined to use more of both.

Finally, technological change not only changes relative prices, it also modifies political relations. By favouring the development of a large, concentrated, agroindustrial sector, new soybean technologies facilitated the creation of powerful interest groups, which successfully lobbied the Brazilian and Bolivian governments to implement policies favourable to the soybean sector.

3. Southern Brazil

Southern Brazil includes Parana, Rio Grande do Sul and Santa Catarina. By 1960, farmers had settled most of this region, except for parts of Parana. Coffee, beans, maize and cassava covered large areas. Small farmers with less than 50 ha of land, many of whom were sharecroppers, tenants or squatters, planted much of that (Stedman, 1996).

3.1. Technologies and policies promoting soybean expansion

Coffee boomed in Parana and the other southern states in the 1950s, expanding from 7% of harvested area to 19% (Stedman, 1996). By 1960, Parana had become Brazil's top coffee-producing state. Soon after, however, low coffee prices, soil erosion, plant diseases and frost caused a crisis in the regional coffee economy (Diegues, 1992). In response, the government introduced a 'coffee eradication programme', designed to replace coffee with traditional food crops, wheat and soybeans (Stedman, 1996).

A handful of farmers in Rio Grande do Sul were already planting soybeans in the early 1960s. But Brazil still had less than 250 ha of soybeans. Yields averaged only 1060 kg ha^{-1} (Kaster and Bonato, 1980; Wilkinson and Sorj, 1992).

Then research centres in São Paulo and Rio Grande do Sul introduced varieties from the USA. Thanks to its similar climate, soils and day length, these varieties adapted easily to southern Brazil. The new varieties permitted average yields to increase 15% between 1960 and 1970 to 1141 kg ha^{-1} (Wilkinson and Sorj, 1992).

Government land, credit and price policies encouraged the spread of soybeans. The 1964 Land Statute gave tenant farmers and sharecroppers greater rights and many large landholders responded by expelling tenants and sharecroppers from their farms. Similarly, landholders reacted to new minimum wage laws by hiring fewer agricultural labourers. One way to achieve that was to plant soybeans and wheat, which required less labour, instead of coffee and traditional food crops. The government further accelerated the shift towards mechanized annual crop production by providing subsidized credit to purchase agricultural machinery (Sanders and Ruttan, 1978). Between 1965 and 1970, the coffee area in the south fell from 1.4 million ha to 1 million ha and, by 1970, farmers were growing more than 1.2 million ha of soybeans (Stedman, 1998).

Then, in 1973, a severe drought in the USA caused international prices to sky-rocket and the USA imposed an embargo on its own soybean exports in response (Smith *et al.*, 1995). This coincided with policies favouring soybeans in Brazil. Exchange rates became less overvalued. Agricultural credit rose almost fivefold between 1970 and 1980 and soybeans received over 20% of that (Skole *et al.*, 1994). The government gave incentives to domestic wheat producers to promote import substitution. This benefited soybean-growers, since farmers frequently rotated wheat with soybeans and the two crops shared the same machinery, equipment and labour (Wilkinson and Sorj, 1992). Rapid urbanization and rising per capita incomes increased domestic demand for soybean products.

Brazil's trade policies had more ambiguous effects. The government used subsidies and fiscal incentives to encourage domestic processing of soybean products, while restricting exports of unprocessed beans. On balance, these policies probably slowed soybean's expansion (Williams and Thompson, 1984).

In 1973, the Federal Government created the Brazilian Agricultural Research Corporation (EMBRAPA). EMBRAPA's National Soybean Research Centre, located in Parana, conducted its own research and coordinated the efforts of various universities, state-level research centres, farmer cooperatives and private companies. The International Soybean Program at the University of Illinois, financed by the US Agency for International Development, helped introduce new technologies from the USA (Wilkinson and Sorj, 1992). By 1981, Brazil had almost 1000 researchers and extension agents working on soybeans (Bojanic and Echeverría, 1990).

Their efforts paid off. By 1980, national breeding programmes had produced 26 of the 48 varieties recommended in Brazil and these varieties provided yields 36–63% higher than their predecessors (Kaster and Bonato, 1980). Researchers also introduced *Rhizobium*-based nitrogen fixation, biological agents for controlling soybean caterpillars, 'no-till' planting, the use of contour bunds across fields and new herbicides and fertilizers (Kaster and Bonato, 1980; Wilkinson and Sorj, 1992). This allowed producers to increase yields, reduce costs and degrade their soils less. *Rhizobium*-based nitrogen fixation alone saved farmers over 5 million t annually of nitrogen fertilizer and reduced fertilizer costs by 80% (Wilksinon and Sorj, 1992). Average yields in 1975 were 1720 kg ha^{-1}, 50% higher than in 1970. Kaster and Bonato (1980) attribute two-thirds of that increase to new varieties and the remainder to improved agronomic practices.

The combination of high international prices, government subsidies and technological progress led to a dramatic rise in the south's soybean area. The cultivated area jumped from 1.2 million ha in 1970 to 5.1 million ha in 1975 and 6.9 million ha in 1980 (Stedman, 1996). Simoes (1985) calculated that, for each percentage increase in expenditure on soybean research between 1973 and 1983, the soybean area grew by 0.28%. Moreover, this calculation ignores how new technologies interacted with other contextual factors, since the sum was certainly greater than the parts.

3.2. Soybean expansion and the loss of natural vegetation (Table 11.1)

We do not know what portion of the soybean expansion in the south directly led to deforestation within the region itself. It was probably less than a third, since the total utilized farmland in the south increased only 1.9 million ha during the 1970s (Stedman, 1996). It is worth noting, however, that virtually all of Parana was originally old-growth forest, with a high prevalence of *Arucaria* trees (M. Faminow, 1998, personal communication).

Table 11.1. Causes of soybean expansion and impact on natural vegetation: Brazil and Bolivia.

	Southern Brazil	Brazilian Cerrado	Frontier areas: Brazilian Cerrado	Bolivian 'expansion zone'
Causes of expansion				
Technology				
Improved varieties/cultural practices: increased productivity	✓	✓	✓	✓
Mechanization/economies of scale leading to:				
• Land concentration/increasing land prices (southern Brazil)		✓	✓	
• Sunk costs in machinery/ infrastructure leading to lobbying by soybean sector			✓	
• Expansion to frontier areas with cheap land		✓	✓	✓
Market conditions				
High international prices (mid-1970s–mid-1980s)	✓	✓		
Andean common market				✓
Government policy				
Subsidized credit	✓	✓		
Development projects		✓		
Land as hedge against inflation		✓	✓	
Structural adjustment				✓
Roads		✓	✓	✓
Private sector				
Settlement schemes/soybean infrastructure			✓	
Political economy: lobbying power				
Mid-1980s: uniform output/fuel prices			✓	
Mid-1990s: tax concessions, credit guarantees, export corridor			✓	
Export of Brazilian capital				✓
Impact on natural vegetation	Migrants to Amazon/ Cerrado	Conversion to agriculture	Conversion to agriculture. Export corridor: Amazonian deforestation?	Conversion to agriculture

General equilibrium effects: the product market

After 1980, soybeans in the south ran out of steam and the area contracted from 6.9 million ha in 1980 to 6.1 million ha in 1990 (Stedman, 1996). Technological change contributed to this process by depressing soybean prices and thus dampening the initial incentive it had provided to increase the area. According to Simoes (1985), falling international prices resulting from Brazil's increased productivity caused soybean farmers in the south and in São Paulo to receive 28% fewer benefits from agricultural research between 1973 and 1983 than they would have if Brazil had been a small player in international soybean markets. Stagnating yields, in part due to growing problems of soil erosion and compaction, the elimination of wheat subsidies and high port costs, also contributed to the decline in soybean area (Wilkinson and Sorj, 1992).

General equilibrium effects: the labour market and farmer incomes

The principal way the advance of soybeans in the south influenced deforestation was through the labour market. The shift to soybeans stimulated land concentration and agricultural mechanization. Many small farmers could not afford the machinery and chemical inputs required for growing soybeans. Rising yields, high soybean prices and subsidized credit pushed up land prices and poor farmers found it increasingly difficult to compete in the land markets (Brandao and Rezende, 1992). Subsidized credit went mostly to large farmers and this accelerated the concentration of landholdings (Goldin and de Rezende, 1993). The number of tractors in Brazil jumped from 134,500 to 545,200 between 1965 and 1980 and southern soybean producers accounted for a lot of this (Stedman, 1996).

As a result of these processes, more than 2.5 million people left rural Parana in the 1970s and the number of farms smaller than 50 ha declined by 109,000 (Diegues, 1992). During the same period, Rio Grande do Sul lost some 300,000 farms (Genetic Resources Action International, 1997).

The majority of migrants moved to urban areas. Nevertheless, a significant number went to the Amazon and cleared forest to grow crops. Sawyer (1990) cites Parana as an important source of migrants to the Amazon in that period.

While the expansion of mechanized agriculture destroyed the livelihoods of many migrants to the Amazon, in other cases soybean and wheat production provided the resources that allowed small farmers to purchase land on the agricultural frontier. Many better-off small farmers who moved to the Cerrado took advantage of land price increases in the south to sell their farms and buy larger areas in the Cerrado, where land was cheap (Coy and Lucker, 1993).

4. The Brazilian Cerrado

Just as soybean production stagnated in the south, it took off in the Cerrado. The term Cerrado refers to a characteristic set of vegetative types, which include natural savannahs and woodlands. This vegetation dominates 1.5 to 2 million km² in Brazil's centre-west states of Mato Grosso, Mato Grosso do Sul, Goias and Tocantins and in parts of Bahia, Maranhao, Minas Gerais and Piaui (Stedman, 1998). In northern Mato Grosso, one finds a transition between Cerrado vegetation and rain forest. The region's soils tend to be highly acidic and deficient in phosphorus (Smith *et al.*, 1998).

4.1. Technologies and policies promoting soybean expansion

Historically, the Cerrado had a low population density and large unoccupied areas, dominated by extensive cattle ranches (Mueller *et al.*, 1992). New soybean technologies, public road construction and subsidized credit, fuel and soybean prices changed that. The total annual crop area in the centre-west rose from 2.3 million ha in 1970 to 7.4 million ha in 1985. The soybean area soared from only 14,000 ha to 2.9 million ha and then reached 3.8 million ha in 1990 (Stedman, 1996). Heavily capitalized farms with between 200 and 10,000 ha grew most of this (Mueller *et al.*, 1992). In 1992, a farmer in Maranhao needed to invest almost $1 million to grow 1000 ha of soybeans (Carvalho and Paludzyszyn Filho, 1993).

Traditionally, the Cerrado's poor and heavy soils and lack of suitable varieties limited intensive crop production in the Cerrado. Farmers solved the first constraint by applying a lot of lime and phosphate and using machinery to plough heavy soils (Sanders and Ruttan, 1978; Goldin and de Rezende, 1993). To overcome the second constraint required local plant breeding. Existing soybean varieties were sensitive to photoperiod and performed poorly in the lower latitudes, where day length is uniform and short. They were also susceptible to aluminium toxicity and required large amounts of calcium (Spehar, 1995).

Beginning in the mid-1970s, the National Soybean Research Centre and other research centres worked to produce varieties adapted to the Cerrado, with the explicit goal of advancing the agricultural frontier (Kueneman and Camacho, 1987). By the early 1980s, they had largely succeeded. Thanks to these efforts, mean yields rose 45% between 1975 and 1983, from 1300 kg ha^{-1} to 1900 kg ha^{-1} (Simoes, 1985). Spehar (1995) estimates that the new varieties increased the annual earnings of soybean producers in the Cerrado by $1 billion.

Without new varieties, soil treatments and machinery, the rapid spread of soybeans into the Cerrado would have been impossible. Nevertheless, other factors also contributed. In particular, as noted earlier, credit subsidies proved an essential precondition for the rapid adoption of agricultural machinery and soil amendments. Between 1975 and 1982, one subsidized credit programme, the Programme for the Development of the Cerrado (POLOCENTRO), gave $577 million in agricultural loans, 88% of which went to farmers with over 200 ha. According to Mueller *et al.* (1992), this was responsible for the conversion of 2.4 million ha of savannah to agriculture. Without government subsidies, soybean production would probably have been restricted to accessible areas with better soils. Subsidies allowed farmers to grow soybeans profitably in more remote areas, such as northern Mato Grosso.

In the 1970s, new roads, such as BR163, which connected Cuiaba and Santarem, and BR158, between Barra do Garcas and Maraba, opened up northern Mato Grosso (Coy and Lucker, 1993). The Brazilian government also made land and credit available to large private companies, which built roads and other infrastructure and then resold part of the land in 50 to 400 ha parcels to enterprising small farmers from the south. By 1986, 104 private colonization schemes covered 2.9 million ha, of which 668,000 ha were planted with annual crops (Mueller *et al.*, 1992). Some large private investor groups used the proceeds from the land sales to grow soybean in the remaining areas and create local infrastructure for storing soybeans and carrying out the initial stages of processing to make vegetable oil. In recent years, they have also collaborated with EMBRAPA to develop improved varieties (Coy, 1992; Franz and Pimenta da Aguiar, 1994).

General equilibrium effects: the product market

From the mid-1980s, several factors turned against soybean production. International soybean prices fell. Higher soybean production in the Cerrado, generated by technological changes, may have contributed to this, but as far as we know no one has studied the issue. The Brazilian government also greatly reduced credit subsidies and real interest rates rose sharply (Goldin and de Rezende, 1993). The 1994 macroeconomic stabilization policy, known as the Real Plan, generated positive real growth rates and radically reduced inflation, both of which stimulated domestic demand for soybeans (Smith *et al.*, 1998). Nevertheless, the exchange rate became progressively overvalued and real interest rates remained high and volatile, causing severe financial stress among indebted soybean farmers (Smith *et al.*, 1999).

General equilibrium effects: the labour market

Thanks to in-migration from the south and the north-east and the limited labour requirements of soybean cultivation, the growth of the soybean area put little upward pressure on wages. Between 1970 and 1985, the area in crops rose by 172%, the cattle herd by 128% and the number of tractors by 660% (Mueller *et al.*, 1992). However, the agricultural labour force in the savannah region grew by only 45%, from 1.4 million to 2 million.

The soybean lobby

Despite low international prices, declining credit subsidies and an overvalued exchange rate, soybean production in the Cerrado has continued to expand, except for a few years in the early 1990s (Stedman, 1996). Soybean exports reached a record 8.3 million t in 1996/97 and the soybean area was projected to reach a record 12.9 million ha in 1997/98 (USDA, 1998).

Powerful interest groups linked to the soybean sector lobbied successfully for compensating government concessions whenever conditions turned unfavourable. This group, which includes processors and exporters, machinery and input manufacturers, investor groups and farmer organizations, has become a potent force in Brazilian politics (Pompermayer, 1984; Coy, 1992). Its great influence appears to be linked to the important contribution made by agricultural exports to meeting balance-of-payments deficits, particularly during the debt crisis of the 1980s and again in the mid-1990s. Between 1994 and 1996, the agricultural sector contributed over $25,000 million to the trade balance, of which soybeans and related products accounted for 26% (USDA, 1998).

To compensate for the decline in subsidized credit and to protect farmers from falling international soybean prices during the mid-1980s, the government purchased large quantities of soybeans from farmers at pre-established prices (Goldin and de Rezende, 1993). Farmers received the same price for their soybeans no matter where they were located, thus encouraging soybean's expansion into remote areas, where high transportation costs might otherwise have impeded commercial production.

The government also established uniform fuel prices, without considering the high cost of transporting fuel to remote areas. This not only made it feasible for farmers to transport their crops long distances to markets, but also lowered fuel costs for the use of agricultural machinery (Mueller *et al.*, 1992).

In the 1990s, the private sector and government agencies initiated several projects designed to reduce the cost of transporting soybeans from the Cerrado to different ports. The US Department of Agriculture (USDA, 1998) reports that a north-west corridor project linking the northern Cerrado to the Amazon will lower soybean transport costs by around $30 t^{-1}, as well as reducing fertilizer costs. In 1990, private companies, banks and government agencies jointly established the northern export corridor initiative to increase soybean production in Tocantins, Maranhao and Piaui, with a goal of 500,000 ha by 1998. The initiative includes fiscal incentives, agricultural research, credit and infrastructure for transporting soybeans to the Amazon River (Carvalho and Paludzyszyn Filho, 1993).

The USDA (1998) also reports other recent policy changes that benefit the soybean sector. Soybean farmers benefited from the 1996 removal of a tax on primary and semi-manufactured exports. In response to the high interest rates of the 1990s, the government provided guarantees to commercial banks to allow exporters to obtain credit at rates similar to those available internationally.

4.2. Soybean expansion and the loss of natural vegetation

Over the last 20 years, soybean and pasture expansion dramatically affected the natural vegetation of the Cerrado. Between 1970 and 1985, conversion of the Cerrado's natural ecosystem was as rapid as in the Amazon. Farmers converted some 2 million ha of natural vegetation to agricultural uses each year, including 350,000–450,000 ha of forest (Smith et al., 1998). Intensive annual crop production accounts for about 20% of this loss (Smith et al., 1998). Only 35% of the Cerrado biome remains in a relatively natural state (Stedman, 1998). Some types of vegetation and fauna, such as mesotrophic woodland and the pampas deer, are becoming rare.

Whether the benefits of this transformation outweighed the costs remains uncertain. On the one hand, the Cerrado probably accounted for almost half of Brazil's $4.4 billion of soybean exports in 1996 (Spehar, 1995; Waino, 1998). On the other hand, the region has one of the richest savannah floras in the world, especially of woody species, and much of this could be lost (Klink et al., 1993). Conversion has brought about large emissions of carbon dioxide into the atmosphere. Agriculture is estimated to be responsible for 50% of the organic matter that enters waterways, and sedimentation could cause serious problems, since the Cerrado forms part of the watershed of major rivers and drains into the Pantanal, one of the world's largest wetlands (Smith et al., 1998).

5. Santa Cruz, Bolivia

Santa Cruz, Bolivia, has many of the same features as the Cerrado. Agricultural research and technology transfer there encouraged the rapid spread of soybeans and this led to large-scale deforestation. Again, however, it was not the only factor.

Bolivian farmers grew only 1000 ha of soybeans in 1970 and even in 1980 still had only 31,000 ha. This grew to 56,000 ha in 1985, 147,000 in 1990 and 470,000 in 1996 (Pacheco, 1998).

Approximately 1900 farmers grew soybeans in 1990 (Bojanic and Echeverría, 1990). Traditionally, Mennonite colonists and, to a lesser extent, Japanese colonists and Bolivian farmers grew most of them. In recent years, large Brazilian farmers have become important and now account for about a quarter of the production (Pacheco, 1998).

5.1. Technologies and policies promoting soybean expansion

As in Brazil, agricultural research and extension helped promote soybean expansion. Local researchers began testing varieties imported from the USA in 1953. Significant research got under way in 1975, when the Tropical

Agricultural Research Centre (CIAT) was established and created a small soybean programme. Through the Cooperative Agricultural Research Programme for the Southern Cone (PROCISUR), CIAT maintained close relations with the soybean researchers at EMBRAPA. All five soybean varieties released in the 1980s came from Brazil. CIAT also tested different products to control weeds, diseases and insects and conducted research on crop rotations, inoculants and fertilizers, direct planting and soil conservation. A local farmer organization (ANAPO), farmer cooperatives and various commercial establishments promoted the results of this research (Bojanic and Echeverría, 1990).

Average summer soybean yields rose from 1333 kg ha^{-1} in 1974–1979 to 1743 kg ha^{-1} in 1980–1984 and 2022 kg ha^{-1} in 1985–1990. Bojanic and Echeverría (1990) attribute between 40% and 60% of that increase to CIAT's research and the private sector's technology transfer efforts. Since the new technology involved only marginal additional costs, the increased yields clearly contributed to the commercial viability of producing soybeans.

The removal of price controls, a currency devaluation, fiscal incentives for exporters, low taxes, road construction and government land grants also contributed to the expansion of soybeans (Kaimowitz et al., 1999). In the mid-1990s, Bolivian producers paid $26.5 t^{-1} less in taxes than their Brazilian counterparts (Monitor Company, 1994). Between 1986 and 1991, the road network in Santa Cruz's so-called 'expansion zone' grew from 430 km to 650 km and in 1989 the World Bank's Eastern Lowlands Project began financing road improvements to facilitate soybean exports (Davies, 1993). The government reversed its policy of allocating land in the expansion zone to small agricultural colonists and began focusing more on large landholders, and this also encouraged soybean production.

Unlike Brazil, credit subsidies had only a minor role in Bolivia's soybean expansion. The government heavily subsidized credit in the early 1980s but eliminated the subsidies in 1985 as part of its structural adjustment programme. Access to capital, however, did not greatly constrain the advance of soybean production, since commercial farmers had large financial reserves and easy access to private credit and Brazilian investors brought additional resources into the area.

Bolivia's entrance into the Andean Common Market also greatly boosted its soybean exports. In 1995, a little more than 80% of Bolivia's soybean and soybean-product exports went to the Andean market, where the country enjoyed a $37.17 t^{-1} tariff advantage over its Brazilian competitors (Monitor Company, 1994).

5.2. Soybean expansion and the loss of natural vegetation

Initially, most soybeans were grown in an area west of the Grande River, near the city of Santa Cruz, known as the 'integrated zone'. That area has been settled for a long time, has moderately high population densities and is dominated by large commercial farmers. Most soybean production there is on land where the natural vegetation had already been removed for other purposes. Although somewhat fragile and susceptible to wind erosion and compaction, the region has much better soils than the Brazilian Cerrado. Most farmers produce soybeans there without fertilizers or soil amendments (Barber, 1995).

Since 1990, most of the soybean growth has been just east of the Grande River in the 'expansion zone'. There, the soybean area rose from 68,000 ha in 1990 to 278,000 ha in 1996 and has continued to expand rapidly since (Pacheco, 1998). Unlike the 'integrated zone', for the most part these lands were directly converted from semi-deciduous forest to grow soybeans and certain areas have climates and soil conditions that are less favourable for soybean production.

Largely as a result of greater soybean production, the annual deforestation rate in the expansion zone in 1989–1992 was 24,207 ha and in 1992–1994 it was 41,604 ha (Morales, 1993, 1996).

As in the Cerrado, we are unable to say whether the benefits of converting forests to soybean fields outweigh the environmental and social costs. Davies and Abelson (1996) attempted such an evaluation and concluded that the financial benefits from soybean production greatly outweigh the costs from reduced carbon sequestration and harvesting of forest products. However, they were unable to assign economic values to the loss of biodiversity and soil erosion and ignored equity. Hecht (1997: 4) has argued that the biodiversity values of these forests are particularly great since 'they embrace Andean, Amazonian, and Chaco biotic elements, and include important (and threatened) centres of diversity for crop plants like peanuts and tomatoes'.

6. Conclusions

The technological changes related to soybean production in Brazil and Bolivia involve a new production system, more profitable production practices and the substitution of capital for labour. These changes directly and indirectly induced the conversion of large areas of natural vegetation to expand annual crop production. In the Cerrado and the Bolivian 'expansion zone', the availability of cheap land in frontier areas particularly favoured production systems characterized by economies of scale.

The low labour requirements of the new soybean technologies led to the displacement of existing agricultural labour in southern Brazil, some of which subsequently moved to the agricultural frontier. In the other regions, they ensured that the growth in soybean production did not put upward pressure on wages and feed back into lower profits.

The high capital requirements of the new technology might have constrained soybean's expansion but it did not, except perhaps for Brazil in the 1990s. During the 1970s and 1980s, the availability of plentiful subsidized credit allowed Brazilian farmers to adopt heavily capital-intensive technologies. Bolivian farmers had ready access to private credit and the Brazilian farmers who moved to Bolivia brought large amounts of money.

The technology involved in the case of the Brazilian Cerrado was specifically suited for the environmental conditions of that region, which was an agricultural frontier area covered with natural vegetation. This undoubtedly increased the environmental impact of the technology's development and dissemination.

A particularly interesting feature of the expansion into the northern Cerrado is the role of political-economy factors. The expansion of soybean after the mid-1980s appears to be closely related to the lobbying power of the soybean sector, which enabled it to wring concessions from the government. By helping to create the soybean sector in the first place, technological developments inadvertently created a strong new political lobby.

Because of the huge production increases made possible by technological change in Brazil and the small size of the Andean market, which buys Bolivian soybeans, in both cases general equilibrium effects in the product markets reduced some of the expansionary impetus created by technological change. These dampening effects were not sufficient to avoid widespread loss of natural vegetation.

Rather than attempting to separate out the relative weight of technology and other factors in the spread of soybeans, we would like to emphasize the interaction between these factors. Changes in production systems of such a large magnitude require both appropriate technologies and favourable policy and market conditions.

Finally, the soybean case highlights the difficulties in determining whether the benefits outweigh the costs in cases where agricultural technology leads to the loss of natural vegetation. Soybeans provide substantial foreign exchange and much more income per hectare than cattle ranching (Davies and Abelson, 1996). The type of natural vegetation they replace typically stores much less carbon per hectare than do rain forests and has less biodiversity. Nevertheless, conversion still involves substantial carbon emissions and biodiversity losses and increases soil erosion. Moreover, both the Brazilian Cerrado and the semi-deciduous forests of Bolivia have richer biodiversity than people often realize (Klink *et al.*, 1993; Hecht, 1997). Mechanized soybean production provides little employment and a small group of wealthy farmers receive most of the income.

References

Barber, R.G. (1995) Soil degradation in the tropical lowlands of Santa Cruz, eastern Bolivia. *Land Degradation and Rehabilitation* 6, 95–107.

Bojanic, A. and Echeverría, R.G. (1990) *Retornos a la Inversión en Investigación Agrícola en Bolivia: El Caso de la Soya.* ISNAR Staff Notes Number 90–94, International Service for National Agricultural Research (ISNAR), The Hague.

Brandao, A.S. and Rezende, G.C. (1992) *Credit Subsidies, Inflation, and the Land Market in Brazil: a Theoretical and Empirical Analysis*. World Bank, Washington, DC.

Carvalho, J.G. and Paludzyszyn Filho, E. (1993) *Diagnóstico do Corredor de Exportaçao Norte*. Companhia Vale do Rio Doce, Brazil.

Coy, M. (1992) Pioneer front and urban development: social and economic differentiation of pioneer towns in northern Mato Grosso (Brazil). *Applied Geography and Development* 39, 7–29.

Coy, M. and Lucker, R. (1993) Mutations dans un espace périphérique en cours de modernisation: espaces sociaux dans le milieu rural du Centro-Oeste brésilien. *Cahiers d'Outre-Mer* 46(182), 153–74.

Davies, D. (1993) Estimations of deforestation east of the Rio Grande, Bolivia, using Landsat satellite imagery. MSc thesis, Silsoe College, Cranfield Institute of Technology.

Davies, P. and Abelson, P. (1996) The value of soils in the tropical lowlands of Eastern Bolivia. In: Abelson, P. (ed.) *Project Appraisal and Valuation of the Environment, General Principles and Six Case Studies in Developing Countries*. Macmillan Press, London, pp. 240–267.

Diegues, A.C. (1992) *The Social Dynamics of Deforestation in the Brazilian Amazon: an Overview*. DP36, United Nations Research Institute for Social Development (UNRISD), Geneva.

Franz, P.R.F. and Pimenta da Aguiar, J.L. (1994) *Characterizacao da Agropecuaria do Estado do Mato Grosso – Sondagem*. Projecto Novas Fronteiras do Cooperativismo (PNFC), Ministerio da Agricultura, Brasilia.

Frechette, D.L. (1997) The dynamics of convenience and the Brazilian soybean boom. *American Journal of Agricultural Economics* 79, 108–118.

Genetic Resources Action International (1997) La industrialización de la soja. *Biodiversidad, Sustento y Culturas* 14, 12–20.

Goldin, I. and de Rezende, G.C. (1993) *A Agricultura Brasileira na Década de 80: Crescimento Numa Economia em Crise*. IPEA 138l, Instituto de Pesquisa Económica Aplicada (IPEA), Rio de Janeiro.

Hecht, S.B. (1997) *Solutions and Drivers: the Dynamics and Implications of Bolivian Lowland Deforestation*. School of Public Policy and Social Research, University of California, Los Angeles.

Kaimowitz, D., Thiele, G. and Pacheco, P. (1999) The effects of structural adjustment policies on deforestation and forest degradation in lowland Bolivia. *World Development* 27(3), 505–520.

Kaster, M. and Bonato, E.R. (1980) Contribução das ciencias agrarias para o desenvolvimento: a pesquisa en soja. *Revista de Economía Rural (Brasilia)* 18(3), 415–434.

Klink, C.A., Moreira, A.G. and Solbrig, O.T. (1993) Ecological impacts of agricultural development in the Brazilian Cerrados. In: Young, M.D. and Solbrig, O.T. (eds) *The World's Savannas – Economic Driving Forces, Ecological Constraints, and Policy Options for Sustainable Land Use*. Man and Biosphere Series, Vol. 12, UNESCO, Paris, pp. 259–282.

Kueneman, E.A. and Camacho, L. (1987) Production and goals for expansion of soybeans in Latin America. In: Singh, S.R., Rachie, K.O. and Dashiell, K.E. (eds) *Soybeans for the Tropics, Research, Production, and Utilization*. John Wiley and Sons, Chichester, pp. 125–134.

Monitor Company (1994) *The Fragile Miracle: Building Competitiveness in Bolivia, Phase One*. Monitor Co., La Paz.

Morales, I. (1993) *Monitoreo del Bosque en el Departamento de Santa Cruz. Periodo 1988/89–1992/3*. Plan de Uso del Suelo, Santa Cruz.
Morales, I. (1996) *Monitoreo del Bosque en el Departamento de Santa Cruz. Periodo 1992/93–1994*. Plan de Uso del Suelo, Santa Cruz.
Mueller, C., Torres, H. and Martine, G. (1992) *An Analysis of Forest Margins and Savanna Agroecosystems in Brazil*. Institute for the Study of Society, Population, and Nature (ISPN), Brasilia, Brazil.
Pacheco, P. (1998) *Estilos de Desarrollo, Deforestación y Degradación de los Bosques en las Tierras Bajas de Bolivia*. CIFOR/CEDLA/TIERRA, La Paz.
Pompermayer, M.J. (1984) Strategies of private capital in the Brazilian Amazon. In: Schmink, M. and Wood, C.H. (eds) *Frontier Expansion in Amazonia*. University of Florida Press, Gainesville, pp. 419–438.
Sanders, J.H. and Ruttan, V.W. (1978) Biased choice of technology in Brazilian agriculture. In: Binswanger, H.P. and Ruttan, V.W. (eds) *Induced Innovation, Technology, Institutions, and Development*. Johns Hopkins University Press, Baltimore, pp. 276–296.
Sawyer, D. (1990) Migration and urban development in the Amazon. mimeo.
Simoes, C.H. (1985) The contribution of agricultural research to soybean productivity in Brazil. PhD thesis, Department of Agricultural Economics, University of Minnesota.
Skole, D.L., Chomentowski, W.H., Salas, W.A. and Nobre, A.D. (1994) Physical and human dimensions of deforestation in Amazonia. *Bioscience* 44(5), 312–322.
Smith, J., Winograd, M., Gallopin, G. and Pachico, D. (1998) Dynamics of the agricultural frontier in the Amazon and savannas of Brazil: analyzing the impact of policy and technology. *Environmental Modelling and Assessment* 3, 31–46.
Smith, J., Cadavid, J.V., Ayarza, M., Pimenta de Aguiar, J.L. and Rosa, R. (1999) Land use change in soybean production systems in the Brazilian savanna: the role of policy and market conditions. *Journal of Sustainable Agriculture* 15, 95–118.
Smith, N.J.H., Serrao, E.A.S., Alvim, P.T. and Falesi, I.C. (1995) *Amazonia – Resiliency and Dynamism of the Land and its People*. United Nations University Press, Tokyo.
Spehar, C.R. (1995) Impact of strategic genes in soybean on agricultural development in the Brazilian tropical savannahs. *Field Crops Research* 4, 141–146.
Stedman, P.A. (1996) Trade and environment: international context, policy response, and land use in Brazil. PhD dissertation, University of Florida, Gainesville.
Stedman, P.A. (1998) *Root Causes of Biodiversity Loss: Case Study of the Brazilian Cerrado*. World Wildlife Fund, Washington, DC.
Taylor, M.Z. (1998) Farming the last frontier. *Farm Journal Today*, 16 November.
USDA (1998) Agricultural Outlook. http://usda2.mannlib.cornell.edu: 70/0/reports/erssor/economics/ao-bb/1998/agricultural_outlook , 9.22.98.
Waino, J. (1998) Brazil's ag sector benefits from economic reform. In: *Agricultural Outlook AO-251, May*. Economic Research Service, United States Department of Agriculture, pp. 37–43.
Wilkinson, J. and Sorj, B. (1992) *Structural Adjustment and the Institutional Dimensions of Agricultural Research and Development in Brazil: Soybeans, Wheat, and Sugar Cane*. OECD/GD (92), OECD Development Centre, Paris.
Williams, G.W. and Thompson, R.L. (1984) The Brazilian soybean policy: the international effects of intervention. *American Journal of Agricultural Economics* 66(4), 488–498.

Kudzu-improved Fallows in the Peruvian Amazon

12

David Yanggen and Thomas Reardon

1. Introduction

Some 200–300 million people practise fallow-based slash-and-burn agriculture on roughly 30% of the world's exploitable soils (Sanchez, 1976; Crutzen and Andreae, 1990). Forest fallow vegetation allows land to recuperate its productive capacity by extracting nutrients from the air and from deep in the soil and returning them to the topsoil through leaf litter when farmers burn it (Sanchez, 1976; Nair, 1993). Fallow vegetation also shades out herbaceous weeds, another major constraint in slash-and-burn systems (de Rouw, 1995).

In regions where low population densities and abundant land permit long fallow periods, slash-and-burn agriculture is a sound way to manage land (Nair, 1993; Kleinman *et al.*, 1995). However, increasing population, shorter fallow periods and the removal of primary forest cover tend to cause declines in agricultural productivity (Nair, 1993; Thiele, 1993). To slow down or reverse this decline, farmers can manage their fallow better by encouraging the natural regeneration of plant species that help the land to recuperate quickly, such as leguminous nitrogen fixers, and by planting those species themselves (Kang and Wilson, 1987). Agronomists call that improved fallowing.

Most research on alternatives to traditional slash-and-burn fallows focuses on the agronomy of how to recuperate soil fertility more quickly during the fallow period. Researchers have devoted less attention to the economics of why farmers manage their fallow-based slash-and-burn agriculture the way they do (Dvorak, 1992). This chapter helps fill that gap by comparing the costs and benefits of traditional and improved fallowing practices and assessing how improved fallow adoption affects deforestation. Specifically, it analyses the impact of kudzu-improved fallow on deforestation in the lowland tropical rain-forest areas surrounding the city of Pucallpa, Peru.

The data for this chapter come from a 1998 survey of 220 farm households, which focused on input use, production and forest clearing, and a smaller survey of 24 households, which looked at farmer use and management of kudzu-improved fallows, traditional forest fallows and primary forest. Section 2 describes the Pucallpa region and the kudzu-improved fallows found there. Section 3 uses the survey data to compare kudzu-improved fallows with traditional slash-and-burn practices in terms of land and labour use and yields. This leads in to hypotheses about how kudzu-improved fallows affect deforestation, discussed in section 4. Section 5 contains an analytical model and two econometric models of how farmer decisions about input use, crop mix and disembodied technologies affect deforestation. Section 6 presents the regression analysis results and interprets them. Then come a summary and some policy recommendations.

2. The Pucallpa Region and Kudzu-improved Fallows

This section presents the key characteristics of the Pucallpa region and of kudzu fallow. An underlying hypothesis of this book is that the impact of technological change on forests depends on both the type of technology introduced and the characteristics of the zone where farmers adopt it.

2.1. Pucallpa

Pucallpa sits 85 km from the Brazilian border in the middle of the lowland rain forest (< 500 m a.s.l.). A paved road that crosses the Andes Mountains connects it to the capital, Lima. Its 200,000 or so people make it the second largest and fastest-growing city in the Peruvian Amazon. Low per capita income, land scarcity and civil unrest in other parts of the country have pushed many people to migrate to the zone. Infrastructure development, cheap abundant land, job opportunities and, until recently, coca production have all been major pull factors (Riesco, 1993; Labarta, 1998).

Pucallpa's economy revolves primarily around timber, gas and oil, livestock and crops. The main semi-subsistence crops are maize, rice, cassava, plantains and beans. Oil palm, palm heart, pineapple, pijauyo (*Bactris gasipaes*), cotton and camu-camu (*Myrciaria dubia*) are being promoted as cash crops but remain of limited importance.

In the early 1990s, the Peruvian government greatly reduced its support for agriculture. In the Amazon, it virtually eliminated subsidized credit and guaranteed minimum prices for crops such as rice and maize. Budget cuts and political violence limited extension services in the zone until recently, but those services are now slowly improving. Outside areas close to the main highway, poor access to markets has kept the prices for semi-subsistence crops low and marketing costs high. To obtain secure land tenure, farmers generally do not need to clear the forest. Occupation of unclaimed land typically suffices.

Low soil fertility, high acidity and aluminium toxicity impede crop growth in the region. The above-ground biomass contains most nutrients. Farmers use few capital goods, such as fertilizers, pesticides and ploughs. Slash-and-burn agriculture dominates. Most farmers clear forest every year. They typically cultivate a parcel for 6 months to 2 years before leaving it in fallow. Average rainfall of around 2000 mm comes in a bimodal pattern, which permits two cropping seasons per year.

Median farm size is 30 ha, of which primary forest accounts for 31%, forest fallow 30%, pasture 25%, annual crops 10% and perennials 4%. Rural population density is low, at 7 persons km^{-2}. Farmland is plentiful and cheap but poorly suited for agriculture. Farmers' capital and labour resources limit their production, while land is relatively abundant.

2.2. Kudzu-improved fallows

According to our survey, a majority (52%) of farmers in the study zone have adopted fallows improved with kudzu (*Pueraria phaseloides*), but few have adopted other improved practices. We selected kudzu fallows for this study because we wanted to understand why farmers have widely adopted them while not adopting other practices, as well as how the adoption of kudzu fallows might affect deforestation.

Kudzu is a leguminous vine that farmers plant or that spontaneously regenerates in areas where annual crops have recently been grown. It spreads rapidly through fallow areas, covering the ground and climbing up bushes and trees. Eventually, farmers once again slash and burn these fallow areas and plant annual crops there.

Research on kudzu in the Peruvian Amazon indicates that it not only fixes nitrogen but also increases soil-available phosphorus, potassium, magnesium and calcium (Wade and Sanchez, 1983). Kudzu's aggressive growth impedes secondary forest regeneration and the spread of herbaceous weeds. Since the soils recuperate more rapidly and the weeds get quickly overwhelmed, farmers can use shorter fallows. More limited regeneration and weeds also lower the labour costs of clearing land and weeding. And kudzu fallows require no capital investment.

Farmers first introduced kudzu about 50 years ago as a cover crop in rubber plantations. When rubber petered out, kudzu was actively promoted as an improved pasture. However, grazing and trampling by livestock rapidly eliminate it, it dries up during the low-rainfall months between June and September and *Brachiaria decumbens*, the dominant improved pasture, tends to outcompete and displace it.

Farmers adapted the use of kudzu to improve their fallows, its current dominant use, largely on their own. Kudzu's informal diffusion led some researchers to question whether farmers were truly 'adopting' kudzu-improved fallows or whether it was simply spreading spontaneously. Survey results, however, indicate that farmers actively manage kudzu in fallow areas. Some agronomists have also expressed concerns about kudzu based on the negative experience in the southern USA, where kudzu has disrupted the ecological balance. In fact, kudzu does not invade primary forest and only temporarily slows secondary forest regeneration, although it does occasionally invade annual and perennial crop parcels. Nevertheless, 94% of farmers expressed a positive opinion of kudzu, largely because of its ability to improve soils and control weeds and its use as fodder.

3. The Economics of Kudzu-improved versus Traditional Fallows

This section compares total land use, labour use and yields of kudzu-improved fallows with those of traditional fallows and primary forest used in extensive slash-and-burn agriculture. Farmers in the zone distinguish between two types of fallows: low and high. The former has secondary forest vegetation below 5 m high, while the latter has higher vegetation. Farmers also have the option of clearing primary forest to grow their crops. In this research, a key emphasis is on farmers' decisions whether to clear primary forest or secondary forest fallow.

Farmers classify kudzu-improved fallows as low fallows. Trees and shrubs grow in these areas, but kudzu vines climb over them and slow their regeneration. The shorter fallow period farmers use when working with kudzu-improved fallow further reduces secondary forest regrowth. Therefore the trees in these fallows may not regenerate enough to be considered full-fledged secondary forests by some classifications. However, they grow in areas that would otherwise become secondary forests and perform many of the same functions as secondary forest in regard to regenerating fertility and controlling weeds. Therefore, for the purposes of this research, we consider kudzu-improved fallow areas to be incipient secondary forest.

3.1. Total land use

Since slash-and-burn agricultural systems require farmers to leave land in fallow, they must clear and cultivate other areas until the fallow areas are ready to cultivate again. We use the term 'total land use' to refer to the entire area cleared and cultivated before the farmer can return to his or her initial parcel and begin the rotation again. One can calculate the total amount of land needed to maintain a given area of annual crops in production based on the average fallow period and the average amount of time that farms cultivate a parcel before leaving it in fallow. The less time that land must be left in fallow to restore its soil fertility, the earlier farmers can return there. Therefore, they need less land to produce crops. Similarly, the longer farmers can cultivate an area before abandoning it, the more time their fallow areas have to recuperate. This also reduces total land needs.

Farmers leave land in high and low forest fallows an average of 6.3 and 2.5 years, respectively. For kudzu-improved fallows, the average is only 1.7 years. The average time farmers plant their fields before leaving them in fallow is 1 year for kudzu fallows and 1.3 and 0.9 years for high and low forest fallows (Table 12.1).

Based on these figures, we can calculate the total amount of land a farmer would need to maintain 3 h in production using each type of fallow. To simplify things we assume that yields remain constant and farmers only use one type of fallow in each scenario. The following formula calculates total land use: TL = HP + [(FY/YC) × HP], where TL = total land use, HP = hectares planted, FY = fallow years and YC = years cultivated. The first HP in the formula represents the original amount of land in production – in our example, 3 ha. The part in brackets is the average fallow duration divided by the average number of years the farmer cultivates his or her field before leaving it as fallow, multiplied by the hectares planted. This gives the amount of additional land that must be kept in fallow until the initial land has regained its fertility and is ready to be cultivated again. The sum of these two gives the total amount of land needed in each type of fallow production system.

This calculation shows that kudzu fallows substantially reduce the amount of land farmers need to support annual crop production, compared with high and low forest fallows. In our simplified model, farmers who use kudzu fallows can clear 116% and 40% less forest than those that use high and low forest fallows, respectively, and still cultivate the same amount of land each year (Table 12.1). Thus kudzu-improved fallows allow farmers to reduce the total amount of land they need for their shifting agricultural production system and to intensify their land use. As long as they cultivate the same area of annuals, this will reduce their forest-clearing needs.

Table 12.1. Comparative analysis of total land use for differing fallow systems.

	Kudzu-improved fallows	High secondary forest fallow	Low secondary forest fallow
Average years of fallow	1.7	6.3	2.5
Average years of cultivation	1.0	1.3	0.9
Total land clearing: 3 ha annual crop system	8.1	17.5	11.3
Comparative land use (kudzu fallow = base 100%)	100%	216%	140%

Can farmers sustain these systems over time? Smith *et al.* (1998) found that fallow vegetation in the Pucallpa region regenerates more slowly after each cropping cycle. Under these circumstances, low forest-fallow systems are likely to rapidly degrade the land. High forest-fallow systems can be sustained for longer periods. The sustainability of kudzu fallow appears relatively promising. Of the 24 farmers we interviewed in our improved-fallow survey, none reported that their productivity had declined over time. While most had used kudzu fallows for less than 10 years, four had used kudzu for 10 years and one each for 12, 21 and 30 years. Thus, at least the medium-term sustainability of kudzu-improved fallows appears likely.

3.2. Labour use

Primary forest and forest fallows require substantial labour to slash-and-burn the land. Since kudzu permits shorter fallow periods and suppresses secondary forest regeneration, it reduces the amount of trees and shrubs in the improved fallow. As a result, farmers need less labour to clear their land (Table 12.2). Further, about half the time kudzu fallows regenerate naturally and the other half farmers broadcast the seed, so seeding requires either no labour or minimal labour.

Kudzu also reduces the amount of labour required for weeding. In general, when more light penetrates the forest canopy, herbaceous weeds develop more. Since little light penetrates the canopy in primary forests, these areas have the fewest weeds. Areas previously under traditional forest fallows have significantly more. Kudzu, however, is an aggressive cover crop, which smothers herbaceous weeds that invade fallow areas. Areas previously in kudzu fallows require less weeding labour than other forest fallows (Table 12.3).

Labour constraints limit production more than available land. On average, farmers only cultivate annual crops on 14% of the land available on their farms (cropland, fallows and primary forest). Forest clearing and weeding are two critical labour bottlenecks in slash-and-burn fallow systems (Thiele, 1993). By reducing the need for labour at these key moments, farmers can put more land into production. This could potentially increase deforestation.

Table 12.2. Use of labour for clearing land (days ha^{-1}).

	Average use	Comparative use (kudzu fallow = base 100%)
Primary forest	26.5	323%
High forest fallow	21.8	266%
Low forest fallow	13.0	159%
Kudzu-improved fallow	8.2	100%

3.3. Yields

In slash-and-burn agricultural systems, yields increase when the amount of biomass burned, and hence the quantity of nutrients made available for crop uptake, is higher. Primary forest has the most biomass. But many of the trees are too large to burn well so their nutrients are not released and the tree-trunks take up space in the field. Therefore, according to farmer interviews, high forest fallows often provide the highest nutrient flush from burning, while low fallows typically provide the lowest.

Weeds are the second principal factor affecting yields. Primary forest has the fewest weeds, followed by kudzu and then other forest fallows.

Kudzu fixes nitrogen and provides other nutrients that accelerate the regeneration of soil fertility. And, as a cover crop, kudzu effectively reduces weed competition. The net result of these two factors is a substantial yield increase over both traditional forest fallows and primary forest (Table 12.4).

4. Hypotheses Concerning Kudzu Fallows and Deforestation

We base our hypotheses concerning how kudzu fallows influence deforestation on an analysis of their key characteristics, identified in section 3: that is, the fact that kudzu-improved fallows reduce fallow periods and labour costs and increase yields. The first two hypothesized impacts are as follows:

Table 12.3. Use of labour for weeding (days ha^{-1}).

	Rice weeding	Comparative use (kudzu fallow = base 100%)	Maize weeding	Comparative use (kudzu fallow = base 100%)
Primary forest	3.6	37%	3.1	24%
High forest fallow	39.7	409%	20.2	158%
Low forest fallow	31.6	326%	23.2	181%
Kudzu-improved fallow	9.7	100%	12.8	100%

Table 12.4. Comparative yields (kudzu fallow = base 100%).

	Rice (t ha^{-1})	Comparison	Maize (t ha^{-1})	Comparison
Primary forest	1.6	76%	1.3	76%
High forest fallow	1.9	90%	1.5	88%
Low forest fallow	1.0	48%	1.4	82%
Kudzu-improved fallow	2.1	100%	1.7	100%

Hypothesis 1: Thanks to its shorter fallows, kudzu decreases deforestation by reducing the total land needed in a shifting slash-and-burn production system.

Hypothesis 2: Kudzu-improved fallows' easing of labour constraints increases deforestation by allowing farmers to put more land into crop production.

These two hypotheses predict that kudzu-improved fallows will have two contradictory effects on deforestation. A priori, therefore, the net impact on deforestation is uncertain. However, if one distinguishes the type of deforestation, a clearer outcome appears likely. We have defined fallows in general as secondary forest areas and kudzu fallows as a type of (emerging) secondary forest. The adoption of kudzu-improved fallows decreases the labour costs for clearing and weeding and increases the benefits (yields) of clearing secondary forest relative to clearing primary forest. This leads to hypothesis 3:

Hypothesis 3: Changes in the relative costs and benefits of secondary versus primary forest clearing due to kudzu fallows increase the amount of secondary forest cleared and decrease the amount of primary forest cleared.

In summary, the labour cost and fallow length consequences of kudzu fallows have opposite effects on total forest clearing, giving an ambiguous net outcome. However, kudzu fallows change relative costs and benefits in a way that encourages clearing secondary forests and discourages clearing primary forest.

5. Models Used

5.1. The analytical model

Farmers practise slash-and-burn agriculture to convert vegetative biomass into ashes that provide nutrients accessible for crop uptake. Deforestation is therefore an outcome of agricultural activities, a means to the end of agricultural production, not an end in itself. That would be the case if farmers clear-cut forests to harvest timber or other forest products. Farmers do selectively harvest and sell high-value hardwood and other species from primary forest areas on their landholdings. However, since harvesting is very selective, the 'residual' primary forest remains relatively intact. Therefore, to understand why the complete removal of forests occurs, it is necessary to understand how farmers make agricultural production decisions.

Our theoretical framework for understanding these decisions is based on a profit-function approach. This approach assumes that farmers choose a combination of variable inputs and outputs to maximize profit, subject to a technology constraint. Farmer decisions about what to produce (outputs) and how to produce it (inputs/technologies) determine the impact of agriculture on deforestation. This basic idea underlies the econometric regression modelling in the following section.

Profit maximization should be understood in the broad sense of farmers using limited resources as efficiently as possible to meet their livelihood (including social) objectives. Furthermore, we do not need to assume that farmers strictly maximize profits to specify our output-supply and factor-demand equations, as long as individual agents behave in a manner that is sufficiently stable over time and can be aggregated over farmers (Sadoulet and de Janvry, 1995).

The profit function for a farm is specified as follows:

$$\pi = pq - wx \qquad (1)$$

where p is a vector of prices for outputs, q is a vector of outputs produced by the farmer, w is a vector of input prices and x is a vector of inputs.

Farmers' profit-maximizing behaviour is constrained by the production function, which describes the technical relationship between the fixed and variable inputs and outputs produced. Quantity of outputs q is a function of variable inputs x and fixed inputs z used in the production process:

$$q = f(x,z) \qquad (2)$$

Variable inputs include those factors the farmer can change in the given amount of time (e.g. fertilizer, seed, labour). Fixed factors are those that farmers cannot adjust in the given time frame. These include private factors (e.g. land, education, family labour), public factors (e.g. infrastructure, credit, extension) and exogenous factors (e.g. soils and market distance).

The farmer chooses the optimal level of inputs x and outputs q to maximize profits. We can write the input-demand and output-supply functions as follows:

$$x = x(p,w,z) \text{ and } q = q(p,w,z) \qquad (3)$$

This indicates that the optimal levels of inputs and outputs are a function of output price, input price and fixed factors. Using the expressions in equation (3), we can rewrite the profit function as:

$$\pi = pq(p,w,z) - wx(p,w,z) \qquad (4)$$

By differentiating (4) in respect of output and input prices, one can derive the output-supply and factor-demand functions:

$$d\pi/dp_i\,(p,w,z) = q_i \quad d\pi/dw_i\,(p,w,z) = -x_i \qquad (5)$$

Rising demand for crop- and pasturelands is the main cause of deforestation. However, not all agricultural products and technologies have the same impact on deforestation. The econometric modelling that follows analyses the determinants of the adoption of kudzu-improved fallow and its impact on deforestation.

5.2. The econometric models

Based on the previous analytical model, we developed two econometric models. The first analyses the determinants of output production and input/technology use. The second focuses on how farmers' decisions about production and input use affect deforestation. We divide outputs into annual crops, perennial crops and livestock. Our inputs and technologies include improved fallows, improved pastures, natural pastures, capital inputs and hired labour. Our dependent variables include hectares of total forest (primary plus secondary forest), primary forest and secondary forest felled in 1998.

To do the input and output regressions, we used ordinary least squares (OLS). When we tried using seemingly unrelated regression equations (SURE), we got similar results but had to greatly limit the number of available observations we could use. For our deforestation regression models, we used two-stage least squares to correct for simultaneity.

The final specification of the econometric model follows. The reader can find the precise definition of each variable in Appendix 1.

$\text{OUTPUT}_{ij} = f(\text{farm size}_i, \text{secondary forest}_i, \text{primary forest}_i, \text{alluvial soils}_i,$
years in lot_i, education_i, age_i, family labour_i, origin_i, off-farm income_i,
credit_i, extension_i, real maize price_i, land tenure_i)

where j = annuals, perennials and livestock and $i = 1, \ldots, n$ observations.

$\text{INPUT}_{ij} = f(\text{farm size}_i, \text{secondary forest}_i, \text{primary forest}_i, \text{alluvial soils}_i,$
sandy soils_i, years in lot_i, education_i, family labour_i, off-farm income_i,
distance social services_i, credit_i, market distance_i, land tenure_i)

where j = kudzu fallows, improved pastures, natural pastures, capital inputs and hired labour and $i = 1, \ldots, n$ observations.

$\text{DEFOR}_{ij} = f(\text{annual}_i, \text{perennial}_i, \text{livestock}_i, \text{kudzu fallows}_i, \text{improved}$
pasture_i, natural pasture_i, capital input use_i, hired labour_i, family labour_i,
land tenure_i, education_i, forest_i, alluvial soils_i, forest product income_i)

where j = total, primary and secondary forest deforestation in 1998 and $i = 1, \ldots, n$ observations.

6. Results and Interpretation

The results presented here focus on the adoption of kudzu-improved fallows and the hypotheses presented in section 4. The four regression model results presented in Table 12.5 are for kudzu fallow adoption, total deforestation, primary forest deforestation and secondary forest deforestation. Due to the cross-sectional nature of the data and the diversity of the agricultural systems, we considered variables significant up to the 0.15 level. We also provide the exact significance levels for the reader's consideration.

Table 12.5. Improved fallow adoption and deforestation regression results.

	Improved fallows R^2 0.33; adj. R^2 0.25			Total deforestation R^2 0.54; R^2 0.50		Primary forest deforestation R^2 0.30; R^2 0.23		Secondary forest deforestation R^2 0.30; R^2 0.23	
Variables	Coefficient[†]	Significance	Variables	Coefficient	Significance	Coefficient	Significance	Coefficient	Significance
Constant		0.01	Constant		0.182		0.74		0.74
Farm size	0.548***	0.00	Annuals	0.490***	0.00	0.323***	0.00	0.301***	0.00
Secondary forest	−0.066	0.44	Perennials	−0.102*	0.10	−0.040	0.60	−0.008	0.92
Primary forest	−0.198*	0.09	Livestock	0.039	0.59	0.020	0.83	0.197**	0.04
Alluvial soils	−0.066	0.48	Improved fallows	0.049	0.41	−0.129*	0.07	0.137*	0.06
Sandy soils	0.168**	0.05	Improved pastures	−0.023	0.75	−0.006	0.95	−0.154*	0.13
Years in lot	0.151*	0.12	Natural pastures	−0.094*	0.13	−0.061	0.45	−0.163**	0.05
Education	0.207***	0.01	Capital inputs	−0.153**	0.03	−0.125	0.16	0.181**	0.04
Family labour	0.091	0.28	Hired labour	0.337***	0.00	0.011	0.89	0.019	0.82
Off-farm income	0.002	0.99	Family labour	0.060	0.31	0.021	0.79	0.202***	0.01
Distant social services	0.140*	0.13	Education	0.129**	0.02	0.003	0.96	0.070	0.33
Credit	−0.151*	0.07	Tenure insecurity	−0.097*	0.09	−0.025	0.75	−0.030	0.69
Market distance	0.189**	0.05	Alluvial soils	0.028	0.65	−0.193***	0.01	0.203***	0.01
Tenure insecurity	−0.010	0.91	Total forest	0.108*	0.09	—	—	—	—
			Secondary forest[‡]	—	—	−0.098	0.16	0.166**	0.02
			Primary forest	—	—	0.254***	0.00	−0.093	0.23
			Primary forest products	—	—	0.168***	0.02	−0.034	0.63

*Significant at the 0.15 level; **significant at the 0.05 level; ***significant at the 0.01 level.
[†] All coefficients are standardized.
[‡] High secondary forest (> 5 m).

6.1. Kudzu-improved fallow adoption

Kudzu fallow adoption is positively and significantly correlated to farm size, distance to social infrastructure (schools and health posts) and distance to markets. Even though kudzu fallows have the effect of increasing the productivity of land, their adoption does not appear to follow a traditional Boserupian scenario of intensification induced by land scarcity, increasing population and proximity to population centres (Boserup, 1965, 1981; Pingali et al., 1987).

Farms with parcels that have been in production longer and with less primary forest have a significantly higher probability of adopting kudzu fallows. Both variables indicate decreasing land quality. The longer a farm has been in production, the more depleted its soil is of nutrients (fallows take longer to regenerate). Less remaining primary forest implies greater depletion of above-ground nutrient stocks contained in the vegetative biomass.

In the land-abundant environment around Pucallpa, we cannot yet talk of a closing of the land frontier in quantitative terms. Farmers typically cultivate only a small proportion of their farms and/or have access to nearby land. However, given that farmers are subjecting the region's fragile tropical soils to slash-and-burn agriculture with declining fallow periods, land quality increasingly constrains production. Much of the adoption of kudzu-improved fallows is associated with the closing of the land 'quality' frontier.

More educated farmers and farmers with sandy soils adopt kudzu fallows significantly more often, while those with credit adopt them less. Education's positive correlation indicates that understanding nitrogen fixation and managing kudzu in an improved fallow system require a relatively high level of knowledge. This is particularly true given that kudzu fallows have tended to spread informally among farmers without much support from extension services. Higher educational levels may also be associated with greater opportunity costs for labour. More educated farmers may adopt labour-saving kudzu fallows to free themselves in order to benefit from increased opportunities to work off-farm.

The negative correlation with credit makes sense in that kudzu-improved fallows enhance productivity without using capital and with lower labour inputs. Thus, they offer an attractive alternative for farmers without access to credit. The positive correlation with sandy soils probably relates to the agronomic conditions in which kudzu flourishes.

The overall significance of our model is reasonably good for this type of cross-sectional adoption analysis ($R^2 = 0.39$, $R^2 = 0.26$). This is particularly true given the multipurpose nature of kudzu in the region. Farmers use kudzu to improve fallows, as pasture/fodder and as a perennial cover crop. Our survey was not always able to distinguish between these three uses clearly and consistently, and this may have reduced the model's predictive power.

6.2. The impact of kudzu-improved fallows on deforestation

To analyse the impact of kudzu fallows on deforestation rates, we examine the sign and significance of the kudzu fallow variable in the three deforestation regression models. First, we briefly review the three hypotheses concerning kudzu fallows:

1. Decreased fallow periods reduce deforestation.
2. Easing labour constraints increases deforestation.
3. Changing the relative costs and benefits of land clearing in favour of secondary forests leads to declining primary forest clearing and increasing secondary forest clearing.

By estimating separate models for total, primary and secondary forest clearing, we can examine the impact of kudzu fallows on each type of deforestation. While we cannot empirically separate out the two opposite effects posited in the first two hypotheses using regression analysis, the total deforestation regression model allows us to estimate which effect is stronger. The primary and secondary forest deforestation models allow us to analyse the third hypothesis directly.

The total deforestation model shows that farmers that use kudzu-improved fallows clear more forest, although the difference is not significant. This may indicate that the labour-saving effect of adopting kudzu fallows has a greater impact on deforestation than its effect on fallow periods. The importance of easing the labour constraint is also reflected in the positive and highly significant correlation of hired labour and the positive correlation of family labour to forest clearing. This implies that, were it not for kudzu-improved fallows' tendency to simultaneously reduce fallow periods, their adoption might result in significantly higher forest clearing.

When we break total forest clearing down into secondary and primary forest deforestation, we find that kudzu-improved fallows are negatively correlated with primary forest clearing and positively correlated with secondary forest clearing, and both results are significant. This supports hypothesis 3. The changes in the relative costs and benefits associated with kudzu fallows lead farmers to reduce primary forest clearing and increase secondary forest clearing.

7. Conclusions and Implications for Policy and Technology Development

This study has analysed one particular technology (kudzu-improved fallows) in one specific setting (the lowland Amazon surrounding Pucallpa, Peru). Perhaps its clearest conclusion is that the impact of technological change on deforestation depends fundamentally on the type of new technology and the biophysical and socio-economic characteristics of the zone where it is introduced.

In our case, we looked at how the technological profile of kudzu-improved fallows has interacted with the site-specific characteristics of Pucallpa to produce a given deforestation outcome. Kudzu-improved fallows shorten fallow periods, increase yields, decrease labour costs and require no capital investment. An acute scarcity of capital and labour and a relative abundance of land characterize agricultural production around Pucallpa. This land, however, is poorly suited for agriculture.

It would be wrong to assume that any and all improved fallows increase total deforestation in the Pucallpa zone. For example, some other improved fallow may require more labour, instead of freeing the labour constraint. Or, if kudzu-improved fallows were introduced in a socio-economic context with no labour constraint, deforestation again might not increase. Hence, any analysis must be both site-specific and technology-specific in order to understand the impact of a technology on deforestation.

This research has also shown that technology and land-clearing patterns may interact in a complex fashion. Kudzu fallows have contradictory effects on deforestation. They simultaneously ease labour constraints (increasing deforestation) and shorten the fallow period required (decreasing deforestation). In addition, they have opposite effects on the clearing of primary and secondary forests. Researchers need to recognize that new technologies may simultaneously affect deforestation in several distinct ways and that they may need to do a more disaggregated analysis of these effects.

The dominant annual crop system in Pucallpa is slash-and-burn agriculture. This system uses lots of land and little labour and capital. It is therefore a rational response by farmers trying to maximize their production with scarce capital and labour resources in a land-abundant environment. Kudzu-improved fallow adoption has been successful precisely because it complements the zone's relative factor scarcities and often provides a superior alternative to the current dominant practice (by reducing labour while increasing yields).

Many other improved production technologies focus on reducing deforestation via soil conservation and/or increasing land productivity. While this approach is not wrong *per se*, it may blind researchers to the fundamental fact that labour and capital, and not land, are typically the main constraints in agricultural frontier contexts. Under such circumstances, attempts to get farmers to conserve land with technologies that require greater use of capital and/or labour are likely to fail. Soil conservation is not a primary objective of farmers in a land-abundant environment. Soil fertility-enhancing technologies will only attract farmers when their costs and benefits are superior to the current practice of extensive slash-and-burn agriculture.

Basic economic theory tells us that farmers maximize the returns to scarce production factors, which, in the case of Pucallpa, typically include labour and capital. Kudzu-improved fallows have been successful precisely because they increase labour productivity. Herein lies a paradox: the reduced labour requirements that encourage adoption of this land-conserving production technique also free labour so that overall deforestation increases. One possible solution might be to introduce high-value crops, such as certain perennial and horticultural crops, which demand a lot of labour but still increase the returns to scarce labour resources.

Improved fallows appear to have important potential for reducing deforestation. In our simple model, kudzu-improved fallows decreased total land clearing needs by 116% and 40% compared with high and low forest fallows, respectively. The challenge is to find creative ways to harness this potential and minimize the negative impacts.

Designing improved fallows that use more labour while increasing returns to labour might also help resolve the 'labour paradox'. Improved fallows that produce useful products at the same time as they help recuperate soil fertility may achieve these dual goals. For example, farmers can plant fast-growing leguminous trees in association with annual crops. For the first year and a half, they weed the trees together with the annual crops. This reduces the weeding labour constraint associated with installing tree plantations. By the time the farmers abandon annual cropping, the trees are developed enough to survive in the fallow without weeding. These trees help land productivity to recuperate while providing useful products that absorb farmer labour. For these systems to succeed, the initial labour and capital cost should be kept low (e.g. by using bare-root-bed tree nurseries, direct seeding, planting in association, etc.) and the secondary tree products must have a high value for home consumption or commercial sale. Research should find ways to reduce the costs of improved fallows and identify tree species that combine soil amelioration with the provision of valuable secondary products.

Turning our attention specifically to kudzu-improved fallows, they seem to increase total forest clearing and secondary forest clearing but reduce primary forest clearing. Although higher total deforestation is not a desired outcome, the reduction of primary forest clearing is clearly a positive environmental impact, since these forests typically provide the greatest amount of environmental services. And, even though secondary forest clearing increases, the easing of the labour constraint allows total production to rise and this helps reduce poverty.

To a certain extent, then, kudzu-improved fallows represent a typical case of trade-offs between goals: primary forest clearing and poverty decreased but secondary forest clearing and total deforestation increased. Under these circumstances, an economist cannot scientifically evaluate which option is better. Economic analysis can, however, help policy-makers understand the nature of the trade-offs and make better-informed decisions. Moreover, this type of analysis can be of use in designing strategies that convert certain trade-offs into the type of 'win–win' situation policy-makers seek.

Kudzu-improved fallows are not associated with the typical intensification scenario of land scarcity nearer to urban centres. Kudzu-improved fallow adoption increases on farms that have been in production longer and on those with less primary forest. It is the land quality constraint and not the land quantity constraint that leads farmers to adopt kudzu fallows. This knowledge can help extension services to save their limited resources by targeting farms for introduction of kudzu and other improved fallows where they are most likely to succeed.

Smith et al.'s (1998) study in Pucallpa and another by Schelhas (1996) in Costa Rica found that farmers in older and more developed frontier zones with less primary forest put a higher value on preserving it. The products and services provided by the remaining primary forest acquire a higher scarcity value. Farmers in these zones are likely to receive improved fallows particularly well, not only because of declining soil fertility, but also because of a stronger desire to preserve remaining primary forest. Thus, improved fallows may be effective in reducing forest clearing in older settlement areas with less primary forest left.

References

Boserup, E. (1965) *The Conditions for Agricultural Growth: the Economics of Agrarian Change under Population Pressure*. Aldine, New York.

Boserup, E. (1981) *Population and Technological Change: a Study of Long Term Trends*. University of Chicago Press, Chicago.

Crutzen, P.J. and Andreae, M.O. (1990) Biomass burning in the tropics: impact on atmospheric chemistry and biogeochemical cycles. *Science* 250, 1669–1678.

de Rouw, A. (1995) The fallow period as a weed-break in shifting cultivation (tropical wet forests). *Agriculture, Ecosystems and Environment* 54, 31–43.

Dvorak, K.A. (1992) Resource management by West African farmers and the economics of shifting agriculture. *American Journal of Agricultural Economics* 74, 809–814.

Kang, B.T. and Wilson, G.F. (1987) The development of alley cropping as a promising agroforestry technology. In: Steppler, H.A. and Nair, P.K.R. (eds) *Agroforestry: a Decade of Development*. ICRAF, Nairobi, Kenya, pp. 227–243.

Kleinman, P.J.A., Pimentel, D. and Bryant, R.B. (1995) The ecological sustainability of slash-and-burn agriculture. *Agriculture, Ecosystems and Environment* 52, 235–249.

Labarta, R. (1998) *Los Productores de la Cuenca Amazonica del Perú y la Dinámica de Uso de la Tierra: Resultados de la Caracterización de Pucallpa y Yurimaguas*. ICRAF Report, Pucallpa, Peru.

Nair, P.K.R. (1993) *An Introduction to Agroforestry*. Kluwer Academic Publishers in cooperation with ICRAF, London.

Pingali, P., Bigot, Y. and Binswanger, H.P. (1987) *Agricultural Mechanization and the Evolution of Farming Systems in sub-Saharan Africa*. Johns Hopkins University Press, Baltimore.

Riesco, A. (1993) Intensificación tecnologica en la selva baja: el caso de Pucallpa. In: Loker W. and Vosti, S. (eds) *Desarrollo Rural en la Amazonia Peruana*. CIAT and IFPRI, Cali, Colombia.

Sadoulet, E. and de Janvry, A. (1995) *Quantitative Development Policy Analysis*. Johns Hopkins University Press, Baltimore.

Sanchez, P.A. (1976) *Properties and Management of Soils in the Tropics*. John Wiley, New York.

Schelhas, J. (1996) Land use choice and change: intensification and diversification in the lowland tropics of Costa Rica. *Human Organization* 55, 298–306.

Smith, J., van de Kop, P., Reategui, K., Lombardi, I., Sabogal, C. and Diaz, A. (1998) *Dynamics of Secondary Forests in Slash-and-burn Farming: Interactions among Land Use Types in the Peruvian Amazon*. Proyecto Manejo de Bosques Secundarios en America Tropical, Convenio CIFOR/CATIE/BID, Pucallpa, Peru.

Thiele, G. (1993) The dynamics of farm development in the Amazon: the barbecho crisis model. *Agricultural Systems* 42, 179–197.

Wade, M.K. and Sanchez, P.A. (1983). Mulching and green manure applications for continuous crop production in the Amazon basin. *Agronomy Journal* 75, 39–45.

Appendix 1: Variable Description

Years in lot	Number of years a farm has been in production (past and current owners)
Sandy soils	Dummy variable, sandy soils dominant on farm
Alluvial soils	Dummy variable, farm located in alluvial soil zone
Farm size	Total hectares of a household's landholdings
Secondary forest	Hectares of secondary forest fallow on a household's landholdings
Primary forest	Hectares of primary forest on a household's landholdings
Forest	Hectares of total forest on a household's landholdings
Family labour	Number of family members >14 years old working on the farm
Education	Dummy variable, household head has secondary education or higher
Origin	Dummy variable, household head not from the jungle region
Forest product income	Value (in soles) of products harvested from the primary forest in the previous year
Off-farm income	Value (in soles) of off-farm income earned by a household in the past year
Age	The age of a household head
Annual	Hectares of annual crops on a household's landholdings
Perennial	Hectares of perennial crops on a household's landholdings
Livestock	Head of cattle owned by a household
Kudzu fallows	Hectares of kudzu-improved fallows on a household's landholdings
Natural pasture	Hectares of natural pasture on a household's landholdings
Improved pasture	Hectares of pasture improved with *Brachiaria* on a household's landholdings
Capital input use	Value (in soles) of capital inputs used in the past year by a household
Hired labour	Days of paid labour used on farm (includes labour exchange – *minga*)
Land tenure	Dummy variable, household head perceives tenure as insecure
Market distance	Household's distance in kilometres to the principal market, Pucallpa
Distant social services	Household's distance in kilometres to nearest (school + health post)
Credit	Dummy variable, household received credit in the last 5 years
Extension	The number of extension visits a household received in the past year
Real maize price	The price for maize received by a household, minus the transportation and labour costs of marketing

Ambiguous Effects of Policy Reforms on Sustainable Agricultural Intensification in Africa

13

Thomas Reardon and Christopher B. Barrett

1. Introduction[1]

African farmers respond to increasing food demand stemming from population and income growth by producing more on existing cropland, extending production into fragile areas with high levels of biodiversity, or both. This chapter's key message is that, to satisfy continued growth in food demand without further degrading the natural environment, African farmers and policy-makers must pursue 'sustainable agricultural intensification' (SAI). This requires the use of capital to maintain and conserve soil fertility while meeting productivity goals. In this context, the term capital includes inorganic fertilizer and organic matter, as well as land improvements, such as water control, erosion prevention and fertility maintenance.

Many African farmers are not following this path. They are either intensifying in an unsustainable fashion – that is, mining their soils and degrading the resource base – or they are extending their production on to additional fragile lands. This is often due to inappropriate policies, which reduce farmers' incentives and capacity to pursue SAI. Economic liberalization measures have reduced government support for farming, thus increasing input prices and market risk, without concomitant public investments in institutional development or physical infrastructure, which might have induced smallholders to intensify in a profitable and sustainable manner. African governments and donors should invest in both institutions and infrastructure and follow a middle path between heavy state involvement and absence of public support.

©CAB International 2001. *Agricultural Technologies and Tropical Deforestation*
(eds A. Angelsen and D. Kaimowitz)

Section 2 presents definitions and a conceptual model. Section 3 discusses macro- and sectoral-level policy reforms. Section 4 offers some examples of both sustainable and unsustainable intensification. Section 5 concludes.

2. Sustainable and Unsustainable Agricultural Intensification: a Conceptual Framework

2.1. Definitions

We use two criteria to define SAI. First, the system adopted must protect or enhance the natural resource base and thus maintain or improve land productivity. Secondly, the system must allow farmers to meet their minimum food and cash requirements and it must be profitable. This implies that total factor productivity, physical yields and output per unit labour time should not decrease over time. It does not necessarily mean that farmers will maintain the full range of existing soil nutrients and biota. At the same time, by our definition conservation technologies that reduce – but do not stop – degradation are not 'sustainable'. They merely limit the rate at which productivity declines.

In practice, satisfaction of the SAI criteria normally requires capital-led intensification. This implies that farmers must use substantial amounts of inputs that enhance soil fertility, such as inorganic and organic fertilizer, and quasi-fixed capital land improvements, such as land and water conservation infrastructure (Reardon et al., 1997; Clay et al., 1998). Of course, since farmers use labour to construct and maintain the latter, capital-led intensification might also be labour-intensive (i.e. require more labour per unit of land).

In contrast, capital-deficient intensification occurs when farmers use too little of these capital inputs. Insufficient use of inorganic fertilizer, organic matter and land improvements, combined with the intensity of land use that characterizes most of the semi-arid and hillside tropics in Africa, leads to soil mining and degradation. Farmers following this path often merely add labour, without investing additional capital. This allows them to crop more densely, weed and harvest more intensively, and so on.

Capital-deficient intensification generally fails to meet the economic criteria for sustainability. Ruttan (1990) estimates that one prominent capital-deficient strategy, low-input sustainable agriculture, has the potential to increase food output by only about 1% a year in Africa, roughly the rate observed over the past 20 years. This is well shy of the expected 3.0–3.5% annual growth in African food demand. Failures to satisfy productivity goals resulting from such capital-deficient strategies may force farmers to extend their crops into fragile forest margins (if available) or lead to additional soil mining.

Soil-science research shows overwhelmingly that inorganic fertilizer is necessary for sustainable growth in productivity, even in fragile soils and low-rainfall zones. In 1995, African farmers used 9 kg ha^{-1}, down from 10 kg in 1993. This was only a fraction of the 1993 developing-country average of 83 kg ha^{-1} (Heisey and Mwangi, 1997; Weight and Kelly, 1998). Rather than a cause for celebration, the low use of chemical fertilizer is a major worry from the perspective of both the environment and food production. Outside Africa, as much as 75% of crop yield increases since the mid-1960s are attributable to fertilizer use (Viyas, 1983). Lower yields mean that farmers require more land to grow the same amount of food and hence threaten forest margins and other fragile areas.

In most places in Africa, it will not be feasible to pursue sustainable intensification using only organic matter. Manure, a key component in most low-input systems, is in short supply in countries such as Rwanda, Malawi and Zimbabwe, because of increasing population pressure. The amount of manure needed to supplant inorganic fertilizer is huge. Weight and Kelly (1998) cite calculations showing that 20 t of manure is needed to replace two 100 kg sacks of nitrogen–phosphorus–potassium (NPK) fertilizer, although this obviously varies across sites. Moreover, collecting and incorporating manure can require substantial amounts of labour. Moser (1999) found that more than 90% of the farmers who adopted a new rice technology in Madagascar have not incorporated organic matter, probably because it demands too much labour and presents health concerns.

The use of compost, mulch and manure may present a serious scaling-up problem. It is certainly feasible for individual farmers to pursue these soil improvement strategies. But in most areas, in order for all farmers to adopt these techniques, livestock densities would have to grow to unsustainable levels. Brown manure is not the most efficient way to convert N or P from biomass into soil nutrients. Moreover, the toxic secondary compounds in many native species (especially woody species) are not healthy for most ruminants. Producing compost or mulch with cuttings from local shrubs/bushes may limit forage availability. It would be extremely ironic if attempts to go organic in fertilization induced a loss of natural vegetation due to excessive harvesting of green manure!

Perennials can contribute to SAI because they trap and form organic-matter deposits, while their roots and stems prevent water and wind erosion. However, perennials largely complement other fertility investments; they do not substitute for them. As Clay *et al.* (1998) observed in Rwanda, incorporating perennials can make the use of other inputs profitable. Perennials can be costly. They often take several years to establish, have high sunk costs and often have risky markets. If the poor are more risk-averse and have higher discount rates, then they would probably be least likely to invest in perennials.

2.2. The conceptual framework

Farmers' choice of agricultural technologies and factor intensities depends fundamentally on the incentives and constraints they face. Policy changes induce changes in market conditions and prices. This affects farmers' decisions, which, in turn, influence environmental outcomes.

Figure 13.1 presents four sets of variables and the linkages between them.[2] The first set is made up of policy variables that are exogenous to the farm communities but influence their incentives and capacity to respond. These include: (i) policy reforms at the macro and sectoral level; (ii) structural changes, such as changes in global markets, urbanization and infrastructure; and (iii) projects, which have elements of (i) and (ii) but apply to a specific area for a limited time.

These variables influence the second set of variables, which includes the incentives facing farmers (input and output price levels and variations) and the capacity of farm households and communities to act on changing incentives. Capacity, in turn, depends on access to public infrastructure, private capital and communally owned capital.

Farmers' incentives and capacity lead them to allocate their labour, land and capital over various on- and off-farm activities and to choose particular technologies, which use land more or less intensively. The farmers' decisions, in turn, have environmental consequences, both on and off the farm.

The key to understanding the link between the form intensification takes at the farm level and its economic and environmental sustainability lies in labour productivity. Labour accounts for a majority of total income in almost all smallholder African households, and, since bringing additional land into production is a very labour-intensive activity, labour use patterns affect environmental sustainability.

Farmers decide whether to enlarge their cultivated area – including whether to clear forest – based on their assessment of what the best way to use their available labour is. Such assessments take into account profitability, risk, transaction costs, etc. What intensification path the farmer follows greatly influences this decision, since the productivity of labour is a function of soil quality, which is itself an increasing function of the quasi-fixed capital investment and inorganic fertilizer application we emphasize.

Market conditions and policy reforms shape non-agricultural wage rates, output and input prices, risk exposure and transactions costs. Thus, they significantly affect how smallholders allocate their labour. This has strong implications for whether the ensuing agricultural growth is sustainable (Barrett, 1999). Households decide how much labour to supply and what to spend their time on in part based on the returns provided by the different options they have available to them. As long as the returns to on-farm and off-farm labour both remain high relative to land-clearing labour, farmers are unlikely to clear additional forest.

Figure 13.2 illustrates the central role of labour productivity in different activities in determining how smallholder households allocate their labour. Utility or profit maximization dictates that households allocate labour to the activity with the highest return. If we assume for the moment that households have no access to off-farm labour markets (an assumption we shall later relax),

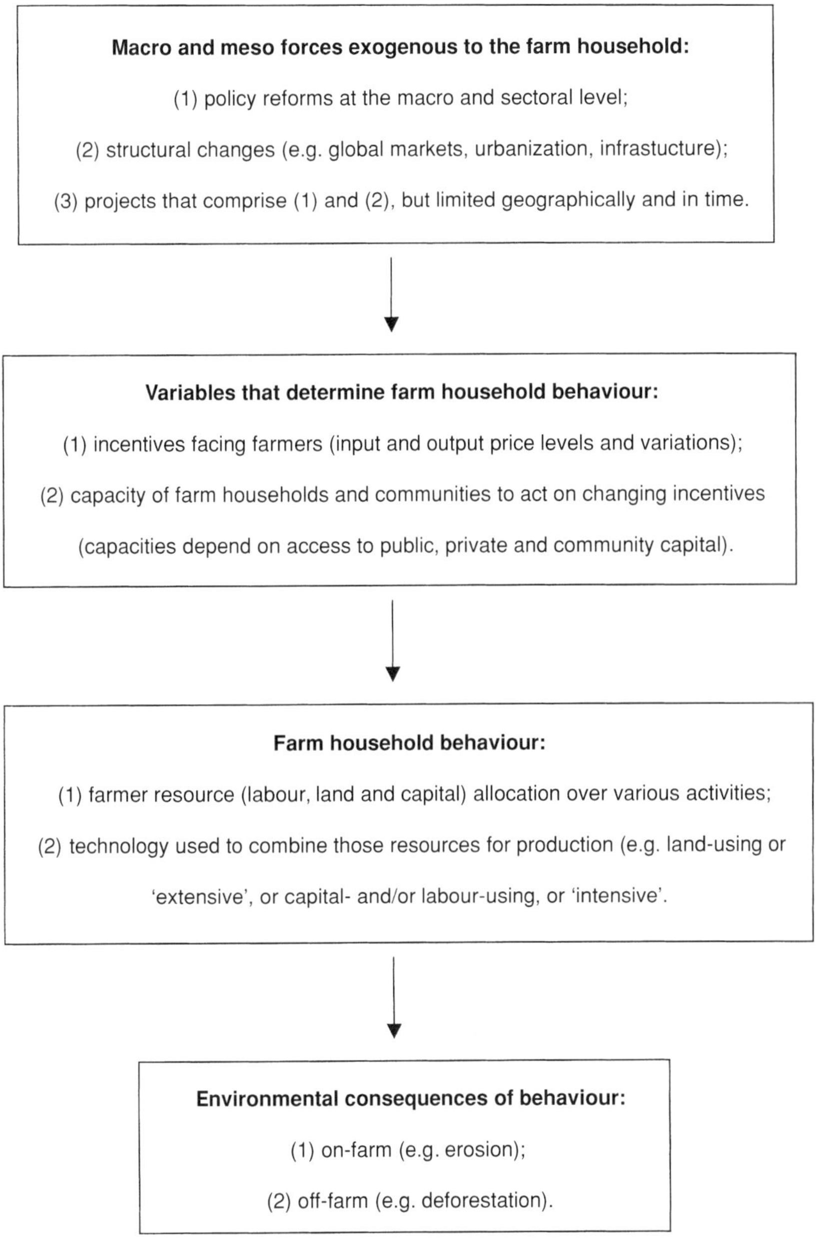

```
┌─────────────────────────────────────────────────────────────┐
│       Macro and meso forces exogenous to the farm household:│
│                                                             │
│       (1) policy reforms at the macro and sectoral level;   │
│                                                             │
│       (2) structural changes (e.g. global markets, urbanization, infrastucture); │
│                                                             │
│       (3) projects that comprise (1) and (2), but limited geographically and in time. │
└─────────────────────────────────────────────────────────────┘
                              │
                              ▼
┌─────────────────────────────────────────────────────────────┐
│       Variables that determine farm household behaviour:    │
│                                                             │
│       (1) incentives facing farmers (input and output price levels and variations); │
│                                                             │
│       (2) capacity of farm households and communities to act on changing incentives │
│                                                             │
│       (capacities depend on access to public, private and community capital). │
└─────────────────────────────────────────────────────────────┘
                              │
                              ▼
┌─────────────────────────────────────────────────────────────┐
│       Farm household behaviour:                             │
│                                                             │
│       (1) farmer resource (labour, land and capital) allocation over various activities; │
│                                                             │
│       (2) technology used to combine those resources for production (e.g. land-using or │
│                                                             │
│       'extensive', or capital- and/or labour-using, or 'intensive'. │
└─────────────────────────────────────────────────────────────┘
                              │
                              ▼
              ┌──────────────────────────────────────┐
              │ Environmental consequences of behaviour: │
              │                                      │
              │     (1) on-farm (e.g. erosion);      │
              │                                      │
              │     (2) off-farm (e.g. deforestation). │
              └──────────────────────────────────────┘
```

Fig. 13.1. The conceptual framework.

households should allocate labour to cultivating existing land up to the point where its productivity equals the productivity of clearing and cultivating new areas. (The marginal revenue product of labour schedules, MRPL and MRPL*,

Fig. 13.2. Farm labour allocation between agricultural uses on existing lands, agricultural extensification and off-farm labour markets.

depicts the productivity of cultivating existing land and cultivating new lands, respectively.)

Several factors determine these marginal labour productivity schedules. They shift upward in response to increases in crop prices and increased use of complementary inputs, such as inorganic fertilizers and land improvements that conserve land or water. They shift down as continued cultivation on a plot reduces soil quality. This gives rise to the traditional cycle of shifting cultivation, wherein forest clearing takes place as labour productivity declines on existing plots after a few plantings. The slope of the two schedules becomes steeper when agricultural products become non-tradable, due to households' preference for home consumption. The total labour allocated decreases as welfare improves and real-income effects induce substitution into leisure.

The limited available evidence from African agriculture suggests that household labour supply is both income- and wage-inelastic (Fafchamps, 1993; Barrett and Sherlund, 2000). This implies that changes in relative returns to labour across different uses will dominate the trade-off between labour and leisure in most settings. Technological advances and investments in quasi-fixed capital (e.g. soil and water conservation) that increase agricultural labour productivity can increase the productivity of labour both in existing fields and in newly opened fields and thereby induce both intensification and extensification. Nevertheless, our informed conjecture is that quasi-fixed capital investments tend to be specific to existing cultivated areas and so, for the most part, they only induce intensification. In this respect, they may differ from crop production technologies.

Our analysis would be incomplete if we failed to recognize the multi-sectoral nature of the African rural economy. If off-farm employment pays more than farm labour and/or helps to reduce overall income risk, then farmers will generally not want to adopt labour-using technologies. Similarly, if off-farm work pays more than farm work at the margin, labour freed from agricultural production on existing lands will generally not be shifted to bringing more land under the plough. The same may apply if on-farm income and non-agricultural income are not highly correlated with each other. This would make it appealing to farmers to diversify their labour allocation across sectors in order to smooth their income stream in a highly stochastic environment.

Figure 13.2 shows some of these aspects clearly. If the market wage rate always exceeds the marginal productivity of forest-clearing labour, then the off-farm market will absorb any labour not employed in agricultural production on existing plots. It does not matter whether technical change saves labour or makes production more labour-intensive. Of course, if real wages fall or crop prices increase, both of which generally occurred in the wake of liberalization measures in Africa, then farmers might both devote more labour to their existing plots, increase their cropped area and allocate less time to leisure and/or off-farm activities. Thus, simply adjusting output prices through measures such as dismantling marketing boards that taxed producer food prices does little to promote intensification and discourage farmers from expanding their agricultural areas. Instead, one needs to shift the MRPL curve more than the MRPL* curve by inducing investment in quasi-fixed factors or the use of agricultural inputs on existing cleared land or some similar measure. Hence, input markets, seasonal financing and rural non-farm labour markets can all play key roles in stemming deforestation.

Non-farm income is important to African farmers. Reardon's (1997) review of 28 field studies in Africa found that, on average, non-farm income represented 45% of total income. However, non-farm activity can be a two-edged sword with respect to intensification. On the one hand, if non-farm activities pay more than farm activities and/or help to reduce income risk, then non-farm activities compete for cash and labour with farming. This also implies that farmers may eschew labour-intensive agricultural technologies even in the face of farm labour surpluses and choose labour-saving technologies if they can afford to adopt them. Low (1986) shows, in a case-study of hybrid maize adoption in Botswana, that farmers deliberately choose labour-saving technologies to free labour for lucrative non-farm work. Where conservation measures require substantial labour, the desire to engage in off-farm activities to diversify income sources might undermine the sustainability strategies. At the same time, to a certain extent, non-farm activities can act as an escape valve from labour/land pressure. Similarly, households do not automatically use labour liberated from the farm to bring more land under the plough.

On the other hand, non-farm income is often a key source of cash for many African rural households. Given weak rural financial systems, non-farm income enhances households' capacity to invest in quasi-fixed agricultural capital (Reardon *et al.*, 1994; Clay *et al.*, 1998). Whether households will invest non-farm income in capital-led intensification is an empirical question. It depends on the characteristics of the local labour market as well as the profitability of alternative investment opportunities.

There is ample evidence, however, that non-farm income is poorly distributed across households. The poorest households have the least access to non-farm opportunities. Since they have little access to non-farm income and lack the means to purchase farm inputs, the poorest households are forced to depend on the land. This often implies either soil mining or clearing additional forested lands. So those with the least capacity to finance investment independently also have the poorest access to non-farm substitutes.

Rising farm productivity induces a disproportionately large demand for non-agricultural goods and services and rural non-farm activities expand as a result. This bids up the price of off-farm labour. Ahmed and Hossain (1990) have demonstrated this effect in the Green-Revolution rice areas of Bangladesh. We conjecture that similar effects occur in African settings where farm productivity increases. Hence, provided that the basic rural market infrastructure is in place, the multiplier effects in general equilibrium probably shield the forests, wetlands and ranges from falling prey to increased agricultural productivity. Historically and worldwide, increased farm productivity leads to reduced employment in agriculture, as industries enjoying higher income elasticities of demand absorb increased labour.

3. Policy Reforms and Incentives for SAI in Africa

The main policy changes in Africa over the past 15 years have been macro and sectoral policy reforms, such as currency devaluation, market liberalization, removal of fertilizer, seed and credit subsidies and removal of marketing subsidies for crop outputs. Too often, policy-makers tend to assume what the effects of changes in output and input prices will be, without empirically testing those assumptions. Policy analysts often claim that 'liberalization will raise farm profitability'. The complex means by which policies actually affect prevailing price distributions, transactions costs and farmer behaviour tend to get lost amid the '*ceteris paribus*' assumptions. We contend that the major macro and sectoral policy changes associated with structural adjustment have had ambiguous and often disappointing effects, generating quite mixed farm-level impacts on intensification patterns.

3.1. Macro-level policy reforms

Macro reforms have analytically indeterminate effects on the incentives facing farmers. Exchange-rate devaluation, for example, could raise the output price of an 'intensification crop', such as rice, maize or cotton, more – or less – than it raises input prices. This depends on how tradable the outputs and inputs are, on the extent to which governments tax away gains from trade rather than passing them on to farmers and on the size of the margins in private commerce. Governments may also take measures to compensate consumers or farmers for the increase in the prices they pay as a result of devaluations, such as when the governments of Mali and Senegal reduced tariffs on fertilizers and rice, respectively, following the 1994 devaluation of the CFA franc. Devaluation can also raise marketing costs and producer price risk, as it did in Madagascar (Barrett, 1999).

The effects of market liberalization can also be ambiguous. Liberalization may reduce commerce margins by stimulating competition, open new output markets and drive down farm-gate input prices, all of which would improve farmers' profits. But market liberalization can also convert interior markets into enclaves, raise transport costs and the prices of imported inputs and increase price risk. The evidence from a number of African rural areas suggests that liberalization favours market concentration and creates barriers to entry in certain markets that tend to produce greater price instability. It also shows that devaluations have ambiguous effects on farm profitability and on input use (Barrett, 1997, 1999).

The limited available evidence suggests that state intervention lowered the mean and variance of agricultural product prices, while liberalization has increased both expected prices and price variability (Barrett, 1997). Price instability can undermine farm investment, even where liberalization raises average medium-term output prices, because price instability discourages investments in quasi-fixed capital (Barrett and Carter, 1999). Price instability also reduces inter- and intrafarm diffusion of yield-increasing technologies, and thus slows down the adoption of technology (Kim et al., 1992).

3.2. Sectoral policy reforms

When not accompanied by macroeconomic stabilization measures, the effects of most sectoral price policies (taxes, subsidies, price controls) on output or input prices are unambiguous. But, when governments implement these policies in the context of macroeconomic stabilization, the effects are uncertain. Sectoral policies can counterbalance macroeconomic reforms, and may even be designed to do so. However, the most recent generation of policy reforms has tended to emphasize macroeconomic policy and to subordinate sectoral policy. To achieve fiscal balance, border parity pricing and similar objectives, governments have tended to limit their sectoral interventions. Nevertheless, sectoral interventions may have important, overlooked, 'crowding-in' effects, encouraging private investment in sustainable technologies. We now consider several specific sectoral policies.

Fertilizer/seed policy

African fertilizer use is the lowest in the world and has declined over the past 15 years. This is not surprising, given that governments have reduced or eliminated their fertilizer, seed and credit subsidies. The effective interest rate for purchasing inputs has risen sharply throughout Africa, as have fertilizer and seed prices. Case-study evidence points to a connection between the rising input and financial service prices and the decline in fertilizer use (Kelly *et al.*, 1996; Maredia and Howard, 1997; Rusike *et al.*, 1997). Moreover, there is growing evidence that private fertilizer and seed merchants have responded much less than expected to the liberalization of input markets resulting from the closure of fertilizer and seed parastatals (Dembele and Savadogo, 1996; Rukuni, 1996; Rusike *et al.*, 1997).

Fertilizer markets in African are plagued by a high-risk, seasonal demand, high transport costs, underdeveloped markets for financial services and cash-constrained farmers. Economies of scale in fertilizer production make domestic production inefficient in most African nations. Consequently, domestic fertilizer prices fluctuate in response to changes in macro, trade and exchange-rate policies and to volatile international fertilizer prices. While fertilizer subsidies and domestic fertilizer production have generally proved ineffective in Africa, it also appears clear that private markets in rural Africa cannot guarantee fertilizer supplies at present. This suggests that some role for government is inevitable in the short to medium term. Given the considerable costs of delivering fertilizer to farmers on time and the restricted availability of fertilizer to most farmers, public investment in improved private marketing infrastructure seems promising (Ahmed *et al.*, 1989; Rusike *et al.*, 1997).

Profitability entails both that an effective market exists and that the ratio of output to input prices is favourable (Dembele and Savadogo, 1996). Only farmers who engage in profitable commercial agriculture will be willing to invest in inorganic fertilizers, animal traction, organic matter and soil conservation. It does not matter in this regard whether they are large or small or whether they grow food crops or other types of crops. For example, in Burkina Faso, farmers use 13 times more manure on cotton and maize, both cash crops, than on sorghum and millet, the main subsistence grains (Savadogo *et al.*, 1998). In Zimbabwe, farmers mainly use improved tillage practices and fertilizers where there are profitable cash crops (Mudimu, 1996). In the highland tropics of Tanzania, farmers confine fertilizer and soil conservation practices to cash crops (Semgalawe, 1998), as they do in Rwanda (Clay *et al.*, 1998). Policy reforms and project-level interventions that render sustainable crops and technologies profitable contribute to environmentally sustainable agricultural intensification.

Financial services policy

Delivery of rural financial services was commonly linked to parastatal distribution of seed and fertilizer and purchase of agricultural outputs. Elimination of public input and output distribution systems not only increased input costs for small farmers in many areas, it also often raised effective interest rates for rural borrowers or eliminated their access to seasonal credit altogether. Many private merchants have found market entry or expansion difficult in the absence of public rural finance schemes, unless they are able to offer consumer credit themselves (Rusike *et al.*, 1997).

Government parastatals were able to establish functioning credit schemes by linking the input markets to the output markets. Under present institutional and legal arrangements, private operators may not be able to do that. In hindsight, the favourable general equilibrium effects of loosened rural liquidity constraints, which resulted from the government's ability to link credit and output markets, appear to have at least partly mitigated the adverse partial equilibrium effects of state monopoly or monopsony.

In the wake of reduced rural credit volumes, smallholders rely increasingly on cash crops and non-farm earnings to finance capital accumulation and to smooth consumption. However, because the biggest farmers earn the most off-farm and from cash crops, since the removal of government credit programmes access to alternative (usually self-) financing has often become quite concentrated. As a result, only larger operators and the most commercial smallholders can afford to follow SAI, leaving the mass of semi-subsistence smallholders to either produce more extensively, intensify unsustainably or exit agriculture (Reardon *et al.*, 1994; Reardon, 1997).

While there is scant empirical evidence on how the demise of public financial services has affected the use of seed and fertilizer, we know even less about its impact on physical capital formation, such as investment in small-scale irrigation, animal traction equipment or postharvest machinery. In theory, increases in effective interest rate should discourage such investment (Lipton, 1991), but few analysts have studied the topic. Physical capital items, such as animal traction equipment, irrigation pumps, spare parts for vehicles, tractors and ploughs, are largely imported in much of Africa. Thus, currency devaluations should drive up their prices. This translates into higher costs for irrigation, transport and land conservation investments. We do not know of any studies of the price elasticity of African farm investment, but it is highly likely that the combination of financial sector retrenchment, contractionary monetary policy and currency devaluation has discouraged investment in agrarian quasi-fixed capital.

Stimulating rural finance is central to promoting capital-led SAI. While some quasi-fixed capital investment requires a considerable amount of labour, e.g. bunding and terracing, it usually also demands a complementary commitment of purchased inputs, e.g. fertilizer and tools. State-directed rural credit schemes were often fiscally unsustainable and ineffective in serving

Africa's most credit-constrained smallholders. None the less, a strong case is to be made for state subsidization of the initial start-up and training costs for self-sustaining rural financial institutions that can mobilize local savings and rotate them within and across communities as loans.

Land policy

Land policy in the past decade has mainly involved titling schemes, gazetting of public areas and some very limited land redistribution. The former would tend to drive up land prices, spurring intensification and long-term investment in land improvements (Place and Hazell, 1993). The latter should increase the marginal value product of land use by raising the ratio of labour to land, as smaller farmers supplant larger farmers (Barrett, 1996). However, the impact of land tenure on technology adoption and investment is ambiguous in sub-Saharan Africa. Migot-Adholla *et al.* (1991) show that many other structural factors, such as rural health, education and infrastructure, blur the impact of land tenure systems.

The past decade's burst of activity in gazetting lands for protected areas increases tenure insecurity for those living in environmentally sensitive areas. If farmers are less certain than before that the state will not appropriate their land for parks and reserves, then they have less incentive to invest in conservation measures required for SAI. The bitter irony is thus that measures designed to achieve environmental conservation may induce environmental degradation by threatening current operators' control over the land.

Back to the future: projects in lieu of policy?

At the same time as governments are dismantling their financial services and input parastatals, public or non-governmental organization (NGO) projects, which are essentially mini-packages of policies that affect smaller groups on a temporary basis, have flourished. These packages basically reproduce a subset of the prestructural adjustment policies – extension services, subsidized 'micro-finance' services, subsidized equipment, inputs and marketing services, etc.

These projects are often presented as 'demonstration projects' in areas where diffusion might eventually have a chance. Good examples include the Sasakawa Global 2000 projects in Ethiopia, Ghana, Mozambique and Tanzania (Putterman, 1995) and a variety of contract farming schemes. Many of these projects have sharply increased yields on participating farms, but only by circumventing the structural obstacles that often impede adoption of SAI methods. The projects have delivered appropriate inputs and financial services directly to farmers in a timely manner and have ensured a market for their output. However, the results often cannot be transferred outside the scheme or sustained after the scheme ends. The schemes themselves may not prove fiscally sustainable on any significant scale.

Such projects demonstrate that African smallholders can achieve higher-yielding and environmentally sustainable agricultural production. They also implicitly demonstrate how weak rural factor and product markets mute both incentives to intensify sustainably and the ability of governments and donors to alter those incentives effectively through macro- or sector-level policy. While macro and sectoral reforms may be necessary to establish a stable macroeconomic environment, they have generally proved insufficient to remedy the underlying structural problems that induce unsustainable intensification and extensification.

4. Illustrative Case-studies

4.1. Examples of SAI

In the cases in this subsection, governments, donors and farmers made investments and took policy measures to redress the problems of risk, high transaction and input costs and low profitability that plague African agriculture. The successes were demand-driven, in the sense that growing demand for the product made agriculture profitable and reduced market-related risk. In each case, profitable intensification that avoids degradation *ceteris paribus* reduced pressure to extend cultivation into the remaining forests, bushlands and wetlands.

Onions and rice in the Office du Niger in Mali[3]

In the 1980s and 1990s, the government of Mali upgraded the irrigation infrastructure to make it easier for farmers to respond to new incentives for rice and onion production. At the same time, it stopped controlling the maintenance of that infrastructure, as well as farm production planning and output and input marketing. This paved the way for private merchants to develop their capacity to react flexibly to new incentives. The 1994 devaluation of the CFA franc provided the necessary incentives. It made rice and onions produced by Office du Niger farmers much more competitive within Mali and in West Africa in general and increased net returns to production.

The new incentives and the improved irrigation infrastructure allowed double cropping (rice followed by onions). This significantly increased farm income and the productivity of the government's infrastructure investments. It also improved farmers' ability to maintain the infrastructure and provided them with funds to purchase fertilizer and farm equipment, and thus to intensify.

Bananas in Rwanda[4]

The emergence of rural towns and rising incomes generated demand for processed products such as banana wine, which, in turn, created a derived demand for bananas. As a result, banana production and area have risen rapidly over the past 20 years. Bananas provide higher earnings than other land uses (except coffee). This, along with high and rising population pressure, which has constrained farmers' access to land, has given farmers an incentive to intensify. Although bananas take a while to get established, food crops can be grown around young bananas, so even the poor can bear the gestation period, which is not true of some other perennials. Bananas also prevent erosion, a major concern in Rwanda.

Cotton/maize zones in Burkina Faso and Mali[5]

Fertilizer and seed subsidies, credit and guaranteed output markets for cotton farmers in these two countries assured profitability and reduced risk. These are administered via vertically integrated, mixed public/private firms linked into the global cotton market. Animal traction equipment programmes helped farmers obtain equipment. This system led to rapid expansion of the area planted in cotton in Mali and Burkina Faso in the 1970s and 1980s.

Farmers reacted to these positive incentives both by increasing their cotton area and by growing cotton more intensively. They tended first to expand production into new areas with high-quality land and then to intensify once no more high-quality land was available. In areas with adequate soils, they generally intensified by using relatively large amounts of fertilizer, organic matter and animal traction on both cotton and the rotation crop, maize. Farmers used the profits from cotton to purchase inputs for maize production and improve its productivity.

4.2. Examples of unsustainable intensification and extensification

Postliberalization African agriculture often lacks the ingredients of successful SAI. There is a dearth of: (i) public and private agrarian capital (e.g. roads, animal traction equipment, irrigation); (ii) affordable inputs; (iii) low-risk output markets; (iv) accessible financial services; and (v) an off-farm labour market to absorb labour from low-productivity farms. In the absence of such conditions, reforms may stimulate unsustainable intensification or extensification. Indeed, considerable evidence suggests that the removal of price stabilization schemes and subsidies to input distribution, marketing and rural credit left a vacuum that was not subsequently filled by the private sector for many African smallholders who produce grains, roots and tubers for domestic markets. The environmental effect, illustrated in the brief case-studies that follow, is induced degradation, either through extensification or through capital-deficient intensification leading to soil nutrient mining.[6]

Rice extensification in Madagascar

The rice sector dominates Madagascar's economy. Market liberalization, currency devaluation and reduced state support for agricultural credit in the 1980s brought higher and more variable rice prices and reduced fertilizer use (Barrett, 1997, 1999). This induced Malagasy rice producers – most of whom are food-insecure net rice buyers – to increase output by expanding the area in cultivation through further shortening of fallow periods and expanding into fragile forest margins (Barrett, 1999). Since they had no new production technologies and minimal access to modern inputs, they had little choice. After liberalization, deforestation appears to have accelerated from the 0.8% annual rate established by aerial imagery for the 1973–1985 period. Smaller farmers with lower rice yields who lived in relatively densely populated areas and households facing greater food insecurity seem to have been responsible for a large portion of the forest loss. These were precisely the farmers most adversely affected by the reform measures (Barrett and Dorosh, 1996; Barrett, 1999).

Liberalization appears to have limited the adoption of a promising new technology package, the system of rice intensification (SRI). Trials on farmers' fields have shown that SRI can deliver lowland rice yields that average 9 t ha^{-1}, compared with roughly 2 t ha^{-1} under traditional methods (Moser, 1999). The key bottlenecks relate to credit for initial investments (especially in water control), for mechanical weeders and to cover seasonal demands for hired labour. Rural credit has largely dried up in Madagascar in the wake of restrictive monetary policies and the withdrawal of state support for smallholder bank credit. Reform-related roll-backs in agricultural extension budgets also pose a serious obstacle to widespread adoption, because SRI is a fairly complicated package of innovations, which few smallholders can pick up without training and follow-up visits. The training and extension record of the indigenous NGO promoting SRI has been reasonably good, but this must be scaled up through national-level investments in agricultural extension.

Cocoa disinvestment in Cameroon[7]

Policy shocks appear to have contributed to deforestation in the rain forest of southern Cameroon. The dismantling of the state's cocoa marketing and price stabilization schemes in the early 1990s led relative crop prices to adjust sharply, favouring plantains and cocoyams over cocoa. This was accompanied by an exacerbation of transport and marketing bottlenecks, due to reduced government investment in rural infrastructure and rising transport costs stemming from the CFA franc devaluation. This increased the price of importing food into the southern forest margins and made it more variable, while also reducing and making more variable producer prices for export crops, such as cocoa and fruit. Producers responded by reallocating labour away from cocoa perennial systems into the slash-and-burn annual systems used in the area for cocoyams, plantains, maize and groundnuts. This occurred in large part because improved technologies do not exist for these crops and input distribution systems work poorly in the forest regions of southern Cameroon. Thus, increased cocoyam or plantain output based on sustainable intensification was not an option.

The policy-induced expansion of annual crop production in low-productivity soils has not only led to deforestation. It has also replaced the previously dominant agroforestry systems, based on cocoa and fruit production, with systems that provide much less biodiversity conservation and carbon sequestration. To get farmers to grow sustainable intensive perennial agroforests in the rain-forest margins of southern Cameroon instead of rotational annual crops will require heavy, renewed emphasis on increased labour and land productivity and on improved interregional food marketing.

Maize in Zambia and Zimbabwe

The maize subsectors of Zambia and Zimbabwe present interesting cases, where prereform policies in the early 1980s fostered smallholder adoption of hybrid maize varieties and fertilizers, practices essential to SAI in the most fragile areas of these countries. However, neither Zambia nor Zimbabwe could afford the public expenditures demanded by depot provision and subsidies to seed, fertilizer and financial services. As a result, they dismantled these services by the early 1990s. Fertilizer use has fallen in both countries, leading to both soil nutrient mining and, where cultivators are near forest margins, extensification through forest clearing (Holden, 1997). Declining real wages in rural labour markets likewise work against sustainable intensification by increasing the relative profitability of extensive agriculture based on labour-intensive clearing (Holden, 1993). Smallholder private markets for outputs and inputs are now slowly coming back. But it is too early to tell whether this will be widespread and successful enough to induce a return to SAI (Eicher, 1995; Howard and Mungoma, 1997; Rusike *et al.*, 1997).

5. Conclusions

The central claim of this chapter is that policy reforms have had ambiguous effects on SAI in Africa, broadly defined as adequate use of inorganic fertilizer, organic matter and agrarian capital, such as soil conservation structures, equipment and irrigation. In an exceedingly capital-constrained continent, SAI is clearly a challenge. At present, most African smallholders appear not to be choosing sustainable paths: hence the interlinked crises of rural poverty, declining per capita agricultural productivity and environmental degradation. None the less, appropriate ancillary investments can reverse the vicious circle. Most needed technologies are already available. The key lies in giving African smallholders the capacity and incentives to choose sustainable expansion paths. Much policy reform has been blind to the net effects on smallholder production incentives, focusing excessively on macro-level reforms, without fully recognizing the underlying structural weaknesses in rural markets.

Recent policy reforms have had mixed effects on African farmers' incentives and capacity to undertake investments needed for SAI. The success stories of SAI appear where necessary investments in farm-level capital had been made in the past or are being made through projects and where market proximity and satisfactory infrastructure enable markets to function reasonably well. Where state or NGO interventions have resolved structural weaknesses in factor or product markets or established an agrarian capital base, farmers enjoy incentives and have the capacity to pursue SAI.

Unfortunately, many of Africa's poorest smallholders live in remote areas, poorly served by infrastructure, financial institutions or public services, and face poor and volatile terms of trade. In their daily struggle against food insecurity and poverty, the capital-led path to SAI remains inaccessible, often leading to a vicious circle of immiseration and environmental degradation. In such settings, liberalization often induces degradation of environmentally fragile areas with high levels of biodiversity.

The pressing issue is how to reverse the decline in conditions for smallholders producing cereals, tubers and roots under rain-fed conditions for local markets. Essential ingredients include policies to spur the private investment necessary for SAI. Heavy-handed state interventions in marketing systems proved fiscally unsustainable failures in most of Africa. But too often economic reform programmes have thrown out necessary state support services for private investment and marketing with the parastatal bath water. The selection of appropriate public investments in physical infrastructure and institutions will need to be made in a country-specific fashion. While there is always a risk that intensification itself can come at a price in terms of deforestation, failure to intensify sustainably is a sure recipe for renewed threats to fragile margins.

Notes

1   This chapter builds on previous work by the authors and collaborators in Reardon et al. (1999, 2000). We are grateful for comments on earlier versions from Arild Angelsen, David Kaimowitz and an anonymous reviewer.
2   The model draws on Reardon and Vosti (1992) and Barrett and Carter (1999).
3   Coulibaly et al. (1995).
4   Byiringiro and Reardon (1996); Clay et al. (1998); Kangasniemi (1998).
5   Dioné (1989); Savadogo et al. (1998).
6   We have omitted other corroborating stories for the purpose of brevity. For example, in Tanzania, Monela (1995) found that fertilizer marketing and price

liberalization provoked increased encroachment on forests, while Angelsen et al. (1999) report that crop price increases stimulate area expansion more than intensification on existing lands.
7   This section draws extensively on Gockowski et al. (2000).

References

Ahmed, J., Falcon, W. and Timmer, P. (1989) *Fertiliser Policy for the 1990s*. Discussion Paper No. 293, Harvard Institute for International Development, Cambridge, Massachusetts.

Ahmed, R. and Hossain, M. (1990) *Developmental Impact of Rural Infrastructure in Bangladesh*. Research Report No. 83, International Food Policy Research Institute, Washington, DC.

Angelsen, A., Shitindi, E.F.K. and Aarrestad, J. (1999) Why do farmers expand their land into forests? Theories and evidence from Tanzania. *Environment and Development Economics* 4(1), 313–331.

Barrett, C.B. (1996) On price risk and the inverse farm size-productivity relationship. *Journal of Development Economics* 51(2), 193–215.

Barrett, C.B. (1997) Liberalization and food price distributions: arch-m evidence from Madagascar. *Food Policy* 22(2), 155–173.

Barrett, C.B. (1999) Stochastic food prices and slash-and-burn agriculture. *Environment and Development Economics* 4(2), 161–176.

Barrett, C.B. and Carter, M.R. (1999) Microeconomically coherent agricultural policy reform in Africa. In: Paulson, J. (ed.) *African Economies in Transition, Vol. 2: The Reform Experiences*. Macmillan, London.

Barrett, C.B. and Dorosh, P.A. (1996) Farmers' welfare and changing food prices: nonparametric evidence from rice in Madagascar. *American Journal of Agricultural Economics* 78(3), 656–669.

Barrett, C.B. and Sherlund, S.M. (2000) Shadow wages, allocative inefficiency, and labour supply in smallholder agriculture. Paper presented to the winter meeting of the Econometric Society.

Byiringiro, F. and Reardon, T. (1996) Farm productivity in Rwanda: effects of farm size, erosion, and soil conservation investment. *Agricultural Economics* 15(2), 127–136.

Clay, D., Reardon, T. and Kangasniemi, J. (1998) Sustainable intensification in the highland tropics: Rwandan farmers' investments in land conservation and soil fertility. *Economic Development and Cultural Change* 46(2), 351–378.

Coulibaly, B.S., Sanogo, O. and Mariko, D. (1995) Cas du coûts de production du paddy à l'Office du Niger. Paper presented at the Atelier Régional du PRISAS sur l'Impact de la Dévaluation du FCFA sur les Revenus et la Sécurité Alimentaire en Afrique de l'Ouest, June 1995, Bamako, Mali.

Dembele, N.N. and Savadogo, K. (1996) The need to link soil fertility management to input/output market developement in West Africa: key issues. Paper presented at the International Fertilizer Development Centre Seminar, 19–22 November, Lomé, Togo.

Dioné, J. (1989) Informing food security policy in Mali: interactions between technology, institutions, and market reforms. PhD thesis, Michigan State University, East Lansing, Michigan.

Eicher, C.K. (1995) Zimbabwe's maize-based Green Revolution: preconditions for replication. *World Development* 23(5), 805–818.

Fafchamps, M. (1993) Sequential labour decisions under uncertainty: an estimable household model of West African farmers. *Econometrica* 61, 1173–1197.

Gockowski, J., Nkamleu, B. and Wendt, J. (2000) Implications of resource use intensification for the environment and sustainable technology systems in the central African rainforest. In: Lee, D.R. and Barrett, C.B. (eds) *Tradeoffs or Synergies? Agricultural Intensification, Economic Development and the Environment in Developing Countries*. CABI, Wallingford, UK, pp. 197–220.

Heisey, P.W. and Mwangi, W. (1997) Fertiliser use and maize production in sub-Saharan Africa. In: Byerlee, D. and Eicher, C.K. (eds) *Africa's Emerging Maize Revolution*. Lynne Reinner, Boulder, Colorado.

Holden, S. (1993) Peasant household modeling: farming systems evolution and sustainability in northern Zambia. *Agricultural Economics* 9(2), 241–267.

Holden, S. (1997) Adjustment policies, peasant household resource allocation and deforestation in northern Zambia: an overview and some policy conclusions. *Forum for Development Studies* 1, 117–134.

Howard, J. and Mungoma, C. (1997) Zambia's stop-and-go revolution: the impact of policies and organizations on the development and spread of maize technology. In: Byerlee, D. and Eicher, C.K. (eds) *Africa's Emerging Maize Revolution*. Lynne Reinner, Boulder, Colorado.

Kangasniemi, J. (1998) People and bananas on steep slopes: agricultural intensification under demographic pressure and environmental degradation in Rwanda. PhD dissertation, Department of Agricultural Economics, Michigan State University, East Lansing, Michigan.

Kelly, V., Diagana, B., Reardon, T., Gaye, M. and Crawford, E. (1996) *Cash Crop and Foodgrain Productivity in Senegal: Historical View, New Survey Evidence, and Policy Implications*. MSU International Development Paper No. 20, Michigan State University, East Lansing, Michigan.

Kim, T.K., Hayes, D.J. and Hallam, A. (1992) Technology adoption under price uncertainty. *Journal of Development Economics* 38(1), 245–253.

Lipton, M. (1991) Market relaxation and agricultural development. In: Colclough, C. and Manor, J. (eds) *States or Markets? Neo-liberalism and the Development Policy Debate*. Clarendon Press, Oxford.

Low, A. (1986) *Agricultural Development in Southern Africa: Farm-household Economics and the Food Crisis*. James Currey, London.

Maredia, M. and Howard, J. (1997) *Facilitating Seed Sector Transformation in Africa: Key Findings from the Literature*. Policy Synthesis No. 33, Department of Agricultural Economics, Michigan State University, East Lansing, Michigan.

Migot-Adholla, S., Hazell, P., Blarel, B. and Place, F. (1991) Indigenous land rights systems in sub-Saharan Africa: a constraint on productivity? *World Bank Economic Review* 5(1), 155–175.

Monela, G.C. (1995) Tropical rainforest deforestation, biodiversity benefits, and sustainable land use: analysis of economic and ecological aspects related to the Nguru Mountains, Tanzania. PhD dissertation, Agricultural University of Norway, Ås.

Moser, C. (1999) Constraints to SRI adoption. Mimeo, Cornell University, Ithaca, New York.

Mudimu, G. (1996) An analysis of incentives for adopting soil and moisture conservation tillage practices by smallholder farmers in Zimbabwe: a case study of Kandeya communal land. Mimeo, Department of Agricultural Economics and Extension, University of Zimbabwe, Harare.

Place, F. and Hazell, P. (1993) Productivity effects of indigenous land tenure systems in sub-Saharan Africa. *American Journal of Agricultural Economics* 75 (February), 10–19.

Putterman, L. (1995) Economic reform and smallholder agriculture in Tanzania: a discussion of recent market liberalization, road rehabilitation, and technology dissemination efforts. *World Development* 23(2), 311–326.

Reardon, T. (1997) Using evidence of household income diversification to inform study of the rural non-farm labour market in Africa. *World Development* 25(5), 735–748.

Reardon, T. and Vosti, S. (1992) Issues in the analysis of the effects of policy on conservation and productivity at the household level in developing countries. *Quarterly Journal of International Agriculture* 31(4), 380–396.

Reardon, T., Crawford, E. and Kelly, V. (1994) Links between non-farm income and farm investment in African households: adding the capital market perspective. *American Journal of Agricultural Economics* 76(5), 1172–1176.

Reardon, T., Kelly, V., Crawford, E., Diagana, B., Dione, J., Savadogo, K. and Boughton, D. (1997) Promoting sustainable intensification and productivity growth in Sahel agriculture after macroeconomic policy reform. *Food Policy* 22(4), 317–328.

Reardon, T., Barrett, C.B., Kelly, V. and Savadogo, K. (1999) Policy reforms and sustainable agricultural intensification in Africa. *Development Policy Review* 17, 4.

Reardon, T., Barrett, C.B., Kelly, V. and Savadogo, K. (2000) Sustainable versus unsustainable agricultural intensification in Africa: focus on policy reforms and market conditions. In: Lee, D.R. and Barrett, C.B. (eds) *Tradeoffs or Synergies? Agricultural Intensification, Economic Development and the Environment in Developing Countries*. CABI, Wallingford, UK, pp. 365–382.

Rukuni, M. (1996) *A Framework for Crafting Demand-driven National Agricultural Research Institutes in Southern Africa*. Department of Agricultural Economics Staff Paper No. 96–76, Michigan State University, East Lansing, Michigan.

Rusike, J., Reardon, T., Howard, J. and Kelly, V. (1997) *Developing Cereal-based Demand for Fertiliser Among Smallholders in Southern Africa: Lessons Learned and Implications for Other African Regions*. Policy Synthesis No. 30, Food Security II Project, Michigan State University, East Lansing, Michigan.

Ruttan, V. (1990) Models of agricultural development. In: Eicher, C. and Staatz, J. (eds) *Agricultural Development in the Third World*, 2nd edn. Johns Hopkins University Press, Baltimore.

Savadogo, K., Reardon, T. and Pietola, K. (1998) Adoption of improved land-use technologies to increase food security in Burkina Faso: relating animal traction, productivity, and non-farm Income. *Agricultural Systems* 58, 441–464.

Semgalawe, Z.M. (1998) Soil conservation and agricultural sustainability: the case of north-eastern mountain slopes, Tanzania. Doctoral dissertation, Department of General Economics, Wageningen University, the Netherlands.

Viyas, V.S. (1983) Asian agriculture: achievements and challenges. *Asian Development Review* 1(1), 27–44.

Weight, D. and Kelly, V. (1998) *Restoring Soil Fertility in sub-Saharan Africa: Technical and Economic Issues*. Policy Synthesis, Food Security II Project, Department of Agricultural Economics, Michigan State University, East Lansing, Michigan.

A Century of Technological Change and Deforestation in the *Miombo* Woodlands of Northern Zambia

Stein Holden

1. Introduction

Agricultural expansion and intensification, driven largely by population growth, migration, technological change and government policies, exposed the *miombo* savannah woodlands in northern Zambia to increasing pressures over the last century. This chapter uses economic theory and agroecosystem analysis to assess the effects of technological changes on deforestation during that period. This provides the basis for drawing wider inferences about the links between agricultural innovation and deforestation in other regions.

The chapter combines historical facts about demographic, policy and technological changes with applied farm household models to illustrate how these changes have affected typical land users in the area. The models use multiobjective programming, which combines lexicographic and weighted-goal programming, and incorporate households' basic needs, evolving cultural preferences, access to technologies, seasonal labour demands and constraints, aversion to drudgery and risk and partial integration into markets (Holden, 1993a). I also draw on my own fieldwork in the area in the 1980s and 1990s.

The chapter highlights two major technological changes: the introduction of cassava during the first half of the century, and the expansion of maize systems involving fertilizer use in the late 1970s. While cassava was labour-intensive, the maize–fertilizer system was capital-intensive. The maize–fertilizer system became more risky after the government introduced structural adjustment policies (SAP) in the 1990s, and in recent years we have witnessed 'technological progress in reverse'.

Section 2 presents the basic environmental characteristics and production systems in the area. Section 3 outlines the theoretical basis for the analysis. Section 4 gives an overview of the historical changes, drawing lessons from economic models to back up the historical facts. Section 5 uses the material from Zambia to evaluate some of the hypotheses discussed in the earlier chapters of this book, followed by a brief conclusion.

2. Zambia's *Miombo* Region and its Production Systems

Zambia is a land-locked country in southern Africa that covers some 753,000 km^2. It rains more in the north than in the south, with an annual rainfall of more than 1000 mm in the high-rainfall areas. The 5-month rainy season extends from November to April. Most of the area is on a plateau about 1200–1500 m above sea level.

Most of northern Zambia has moderately to severely leached and acid soils, many of which are ultisols and oxisols. According to Lal (1987), currently only 21% of Africa's potential cultivable land area is cultivated. Sanches and Salinas (1981) have noted that the greatest potential for expanding the agricultural frontier lies precisely in areas with soils similar to those found in northern Zambia, where, to date, low natural soil fertility and low population density with limited market infrastructure have constrained their utilization.

Ultisols and oxisols are not suited for continuous cultivation. Using low-input systems, farmers can only cultivate them 1 out of every 4 years. With medium and high input levels, this rises to 2 and 3 out of every 4 years, respectively. The rest of the time farmers must leave the land in fallow to restore its nutrients, organic matter and soil structure and to control weeds, pests and diseases (Young and Wright, 1980). Shortening fallow periods can negatively affect short- and long-term soil productivity.

Zambia's vegetation map classifies 66% of Northern Province as *miombo*, an open woodland dominated by genera such as *Brachystegia, Julbernardia* and *Isoberlinia* (NORAGRIC, 1990). In more populated areas, farmers cut down the trees and have fallow periods that are too short for them to regenerate, so grasslands are gradually replacing them. Uncontrolled fires are common and affect the regrowth of woody vegetation and species composition.

Early in the century, the *chitemene* shifting cultivation system dominated most of northern Zambia. In this system, farmers chop down a large area of trees, pile the trunks on to a smaller area and burn them. Then they grow crops in the ash for a few years. The fire releases the nutrients in the woody biomass and makes them available for crops and provides a seed-bed free of weeds. The heat also affects the soil structure, leaving a fine seed-bed for finger millet (*Eleusine coracana*), the first crop. Different ethnic groups practise various forms of *chitemene* and the system has evolved over time.

Towards the border with Tanzania, in the far north-east, where the soils are more fertile and population densities were higher, farmers practised another system, the grass-mound (*fundikila*) system. Producers there incorporated grass turf into mounds, waited for the organic matter to decompose and then spread the mounds out before planting finger millet and then beans. The *fundikila* system later spread to other areas and, like the *chitemene* system, underwent various changes.

The British introduced cassava in the first half of the century. Cassava cultivation was labour-intensive and greatly improved food security. Farmers typically planted cassava on ridges or mounds as the main crop in so-called 'cassava gardens', often alongside other crops during the first year.

Beginning in the late 1970s, infrastructure improvements, market integration and investments in research and extension facilitated the spread of maize production, associated with the use of fertilizers. The maize cropping system is often referred to as 'permanent maize production'. A more appropriate term might be 'high-external-input shifting cultivation', since the fertilizers acidify the soils, leading productivity to decline over the long term.

3. Conceptual Framework

Peasant farm households dominate Africa's humid tropics. They typically employ extensive agricultural practices and many analysts consider them the main sources of deforestation. These households are simultaneously producers and consumers and generally behave rationally, given their preferences, resource constraints and limited access to information and the imperfect markets they face. Low population densities and poor infrastructure contribute to weak communications systems and high transaction costs. This leads to pervasive market imperfections (Binswanger and McIntire, 1987). Since land is abundant, often no land market exists. Similarly, credit, labour, input and output markets may be absent or very imperfect. For example, farmers may only be able to find employment during certain seasons or may have difficulty obtaining credit. This affects household behaviour, including the technologies households choose and their decisions about whether to clear forest. Under such circumstances, household decisions about production and consumption depend on one another and specific technology, market and household characteristics determine many of the outcomes. Missing credit markets, combined with poverty, induce farm households to heavily discount future income, which may lead them to ignore the long-term effects of their land management decisions (Holden *et al.*, 1998a).

In such a context, it is realistic to model farmers' decisions using static household models that analyse production and consumption decisions simultaneously and incorporate market imperfections. Chayanov (1966) developed the first farm household model early in the 20th century to analyse typical farm households in Russia, where there were practically no land and labour markets and households varied the amount of land they farmed in response to changes in the age and numbers of their members. Hunt (1979),

Holden (1991, 1993a) and Low (1986) have shown that Chayanov's model has wide relevance in many African contexts.

The literature on the 'economics of rural organization' shows that imperfect information leads farmers to consider hired labour a poor substitute for family labour (Feder, 1985). This partially explains the frequently found inverse relationship between farm size and efficiency (Berry and Cline, 1979; Feder, 1985; Heltberg, 1998). Moral-hazard situations due to imperfect information cause rationing in credit markets (Stiglitz and Weiss, 1981). Cash and credit constraints also contribute to labour market imperfections, which explains why land-abundant agricultural economies have many Chayanovian features. The characteristics of the outputs involved and market access influence how product markets function. Some outputs have no markets or only local markets. This implies that each household or village faces a distinct set of shadow or market prices and that supply fluctuations may have large effects on those prices. If the local demand is inelastic, technological change in the production of tradable and non-tradable products may have opposite effects on deforestation.

Boserup's (1965) theory of the evolution of agricultural development describes a general tendency for production to become more intensive as labour productivity falls in response to greater population pressures. As population rises, farmers turn from shifting cultivation to long fallow, short fallow, permanent and multiple cropping systems. Similarly, Ruthenberg (1980) concluded that shifting cultivators generally do not experience labour shortages, nor do they work as hard as they can because they have no need or incentive to do so. However, when farmers adopt fallow systems or permanent upland cultivation, labour becomes a limiting factor for production. The demand for labour in these systems has more pronounced seasonal peaks, particularly for land preparation and weeding, and the physical yields per unit of cropped area (excluding fallow) are lower than in shifting-cultivation systems. This implies that shifting cultivation offers many desirable features for farm households in sparsely populated areas.

Farmers often continue practising shifting cultivation even after they have exceeded an area's carrying capacity. As a result, fallow periods decline and the system becomes unsustainable. Holden (1998) has shown that, under such conditions, the high discount rates of poor shifting cultivators will typically lead them to have a short-term view and disregard the future benefits of forest regeneration. Others have blamed such situations on open access, tenure insecurity or the failure of collective action. However, empirical evidence indicates that poor peasant farm households' discount rates are much higher than the rate of forest regrowth (Holden, 1997, 1998; Holden et al., 1998a). Under such circumstances, tenure security becomes irrelevant.

Lexicographic and weighted-goal programming is used to represent the basic goals of farm households, such as food requirements (energy, protein, taste), housing requirements, energy requirements, water, market-purchased goods, security (risk avoidance), social obligations and needs and leisure (drudgery aversion). It is assumed that they want to satisfy their basic needs with the lowest possible work effort. Beyond this, it is assumed that drudgery is minimized subject to an income constraint or a weighted income–leisure goal is maximized.

The optimization is subject to a production possibility set that is changed according to availability of technologies and resources. Risk is modelled by the introduction of a set of safety-first constraints, such that basic needs are met even in years with a late start of the rain and non-availability of fertilizer. Seasonality was modelled by dividing the year in 11 different periods to obtain the typical seasonal variation in labour supply, demand and shadow wage rates.

4. Historical Facts and Models of the Evolution of Northern Zambia's Agriculture

4.1. The period before cassava

In the early 20th century, the area currently known as northern Zambia had a very low population density. People practised the *chitemene* and *fundikila* systems and had not yet begun growing cassava. The British established a research station with long-term cropping trials and studied the indigenous land-use systems. In 1938, they concluded that the only way to control deforestation was by keeping farmers from practising the *chitemene* system, but that it was practically impossible to do so as long as there was still forest available to clear. They also found that the *fundikila* grass-mound system required 3 years of fallow for each cropping year (Boyd, 1959).

The first set of household models was constructed to simulate the choice farmers faced during this period between the *chitemene* shifting-cultivation and *fundikila* grass-mound systems (Table 14.1). At the time, the population density was low (land was abundant) and pest problems (locusts and wild animals) made production risky. Households could sell their crops but not their labour. Their basic goals were simply to grow enough food to meet their subsistence requirements and to work as little as possible. How much area they needed to cultivate to achieve this depended on their cultural preferences, production technologies, market access, seasonal labour demands and other factors. In general, however, they had little incentive to produce more once they met their subsistence requirements, so any improvement in land productivity tended to reduce the total area they cultivated.

The model shows that, given the prevailing yields and labour requirements at the time, farm households should have strongly preferred the *chitemene* system, because the grass-mound system provided much lower returns to labour on the infertile soils of the central plateau. This may explain why population densities remained low in these areas, as the grass-mound system was too unproductive. The *chitemene* system allowed people to produce

Table 14.1. Simulations of the situation before the introduction of cassava.

	Net income constraint (kwacha)		
	0	200	300
Labour requirement in production (h year^{-1})	2303	2653	Infeasible
Annual cropped area (ha per household)			
Total	1.45	1.72	
Chitemene system	1.45	1.72	
Grass-mound system	0	0	
Carrying capacity (persons km^{-2})	2.46	2.2	
Consumption activities (kg consumed year^{-1})			
Finger millet	996	1108	
Groundnuts	305	231	
Beans	67	67	
Selling activities (kg sold year^{-1})			
Groundnuts	0	74	
Beans	0	38	

a small surplus of groundnuts and beans, which they could sell. The carrying capacity of the land use was only 2.2–2.5 persons km^{-2}.

4.2. The introduction of cassava

The British actively controlled population densities by moving people to less populated areas. They also forced the native population to grow cassava to ensure food security, arguing that, when locusts devastated other crops, farmers would be able to fall back on cassava, since its roots were protected underground. Initially, people disliked the new crop, but they soon discovered some of its advantages and began to adopt it widely. Eventually, it became most people's main staple. Cassava was gradually adapted into the *chitemene* and grass-mound systems by intercropping and it was also grown in separate cassava gardens. It produced a good yield even in the poor soils where most other crops failed.

Table 14.2 simulates some of the effects of cassava's introduction. It clearly illustrates that cassava was both yield-increasing (land-saving) and labour-saving, besides enhancing food security by resisting locust attacks. Cassava boosted the carrying capacity of the *chitemene* system two to six times and made it possible for households that adopted cassava as their main staple, instead of finger millet, to meet their food requirements with 40% less labour input. This also enabled them to produce much larger surpluses for sale. Rural households were particularly interested in reducing their labour requirements, because a large portion of northern Zambia's males migrated to work in the

Table 14.2. Simulation of the effect of introducing cassava in a low-population context.

	Net income constraint (kwacha)		
	0	300	1000
Labour requirement in production (h year^{-1})	1374	1762	3176
Annual cropped area (ha per household)			
Total	0.64	0.88	1.93
Chitemene	0.51	0.88	1.82
Grass-mound system	0	0	0
Cassava garden	0.13	0	0.1
Carrying capacity (persons km^{-2})	14.8	7.8	3.4
Consumption activities (kg consumed year^{-1})			
Finger millet	143	100	379
Groundnuts	44	44	44
Beans	67	67	67
Cassava	3046	3151	2470
Selling activities (kg sold year^{-1})			
Groundnuts	0	52	211
Beans	0	0	147
Beer sale	0	190	333

rapidly expanding copper industry. The cassava technology made female-headed households less dependent on male labour, because cassava could be grown without climbing and cutting trees, which were strictly male tasks.

The introduction of cassava reduced population pressure and deforestation both directly, by increasing land productivity in a context where farmers only sought to meet their subsistence requirements, and indirectly, by stimulating out-migration. However, it also facilitated the concentration of population practising intensive cassava systems along lakes and near towns and roads, which provoked greater deforestation in these locations, as illustrated in Table 14.3, but also reduced deforestation in other locations. By incorporating cassava into their systems, farmers could replace the *chitemene* system with grass-mound and cassava-garden systems, which made it possible to maintain population densities ten to 15 times higher than those that could have been sustained previously. The carrying capacity of the *chitemene* system was no longer a binding constraint to population growth. Consequently, even though the short-run effect of introducing cassava was to reduce deforestation, the long-run effect may actually have been to increase it.

Growing of cassava reduced production risk, which also reduced deforestation in the short run. Another advantage was that cassava evened out farmers' seasonal labour requirements, since they could plant it at any time over an extended period, from the start of the rains in November till early

Table 14.3. Simulation of household cropping systems in high-population (deforested) areas with cassava.

	Net income constraint (kwacha)		
	0	300	400
Labour requirement in production (h year^{-1})	1570	2375	2697
Annual cropped area (ha per household)			
Total	0.9	1.38	1.52
Chitemene	0	0	0
Grass-mound system	0.69	1.21	1.35
Cassava garden	0.22	0.17	0.17
Carrying capacity (persons km^{-2})	57.4	28.6	24.6
Consumption activities (kg consumed year^{-1})			
Finger millet	128	100	100
Beans	95	103	103
Cassava	3183	3233	3233
Selling activities (kg sold year^{-1})			
Beans	0	15	9
Beer	0	240	341
Cassava, pounded	0	50	50

March, and harvest it throughout the year. Farmers preferred bitter types of cassava, which were less vulnerable to damage by wild pigs, which were initially a major threat to the production of the other varieties, particularly in the distant *chitemene* fields, which had to be fenced for that reason. Over time, hunting reduced the wild pig population and the problems they caused, increased average yields and allowed farmers to stop fencing in their parcels of land.

During the colonial period, the British resettlement programmes moved approximately 160,000 people from densely populated areas of Zambia to underpopulated areas in order to stop the degradation in the overpopulated areas. They also introduced new agricultural practices, such as anti-erosion measures, planting of fruit trees, improved seeds and early burning. Allan (1967) reports that, by the time he visited these areas 15 years after resettlement, the ecological balance had been restored in most of the region. By then, most villages had changed sites once or twice.

On the other hand, Allan also found no serious degradation in Serenje, an area that had also been declared overpopulated in 1945 but where no resettlement occurred. In that area, the local Lala population practised small-circle *chitemene*. Instead, when he visited that area 15 years later, he observed a spontaneous shift away from *chitemene* cultivation where the tree vegetation had been depleted and towards mound gardens with cassava, sorghum and maize.

In the grass-mound system, fallow periods became shorter as a consequence of population growth. Boyd (1959) reports that average fallow periods were around 18 years in the 1930s. Alder (1960) found practically the same average duration 30 years later, but noted that fallow lengths in more populated areas were much shorter. He describes the system's evolution as 'a change from an ingenious rotation of cereals, legumes and fallows to the monoculture of fingermillet with quite obviously decreasing periods of rest between cultivation sequences'.

Alder also noted that, due to the increasing labour shortages resulting from migration to towns, farmers had begun to leave fields fallow in the mound stage rather than the flattened stage, to save time. Previously, it had been customary to leave the field fallow in the flat stage after a millet crop, since gardens abandoned in the mounded stage take longer to regenerate.

Cassava became an important crop in the grass-mound system in the 1960s, as a result of land shortage, and became the main staple in the northeastern Mambwe area close to the Tanzanian border (Pottier, 1983, 1988).

Adoption of cassava was the most significant technical change in northern Zambia in the 20th century. It greatly increased agricultural productivity and land-carrying capacities. As a consequence, deforestation was reduced in the short run. However, cassava increased labour productivities considerably and made short rotation systems feasible alternatives to *chitemene*. The long-run effect of the introduction of cassava may therefore have been increased deforestation.

4.3. Market integration and technical change since the 1970s

Hybrid maize, variety SR 52, was introduced and spread in the early 1970s. It partly replaced finger millet, because it required less labour on short-fallow land, where weeding of millet demanded a lot of labour (Bury, 1983; Pottier, 1988). Shorter fallows had led to a greater prevalence of the weed *Eleusine indica*, which made the weeding of finger millet (*E. coracana*) much more difficult, due to its similar appearance during the early growth stage. Some farmers tried to reduce the weed problem by increasing the size of the mounds and by burying the weeds under more soil. Others switched to hybrid maize and cassava to reduce their labour requirements.

Fallow periods in the Mambwe area were particularly low and this depleted the region's soils (Pottier, 1983). In one village in this area, Watson (1958, cited in Sano, 1989) had reported fallow periods in the grass-mound system of 5–6 years in the late 1950s, but by 1988 Sano (1989) found that fallow periods had fallen to only 2–4 years.

As a result of declining soil fertility, mounding also became more common in older *chitemene* gardens. These mounds can be thought of as concentrated topsoil and their soils generally have a higher pH and more nutrients. Groundnuts became less common in grass-mound fields, due to empty pods on acid soils, while bean production for sale expanded. To find land for new *chitemene* plots, farmers had to go further away from their villages and the time required to walk back and forth from the fields consequently increased. Farmers began to grow maize on the nearby fields and cassava gardens in the intermediate zone. By then, it had become common to harvest one cassava garden each year. Cassava demanded less labour than millet and sorghum, performed well on poor soils and gave high yields (Sano, 1989).

Peasant agriculture in northern Zambia has changed dramatically since Pottier studied the area in the late 1970s. Improved infrastructure, subsidized inputs and subsidized transport (pan-territorial pricing) allowed maize production to spread to remote areas. This happened first in the Serenje, Mpika and Chinsali Districts in response to the arrival of an integrated rural development project (IRDP) in this area, but later spread to other areas as well. To produce maize, farmers depended critically on access to external inputs and cash or credit. The introduction of maize tended to increase social differentiation (Sano, 1989). Peasants welcomed the new crop mostly because they were interested in increasing their cash income, although in the southern parts of the region, near towns and in more wealthy households farmers also increasingly grew maize for household consumption.

Peasants continued using their traditional systems of cultivation alongside the maize system, and several household surveys found a positive correlation between maize area and area under *chitemene* or other systems (Sano, 1989; Holden *et al.*, 1994). This indicates that the systems complemented each other, but not that increased maize production caused an increase in *chitemene*. The households that could grow large areas of maize also had the means to put large areas under *chitemene*, as the labour peaks in maize production did not severely conflict with the labour requirements for *chitemene* production.

Maize was frequently found in grass-mound fields near the densely populated Kasama area in the late 1980s (Holden, 1988). In the less populated Chimbola area, maize replaced cassava on the nearby more permanent fields, while farmers continued to plant cassava in *chitemene* gardens. Price policies, subsidized credit, fertilizer and hybrid-seed distribution, output marketing by parastatal organizations and public extension programmes all greatly encouraged 'permanent' maize production. However, continuous monocropping of maize led to rapid yield declines, due to the acidifying effects of nitrogen fertilizers and increasing problems with aluminium toxicity and micronutrient deficiencies. Farmers had no access to lime, nor would it have been profitable to use if they had (Øygard, 1987). Thus, they had to abandon their fields after a few years and wait for a very long time for their fertility to be restored. Farmers' perceptions were that the soils had become addicted to fertilizer.

Table 14.4 illustrates the effect of introducing maize in low-population-density areas, with results of household simulation models for typical male- and female-headed households. In these models, it is assumed that maize is a pure cash crop. The models assume that the household's objective is to

Table 14.4. Effects of introduction of maize in low-population-density areas.

	Household type			
	Male-headed household		Female-headed household	
Produces maize and has access to input markets	No	Yes	No	Yes
Objective function (utility)	870	1769	16	81
Net income	2334	3417	351	473
Labour requirement (h year^{-1})	6387	6910	1597	1765
Annual cropped area (ha)				
Total	4.62	4.8	0.87	0.99
Chitemene	4.51	3.88	0.5	0.5
Cassava garden	0.11	0.15	0.01	0.01
Maize	0	0.77	0	0.05
Carrying capacity (persons km^{-2})	1.69	1.82	9.09	8.33
Input use (kg year^{-1})				
Fertilizer	0	294	0	20
Maize seeds	0	34	0	2
Consumption (kg year^{-1})				
Finger millet	1374	1181	60	60
Groundnuts	55	55	26	26
Beans	83	83	40	40
Cassava	969	1355	1883	1920
Selling activities (kg sold year^{-1})				
Groundnuts	650	574	16	16
Beans	458	375	31	51
Beer	333	333	245	198
Maize	0	2772	0	135

maximize a weighted combination of income and leisure. Females were particularly involved in the brewing and selling of beer made from maize or finger millet, which was considered to be a typical female activity. Such beers had a good market in the 1980s.

According to Table 14.4, the introduction of maize and fertilizer technologies affected *chitemene* production relatively little, but this was due in part to the unreliability of farmers' access to credit, input and output markets for maize. The parastatals were poorly managed. Thus farmers faced the risk of receiving their credit or fertilizers too late, or not at all, and of not having their output collected or paid for. This kept them from switching more from *chitemene* to maize production. These problems were less severe in areas close to marketing depots, which allowed those areas to concentrate more on maize production (Holden *et al.*, 1994). Thus, the introduction of maize reduced deforestation, at least in the short run, but less than one might have hoped. Because of the risks involved in the maize system, it could not replace the *chitemene* system.

Table 14.4 also shows that female-headed households, which were labour-poor, farmed much smaller areas and therefore deforested much less than male-headed households. Female-headed households were also less able to benefit from maize production, because of its high seasonal demand for labour. This implies that the spread of maize reduced deforestation more in male-headed households than in female-headed households.

Agricultural labour demand was highly seasonal. Labour supply was lowest when demand was highest, since there were no landless households in this relatively land-abundant area. In more densely populated areas, households with access to off-farm employment typically had much higher incomes. Female-headed households often engaged in business activities in these areas (Holden, 1988). Table 14.5 illustrates the effects of access to off-farm employment and business activities on typical male- and female-headed households in a densely populated area (Holden, 1991). Both lead households to become less involved in farming, because the other activities compete for family labour and hired labour is not a perfect substitute for family labour. These results are consistent with findings in other econometric studies (Holden, 1991; Holden et al., 1994). Hence, we can expect greater access to off-farm income to reduce deforestation.

Economic crisis caused the government to implement adjustment policies from the late 1980s. It removed subsidies on farm inputs, credit and transportation and took steps towards privatizing input and credit supply and the marketing of maize. This led to higher fertilizer and seed prices and interest rates and a contraction in the credit supply (Holden, 1997). Maize production fell drastically as a result. Many farmers increased or switched back to *chitemene* production. In some locations, the change also destabilized village structures, as many households moved away to settle in forested areas (Holden et al., 1994). Overall, SAP reduced market integration and increased deforestation in northern Zambia, as technological change and development went 'in reverse'.

5. What the Zambian Experience Tells Us

This section attempts to evaluate some of the general hypotheses developed in this book's initial chapters in the light of the Zambian experience. To make this discussion easier to follow, Table 14.6 pulls together some of the key findings from the discussion in the previous section. It shows how the main production systems evolved over time, the development of distinct types of markets, the changes in policies, the population densities and carrying capacities associated with each production system and their short- and long-term effects on deforestation.

Table 14.5. Simulation of effects of access to off-farm income in high-population areas.

	Household type			
	Male-headed household		Female-headed household	
Employment/business activity	No	Yes	No	Yes
Objective function (utility)	1466	4030	122	824
Net income	2995	5678	563	1265
Labour requirement (h year^{-1})	6849	6974	1890	1878
Annual cropped area (ha)				
Total	2.95	2.06	0.69	0.54
Grass-mound system	0.05	0	0.13	0.06
Cassava garden	0.83	0.84	0.45	0.23
Maize	1.06	1.22	0.1	0.25
Carrying capacity (persons km^{-2})	13.7	18.2	38.5	40
Input use (kg year^{-1})				
Fertilizer	425	469	39	110
Maize seeds	49	54	5	11
Consumption (kg year^{-1})				
Finger millet	125	125	60	60
Beans	83	83	38	38
Cassava	2507	2570	1855	1855
Maize	663	600	48	47
Selling activities (kg sold year^{-1})				
Beans	58	7	0	0
Beer	1000	275	0	0
Maize	1923	3213	205	646
Cassava	5000	5000	2226	221

5.1. The effect of the type of technology on deforestation: cassava

The cassava technology introduced in Zambia saved both labour and land and reduced risk. On balance, it lowered the pressure on forests both directly and indirectly (by facilitating out-migration of male labour), although in certain locations (around lakes and near towns) it increased the pressure by allowing the population to rise. Prior to the arrival of cassava, the *chitemene* system's limited carrying capacity had constrained population growth (Table 14.6). In these locations, over the long term, the population growth facilitated by the adoption of cassava production caused a shift from forest/bush fallow to grass fallow and increased deforestation and soil degradation.

Table 14.6. The evolution of the agricultural production systems in northern Zambia during the 20th century.

Period	Dominant production system	Market development and policy changes	Population density (exogenous)	Carrying capacity (persons km^{-2}) (endogenous)	Impact on deforestation Short run	Impact on deforestation Long run
Early 20th century	*Chitemene*	No markets	Low	2.5		
	Chitemene	Output markets	Low	2.2	More	Stable
1930–1980	*Chitemene* + cassava	Output markets	Low	3.5–8	Reduced	Increased
	Grass-mound + cassava	Output markets	High	25–29	Reduced	Increased
1980s	*Chitemene* + maize	Output, input and credit markets	Low	2–9	Reduced	?
	Grass-mound, cassava + maize	Output, input and credit markets	High	14–38	Reduced	?
1990s	*Chitemene* + maize	Removal of subsidies and pan-territorial pricing	Low	2–5	Increased	Increased
	Grass-mound, cassava + maize	Removal of subsidies and pan-territorial pricing	High	15–40	Increased	Increased

5.2. The effect of the type of technology on deforestation: maize and fertilizer

The more capital-intensive maize–fertilizer technology generally increased the aggregate demand for labour and discouraged extensive shifting cultivation. This reduced deforestation. Farmers did not adopt labour-intensive technologies, such as alley cropping and planted fallows, due to their limited potential on the infertile and acid ultisols and oxisols in the area (Holden, 1991, 1993b). Such technologies are more suitable in densely populated areas with more fertile soils.

5.3. Farm household characteristics: household labour supply and gender

Labour-poor households were less able to generate a surplus for sale. As a result, they were less integrated into markets, were more subsistence-orientated and cleared less forest than labour-rich households. Poor households that adopted cassava as their main staple food found the cassava technology particularly useful. The gender division of labour kept female-headed households from clearing much forest, because females were not supposed to climb trees to cut branches for *chitemene*.

5.4. Output-market characteristics: pan-territorial prices

Market characteristics, rather than household preferences, influence whether technological change leads to more or less deforestation. Pan-territorial prices improved farmers' access to markets. This favoured the adoption of maize production and reduced deforestation. When households can choose between intensive and extensive production systems and the crop associated with the intensive system becomes easier to market or receives a higher price, farmers shift resources towards the intensive system. This reduces the cultivation of the extensively cultivated crop and associated deforestation (Holden *et al.*, 1998b) (it is assumed that farm households sell both the extensively and intensively grown crops, prefer more leisure and face imperfect credit and labour markets). Removal of pan-territorial pricing, as the Zambian government did in the early 1990s (Table 14.6), implies privatization of transportation costs and this leads farmers in remote locations to revert to less capital-intensive and more land-extensive technologies and hence to clear more forest.

5.5. Imperfect labour markets and population growth

Low-population-density areas tend to have imperfect labour markets, due to high transaction costs, abundance of land, seasonal labour demand in agriculture and the difficulty of supervising hired agricultural labour. Family labour therefore dominates and this may constrain the expansion of production, even though families usually do not use all of their available labour, since the demand for labour is seasonal and they desire leisure. Labour market imperfections cause households with off-farm income opportunities to reduce their farming and clear less forest. Population growth increases labour supply and this increases deforestation. Emigration has the opposite effect.

5.6. Credit

Subsidized government credit stimulated capital-intensive maize production, which reduced deforestation. This could not be sustained once the government removed the subsidies in the 1990s (Table 14.6). Farm households facing cash and credit constraints find it harder to hire labour. Hence, their available household labour's access to forested land largely determines how much forest they clear.

5.7. Discount rates and property regimes

Empirical studies have shown that farm households' discount rates in northern Zambia are much higher than the physical rate of growth of *miombo* woodlands, so households largely ignore the long-term benefits of leaving land fallow. The main reason they leave land fallow is that, in the short term, it costs them more to use it than they get in return in the short run (Holden, 1991, 1998; Holden *et al.*, 1998a). This means that we can simulate household behaviour using static and/or time-recursive farm household models. These predict that households will practise the *chitemene* system in an unsustainable fashion until all the trees are finally removed.

The property regime appears to have little effect on deforestation in this context. Tenure insecurity is not a major problem and households have specific use rights to land, including fallow land (Sjaastad, 1998). High discount rates lead to low investment in intensification and the failure to take intertemporal externalities into account, not tenure insecurity.

6. Conclusions

The most significant technological change in northern Zambia during the 20th century was the introduction of cassava. It represented a labour- and land-saving technological change, which also made production less risky (the initial reason the British introduced it). It reduced short-run deforestation but at the same time facilitated population growth and concentration, which led to localized deforestation near lakes and towns. This resulted in a more complete removal of trees and tree roots than under the *chitemene* shifting cultivation system. Hence, the cassava technology paved the way for much higher population densities in the future.

Policy-makers and researchers focused their technological efforts on introducing capital-intensive maize production from the late 1970s. This system temporarily reduced deforestation, but much of this effect disappeared after the government eliminated the subsidies that stimulated maize production as part of its SAP (Holden *et al.*, 1998b).

Market imperfections continue to greatly influence farm households' decisions regarding forest clearing in northern Zambia. Credit and labour market imperfections shape household responses and SAP have strengthened these imperfections. Population growth is likely to be the main driving force behind deforestation in the future, unless new technologies and/or policies come into play.

References

Alder, J. (1960) *A Report on the Land-usage of the Aisa Mambwe in Reserve 4, Abercorn District.* Department of Agriculture, Abercorn.

Allan, W. (1967) *The African Husbandman.* Oliver & Boyd, Edinburgh.

Berry, R.A. and Cline, W.R. (1979) *Agrarian Structure and Productivity in Developing Countries.* International Labour Organization, Geneva.

Binswanger, H.P. and McIntire, J. (1987) Behavioural and material determinants of production relations in land abundant tropical agriculture. *Economic Development and Cultural Change* 22, 75–99.

Boserup, E. (1965) *The Conditions of Agricultural Growth: the Economics of Agrarian Change under Population Pressure.* George Allen & Unwin, London.

Boyd, W.J.R. (1959) *A Report in the Lunzuwa Agricultural Station: its History, its Development and its Work.* Publication No. 1, Misamfu Agricultural Station, Kasama.

Bury, B. (1983) The human ecology and political economy of agricultural production on the Ufipa plateau, Tanzania: 1945–81. DPhil thesis, Columbia University, New York.

Chayanov, A.V. (1966) *The Theory of Peasant Economy.* Edited by Thorner, D., Kerblay, B. and Smith, R.E.F. Irwin, Homewood, Illinois.

Feder, G. (1985) The relation between farm size and farm productivity: the role of family labour, supervision, and credit constraints. *Journal of Development Economics* 18, 297–313.

Heltberg, R. (1998) Rural market imperfections and the farm size–productivity relationship: evidence from Pakistan. *World Development* 26(10), 1807–1826.

Holden, S.T. (1988) *Farming Systems and Household Economy in New Chambeshi, Old Chambeshi and Yunge Villages near Kasama, Northern Province, Zambia. An Agroforestry Baseline Study.* Zambian SPRP Studies No. 9, NORAGRIC, Ås, Norway.

Holden, S.T. (1991) Peasants and sustainable development – the *chitemene* region of Zambia – theory, evidence and models. PhD thesis, Department of Economics and Social Sciences, Agricultural University of Norway, Ås.

Holden, S.T. (1993a) Peasant household modeling: farming systems evolution and sustainability in northern Zambia. *Agricultural Economics* 9, 241–267.

Holden, S.T. (1993b) The potential of agroforestry in the high rainfall areas of Zambia: a peasant programming model approach. *Agroforestry Systems* 24, 39–55.

Holden, S.T. (1997) Adjustment policies, peasant household resource allocation and deforestation in northern Zambia: an overview and some policy conclusions. *Forum for Development Studies* 1, 117–134.

Holden, S.T. (1998) *Shifting Cultivation and Deforestation: a Farm Household Model.* Discussion Paper No. D-08/1998, Department of Economics and Social Sciences, Agricultural University of Norway, Ås.

Holden, S.T., Hvoslef, H. and Sankhayan, P.L. (1994) *Impact of Structural Adjustment Programs on Peasant Households and their Environment in Northern Zambia.* Ecology and Development Paper No. 11, Ecology and Development Programme, Agricultural University of Norway, Ås.

Holden, S.T., Shiferaw, B. and Wik, M. (1998a) Poverty, market imperfections, and time preferences: of relevance for environmental policy? *Environment and Development Economics* 3, 105–130.

Holden, S.T., Taylor, J.E. and Hampton, S. (1998b) Structural adjustment and market imperfections: a stylized village economy-wide model with non-separable farm households. *Environment and Development Economics* 4, 69–87.

Hunt, D. (1979) Chayanov's model of peasant household resource allocation. *Journal of Peasant Studies* 6(3), 247–285.

Lal, R. (1987) Surface soil degradation and management strategies for sustained productivity in the tropics. In: Sanchez, P.A., Stoner, E.R. and Pushparajah, E. (eds) *Management of Acid Tropical Soils for Sustainable Agriculture. Proceedings of an IBSRAM Inaugural Workshop.* IBSRAM Proceedings No. 2, IBSRAM, Bangkok, Thailand.

Low, A. (1986) *Agricultural Development in Southern Africa. Farm Household-Economics and the Food Crisis.* James Currey, London.

NORAGRIC (1990) Environmental effects of agricultural change and development in the Northern Province, Zambia. Unpublished draft report, NORAGRIC, Ås, Norway.

Øygard, R. (1987) *Economic Aspects of Agricultural Liming in Zambia.* Zambian SPRP Studies No. 7, NORAGRIC, Ås.

Pottier, J. (1983) Defunct labour reserve? Mambwe villages in the post-migration economy. *Africa* 53(2), 2–23.

Pottier, J. (1988) *Migrants No More: Settlement and Survival in Mambwe Villages, Zambia.* International African Library, Manchester University Press, Manchester.

Ruthenberg, H. (1980) *Farming Systems in the Tropics,* 3rd edn. Clarendon Press, Oxford.

Sanches, P.A. and Salinas, J.G. (1981) Low-input technology for managing oxisols and ultisols in tropical America. *Advances in Agronomy* 34, 279–406.

Sano, H.O. (1989) *From Labour Reserve to Maize Reserve: the Maize Boom in the Northern Province in Zambia.* CDR Working Paper 89.3, Centre for Development Research, Copenhagen.

Sjaastad, E. (1998) Land tenure and land use in Zambia: cases from the Northern and Southern Provinces. PhD thesis, Department of Forestry, Agricultural University of Norway, Ås.

Stiglitz, J.E. and Weiss, A. (1981) Credit rationing in markets with imperfect information. *American Economic Review* 71(3), 393–410.

Watson, W. (1958) *Tribal Cohesion in a Money Economy: a Study of the Mambwe People of Zambia*. Manchester University Press, Manchester.

Young, A. and Wright, A.C.S. (1980) Rest period requirements of tropical and subtropical soils under annual crops. In: *Report on the Second FAO/UNFPA Expert Consultation on Land Resources for Populations of the Future*. FAO, Rome.

Livestock Disease Control and the Changing Landscapes of South-west Ethiopia

Robin S. Reid, Philip K. Thornton and Russell L. Kruska

1. Introduction[1]

New technologies for controlling trypanosomosis may strongly affect agricultural expansion in Africa. The bloodsucking tsetse-fly (*Glossina* spp.) transmits trypanosomosis, which causes morbidity and mortality in people and livestock (Jordan, 1986). The human disease is restricted to small foci (de Raadt and Seed, 1977), while the livestock disease is spread over 10 million km² of Africa (Jahnke *et al.*, 1988). Control of livestock trypanosomosis will reduce mortality, increase productivity per animal and cause livestock populations to grow. Farmers with more and healthier oxen will be able to plough more land more effectively, possibly allowing them to cultivate larger areas and work the land more intensively. Greater consumption of livestock products may improve human nutrition (Huss-Ashmore, 1992; Nicholson *et al.*, 1999) and human health. Areas freed from the disease may attract migrants. More people and livestock will require greater quantities of fuel wood, wild foods, forage and water. Thus, the disease not only affects livestock populations directly, but can also indirectly affect human populations, the extent of cropland and the effectiveness of cropped agriculture.

This chapter examines how controlling livestock trypanosomosis affects the rate and location of agricultural expansion. We look at agricultural expansion in general, rather than solely deforestation, because lightly wooded savannah, woodland and wooded grassland cover much of Africa. Furthermore, trypanosomosis is a problem that crosses ecological zones (semi-arid to humid) and vegetation types (open grassland to humid forests); several species can even survive in peri-urban settings (Okoth, 1982). Thus, control of trypanosomosis may affect agricultural expansion in a wide variety of African landscapes.

The Ethiopian data presented in this chapter and evidence from other African studies show that controlling trypanosomosis can encourage agricultural expansion. In some places, the disease is only one of several factors that affect expansion; in others, its effect dominates. Six broad factors determine whether trypanosomosis control is likely to encourage agricultural expansion: the agroecological conditions and strength of the disease constraint, land availability, the type of technology and its likelihood of adoption, accessibility and functioning of markets, farmer characteristics and culture.

We first review the role of trypanosomosis in shaping African landscapes in section 2. Section 3 examines what we know about how the use of trypanosomosis control technologies affects agricultural expansion. In section 4, we describe the Ghibe Valley study area in Ethiopia and the control technology. Section 5 presents the results of a land-use model we developed to simulate the effects of trypanosomosis control on agricultural expansion at the landscape scale. The last section discusses how different factors condition the impact of disease control and the policy implications of our work.

2. How Trypanosomosis Shapes African Landscapes

Most of the 23 species of tsetse-fly live in areas with more than 500 mm of rainfall that lie below 1800 m, although there are a few exceptions (Jordan, 1986). The area infested by the flies expands during the wet season and in wetter years and contracts when there is less rainfall. Infested area is also influenced by removing the flies' favoured habitats (woody thickets), eliminating wild hosts and killing flies directly.

The constant shifts in the tsetse-fly's distribution have affected the spatial patterns of cultivation and cattle raising in Africa for decades, perhaps centuries. In precolonial times, farmers controlled *Glossina* by burning bush late in the growing season, living in dense settlements that had few or no flies and leaving the rest of the land infested (Ford, 1971; Giblin, 1990). Despite these efforts, tsetse often reinfested settlements during wet periods, sometimes forcing farmers to move quickly to areas with no flies (Ethiopia: Turton, 1988; Nigeria: Kalu, 1991; Tanzania: Ford, 1971; Giblin, 1990). Even today pastoralists alter their grazing patterns to avoid tsetse (Jordan, 1986; Dransfield *et al.*, 1991; Roderick *et al.*, 1997).

The strategies used to avoid tsetse have kept human and livestock populations low in tsetse-infested areas. Bourn (1978) and Jahnke *et al.* (1988) found that tsetse-infested areas had fewer cattle and tropical livestock units (TLUs) than tsetse-free areas in the same rainfall zones across several African countries. Almost all of Tanzania's cattle live in places with no tsetse. Farmers and pastoralists avoid the large tsetse belts that cover the country's central areas (Jordan, 1986). Zambian farmers cultivate 16% of the land in tsetse-free areas, but only 9% in infested areas (Reid *et al.*, 1996). In Nigeria, Rogers *et al.* (1996) show that areas infested with the tsetse species *Glossina morsitans* have lower cattle population densities and less agricultural expansion than areas with no flies. At the landscape scale, farmers avoid the more heavily infested low-lying areas and riverine forests in the Ghibe Valley, Ethiopia (Reid *et al.*, 2000).

3. How Controlling Trypanosomosis Affects Agricultural Expansion[2]

Many authors have speculated about the impacts of controlling trypanosomosis. Ormerod suggests that tsetse control will exacerbate overgrazing in the Sahel, which will accelerate desertification and climatic change (Ormerod, 1978, 1986, 1990). In contrast, Jordan (1979, 1986, 1992) argues that human population pressure influences the growth of agricultural land much more than trypanosomosis.

Recent studies have used remote sensing (aerial photographs and satellite images) or oral history/farmer recall to measure or describe how controlling tsetse and trypanosomosis influences the extent of cropping and grazing. Remote-sensing studies measure land-use changes at broad scales (tens of kilometres), while oral history/recall studies describe household-level changes and give more insights into farmers' decisions. The best studies integrate both approaches, although this is rare.

3.1. Remote-sensing studies

Bourn (1983) conducted the first remote-sensing study in Lafia District, Nigeria. He compared human populations, cattle populations and land use in areas with and without tsetse over time. He found that the amount of land farmers cultivated did not significantly change in either area. Surprisingly, the cattle population increased more in the area with tsetse than in that without. Bourn concluded that human population growth and other socio-economic factors probably influenced land use more than tsetse control.

Several other studies also found that tsetse/trypanosomosis control was not the most important factor explaining land use. Human population growth appeared more important than disease control in northern and central Côte d'Ivoire (Erdelen *et al.*, 1994; Nagel, 1994). Similarly, Bourn and Wilson (1997) and Oloo (1997) attributed the strong changes in land cover on Galana Ranch and Nguruman, Kenya, between the 1950s and 1980s to factors other than trypanosomosis. Mills (1995) and Pender *et al.* (1997) concluded that land-use change bore little relation to tsetse control operations in the mid-Zambezi Valley. Farmland expanded rapidly in Kanyati communal land in Zimbabwe between 1984 and 1993, despite the presence of small tsetse populations (Wangui *et al.*, 1997). In Kenya, agriculture and human populations also expanded over the last few decades in Busia District (Rutto, 1997) and the Lambwe Valley, even though those areas were still infested (Muriuki, 1997).

Only one remote-sensing study distinguished the effect of tsetse control from other factors driving land-use change. It used an integrated cross-sectional and longitudinal study design to analyse several areas with and without tsetse control in the Ghibe Valley of Ethiopia to control for different factors. This study showed that changes in trypanosomosis severity could lead to rapid changes in agricultural expansion (Reid et al., 2000). Trypanosomosis in the Ghibe Valley became much more severe in the early 1980s, causing farmland to contract by 30% over 5 years. Other effects of trypanosomosis included a decline in human populations (from out-migration) and livestock populations and a decreased ability by farmers to plough the land.

3.2. Oral history/recall studies

The best studies of this type come from Ethiopia (Kriesel and Lemma, 1989; Slingenburgh, 1992; Reid et al., 1997) and Burkina Faso (Kamuanga et al., 1998). A group of 63 households interviewed in the Didessa Valley in Ethiopia reported that a recent increase in disease severity caused sharp declines in cattle populations, cultivated field areas and milk production (Kriesel and Lemma, 1989; Slingenburgh, 1992). Farmers in the nearby Ghibe Valley mentioned similar effects (Reid et al., 1997).

On average, Ghibe Valley households living in an area with a low incidence of trypanosomosis were able to plough 50% more land than households living in an area of high incidence. Each ox in the tsetse-free area could apparently plough twice as much land as an ox in tsetse-infested areas. This supports the hypothesis that healthy oxen plough more land than unhealthy oxen. However, since only one area was studied in each case, the researchers were not able to control for other factors influencing farmers' land-use decisions (Swallow et al., 1998). Even though the pour-on treatment benefits nearby untreated cattle by reducing tsetse populations, farmers with treated oxen ploughed 22% more land than nearby households with untreated oxen. In south-western Burkina Faso, farmers used oxen for ploughing more frequently after tsetse/trypanosomosis control than before control (Kamuanga et al., 1998). However, in Somalia, tsetse eradication apparently had no effect on farmers' use of draught power (Hanks and Hogg, 1992).

The studies also highlight the effects of trypanosomosis control on livestock populations. In the Ghibe Valley, farmer households in the areas with low trypanosomosis incidence hold more adult cattle and goats than those where incidence is higher. In The Gambia, farmers also have more cattle in areas with a moderate risk of trypanosomosis than in areas with a high risk (Mugalla et al., 1997). In Burkina Faso, the mean number of cattle owned increased 97% over 10 years after tsetse control began in one region, and only

38% in a nearby region still infested with tsetse over the same period (Kamuanga *et al.*, 1998).

Swallow *et al.* (1998) hypothesize that tsetse control will increase farmers' future expected earnings by reducing livestock mortality and improving reproductive capacity and that, in turn, will encourage in-migration. Their data from the Ghibe weakly support this hypothesis for the period 1993–1996. The evidence from other countries is mixed: tsetse control seemed to attract migrants in north-western Zimbabwe (Govereh and Swallow, 1998) but not in south-western Burkina Faso (CIRDES/ITC/ILRI, 1997).

Most authors have found it hard to separate the effect of trypanosomosis control from other factors that drive agricultural expansion. This distinction is essential in order to understand how the technology affects land use. Our research team developed the method described above, which allowed us to separate the different effects. We have also tracked the different forces driving land-use change over half a century in the Ghibe Valley (Reid *et al.*, 2000). Drought and migration, changes in settlement and land-tenure policy and changes in the severity of trypanosomosis all contributed to rapid changes in cultivated area. During a 13-year period of increasing trypanosomosis severity, the disease caused a 25% decrease in cropped area in the Gullele study site. The scale of the causes of agricultural expansion varied from local to regional to international. At the landscape scale, each cause affected the location and pattern of cropping differently. After control of the fly in 1991, cropping began to expand again.

In summary, these studies demonstrate that control of trypanosomosis can cause agricultural expansion. In some places, trypanosomosis appears to be only one of several factors affecting expansion, while, in others, it dominates. Unfortunately, problems with study design make these conclusions tentative.

4. Technological Change in the Ghibe Valley, Ethiopia

The rest of this chapter focuses on the effect of introducing trypanosomosis control in the Ghibe Valley, Ethiopia. The Ghibe Valley is located 180 km south-west of the Ethiopian capital, Addis Ababa, where the road to Jimma first crosses the Ghibe River (Fig. 15.1). Average rainfall from 1986 to 1993 was 838 mm year^{-1}, with a pronounced dry season from December to March (Reid *et al.*, 1997). Unlike much of Ethiopia, precipitation fluctuates little from year to year (EMA, 1988).

The Ghibe River cuts across the study area from north to south. The landscape is dominated by the heavily forested Boter Becho mountains (2300 m) to the west, rising out of a 1600 m plateau, which is deeply incised by the river in places, forming rocky canyons, which are uninhabited and unsuitable for crops. Wooded grasslands cover the rest of the landscape, with thick riverine forests along watercourses (Reid *et al.*, 1997).

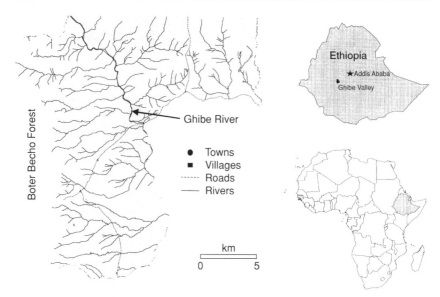

Fig. 15.1. The Gullele study site, showing the Boter Becho forest, roads, rivers, towns and villages in 1993.

Compared with the nearby highlands, the study area supports relatively little agriculture. A recent livestock census in the study area (B.M. Swallow, unpublished data) showed that the Ghibe's cattle population densities were about 80% lower than those of the nearby highlands (9.3 km^{-2} vs. 52 km^{-2}). In 1993, farmers used about a quarter of the arable land in the Ghibe Valley for cropped agriculture, with 97% of the valley farmed by smallholders and about 2–3% by large landholders (Reid *et al.*, 2000). The few roads in the area are generally of poor quality.

Most Ghibe smallholders plough with either hand-hoes or ox-drawn ploughs, although a few use tractors. Smallholders cultivate teff, maize, sorghum, noug or niger-seed, false banana, hot peppers, wheat and groundnuts. Large farmers use tractors to cultivate a less diverse range of crops, principally cereals and legumes. On smallholder farms, total cultivated field area (= farm size) varies from 0.05 to 9 ha. Large farmers claim to cultivate up to 480 ha (Reid *et al.*, 1997). The only government policies that have significantly touched the lives of the valley's farmers have been some major changes in land tenure and settlement policies (Reid *et al.*, 2000). Peasant associations allocate land and newcomers must receive approval by the association leadership before they can have access to land.

Three tsetse-fly species infest the Ghibe Valley, *G. morsitans submorsitans* Newstead, *Glossina pallidipes* Austen and *Glossina fuscipes fuscipes* Newstead. The first tsetse control trial began in 1990 in a 125 km^2 portion of the study area called Ghibe-Abelti, where cloth 'targets' were sprayed with insecticide and then strung on eucalyptus poles (Mulatu *et al.*, 1993). In January 1991, the International Livestock Research Institute (ILRI) scientists began controlling the fly in the Gullele area of the Ghibe Valley, using a cypermethrin 'high-*cis*' compound (Leak *et al.*, 1995). Researchers originally developed this insecticide for tick control, but it was modified as a pour-on for tsetse control for this trial. The pour-on is applied to the backs of the cattle at monthly intervals at seven crushes spread from the town of Toley in the south-west to the Gullele area in the north-east. About 1 ml of the pour-on is applied for each 10 kg of body weight. When flies land on the cattle, they contact the insecticide while they bite and then die. The insecticide may also kill the flies before they bite, but this is not certain (Swallow *et al.*, 1995).

In general, tsetse control is a labour-saving, capital-intensive, embodied technology, which may increase output per hectare in the Ghibe Valley. More importantly, tsetse control allows farmers to plough larger tracts of land. Increased access to healthy oxen substitutes for human labour and reduces the amount of cultivation, weeding and ploughing by hand. The tsetse control itself is relatively expensive, and hence capital-intensive. As livestock become healthier, livestock production increases. So do crop yields, since greater draught power allows farmers to prepare the land more effectively.

Scientists began using biconical traps to monitor tsetse populations and to track cattle productivity well before the control began (1986) and continued throughout the trial (Leak *et al.*, 1995). From 1991 to 1992/93, populations of *G. pallidipes* and *G. m. submorsitans* decreased 72% and 93%, respectively (Leak *et al.*, 1995). Fly populations have remained low since then in areas where cattle graze and thus the pour-on affects fly populations. Throughout the experiment, researchers observed low populations of *G. fuscipes*, although this species is not considered an economically important vector of trypanosomosis (Leak *et al.*, 1993).

Tsetse control with pour-on has both private and public benefits to farmers (Swallow *et al.*, 1995). Private benefits include tick control and reduced tsetse bites, if the flies die before they bite. Other benefits accrue to all livestock owners in the area because the pour-on reduces fly populations wherever there are treated cattle. Thus, one can also consider tsetse control a public good. The size of the area that benefits from tsetse control depends on the fly species involved, the number of cattle treated in an area and the fly population density.

Treatments were given free of charge from January 1991 until December 1992, when a cost recovery programme was initiated (Swallow *et al.*, 1995). Even very poor farmers participated in the programme, despite the fact that costs were relatively high (Swallow *et al.*, 1998).

Farmers claim that trypanosomosis attacked their cattle for the first time in the early 1980s, a decade before tsetse control began (Reid *et al.*, 2000). Analysis of aerial photography shows that the cultivated area decreased by at least 50% in many areas after trypanosomosis appeared. Policy changes also affected this decrease, but the increase in disease incidence caused about half of the change (a 25% decrease in cultivated area). Thus, the disease not only reduced the need to clear new land, it also reduced the area of land under agriculture.

Although tsetse control began in 1991, the effects of the control did not become apparent until 3–4 years later. When we first used satellite imagery to quantify how much land farmers cultivated in 1993, we could not see the effects of control on land use (Reid et al., 1997, 2000). But, by 1997, the expansion of cultivation following tsetse control was clearly visible on the LANDSAT-TM images.

Most of the additional cropland replaced wooded grassland, the dominant land cover. In previous work, we have shown that converting wooded grassland to cropland has little impact on either tree cover or the number of tree species (Reid et al., 1997). When farmers clear wooded grasslands to cultivate, they often do not have to reduce tree cover, because it is already sparse. Over time, farmers actually increase tree cover slightly by planting small woodlots and hedgerows around their fields and houses.

However, analysis of the 1997 imagery shows that farmers are cultivating much closer to the riparian forests than they did in 1993. When cultivation expands into these biologically rich riparian forests, we expect the impacts to be much greater than in the wooded grasslands. Although these forests cover only 4–8% of the Ghibe landscape, 20–30% of the species live in these forests and nowhere else (Reid et al., 1996, 1997; Wilson et al., 1997). Thus, removal of the forest for cultivation is likely to have strong ecological effects.

5. Predicting Agricultural Expansion: a Simple Modelling Approach

Researchers have used various methods to model land use (Lambin, 1994; Kaimowitz and Angelsen, 1998). In our case, we applied a non-stationary Markov transformation matrix, governed by a few simple decision rules, to model land-use changes in the Ghibe over time, an approach that grew out of Thornton and Jones (1997a, b). We wanted a simple model focusing on only a few processes that could still be useful for statistical analyses of regional land-use changes. If we managed to validate a simple model like this, it could provide a lot of valuable information. On the other hand, if the model did not work or failed to provide useful insights, we could abandon it with little loss of time and effort.

The model uses farmers' preferences for different land-use alternatives and the expected economic returns from each alternative to predict land use at any given time. Net returns depend on production (itself a function of land quality and previous land use), input costs and location. Locations that have poorer access to roads and are farther from markets have higher costs. We used a first-order Markov model to take into account the effects of soil fertility changes on productivity over time.

We applied our approach to modelling land-use change in the Ghibe in three stages. We first attempted to model land use in the area in 1973. Then we combined the results from this first exercise with several changes in the model parameters, which reflect important shifts in government policy, to predict land use in 1993. Finally, we simulated land-use change to 1997, after the successful control of trypanosomosis.

5.1. The study area and decision rules

The model encompasses the Gullele/Toley area, where tsetse control was applied in 1991. The study area's western boundary is the 1800 m elevation line. This more or less coincides with the edge of the Boter Becho forest, where settlement is prohibited by government policy, as well as the cut-off point for the presence of tsetse-flies. We assumed that the reach of the tsetse control programme determined the area's northern and north-eastern boundaries, and the edge of the river gorge defines its southern and south-eastern limits. Farmers claim that areas within the river gorge are unsuited for agriculture and our field observations support this. The northern edge of a military camp constitutes the study area's south-western boundary.

We developed the rules used in the model from group and key informant interviews conducted throughout the study area (Reid *et al.*, 2000). We asked long-term residents in 11 villages why they lived where they did, rather than in other, uninhabited, parts of the landscape. The most important factor cited in settlement and land-use decisions was the desire to live at an elevation above the area infested by tsetse-flies and malaria-carrying mosquitoes. Farmers admitted that the lower areas were more fertile, but they were more worried about disease in the lowlands than lower soil fertility in the uplands. After 1986, the Ethiopian government instituted a programme that relocated farmers together in a small number of villages. This was the second most important factor influencing settlement location. At the same time, government officials moved farmers farther from the Boter Becho forest. Finally, farmers said they preferred to cultivate on slightly sloping ground, particularly before they had access to healthy oxen, because the soil was easier to work by hand. The model's decision rules reflect these preferences and constraints.

5.2. Stage 1: modelling land up to 1973

We hypothesized that we could explain the spatial distribution of land use in Gullele in terms of the physical nature of the landscape. We amalgamated all our available land-use data into three categories: upland woodland and grassland (WG), riparian forest (RF) and smallholder cultivation (SC) (Reid *et al.*, 1997).

Starting with a map of the area under upland woodland and grassland at some moment long before 1973, we applied two rules. We gave pixels within 100 m of a major river or within 50 m of a minor river a high probability of being transformed into riparian forest. Secondly, the probability that a pixel is converted to smallholder cultivation from upland woodland–grassland is related to its proximity to a road or to a town and to the elevation of the pixel. Thornton et al. (unpublished) gives full details of the rules model. We adjusted these rules until the simulated proportion of pixels in each land-use category was similar to the observed proportions. Figure 15.2, which compares the observed and simulated land uses for Gullele in 1973, presents the results. Table 15.1 summarizes the percentage land use in each category and shows that the percentage of all pixels classified correctly was 69%.

5.3. Stage 2: modelling land use from 1973 to 1993

Many changes occurred in Ethiopia between 1973 and 1993 (Reid et al., 2000). The policy decision, put into practice in 1986/87, to assemble smallholders into villages of the government's choosing drove a large proportion of the change in land use during this period. One effect of this policy was to keep

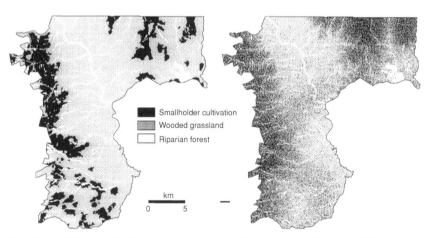

Fig. 15.2. Observed (left) and simulated (right) land use in Gullele, 1973.

Table 15.1. Observed and simulated land use in Gullele, 1973/1993/1997.

	Woodland–grassland	Riparian forest	Smallholder cultivation
Observed (%)	72/76/74	8/4/4	20/20/22
Simulated (%)	72/73/74	8/6/5	20/21/22
Percentage of pixel correctly classified	79/80/76	49/39/35	38/31/30

Total pixels correctly classified (%) = 69/67/64.

smallholder cultivation out of the forests and forest margins. In some cases, farmers described whole villages being removed from the forest and forest margins and relocated elsewhere. For further iterations of the model after 1973, we assume that farmers did not convert riparian forest to smallholder cultivation, since farmers told us that government policy during this period prohibited cultivation within 50–100 m of the river at this time. When farmers were forced out of the forest margins, some areas were transformed from smallholder cultivation to wooded grassland. To model this, we assumed that within 1 km of the forest, a pixel had no chance of being in smallholder cultivation, but the probability of smallholder cultivation increases linearly to a distance of 4 km from the forest boundary. If the pixel is close to a river, then we assume it reverts to riparian forest. We assumed that some land close to roads, towns and villages was converted from riparian forest to woodlands and grasslands. Similarly, we modelled the conversion from woodlands and grasslands to smallholder cultivation as a function of elevation and proximity to roads and towns.

We then used the 1973 simulated land-use coverage as an input to the stage 2 model and applied transformation probabilities, based on the decision rules, to simulate land use in 1993. This is shown in Fig. 15.3, together with actual 1993 land use. Table 15.1 shows the observed and simulated percentage of each land-use category and the classification success percentage.

5.4. Stage 3: modelling land use from 1993 to 1997

For stage 3, we modified some of the relations to take account of the changes that occurred in Gullele in the 1990s. We allowed smallholders to cultivate plots further away from the villages and nearer the forest boundary. We also relaxed the elevation rule, since the introduction of tsetse control allowed

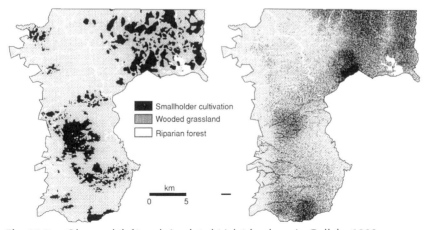

Fig. 15.3. Observed (left) and simulated (right) land use in Gullele, 1993.

smallholders to cultivate at lower elevations. Otherwise, we used the same transformations in the model as for stage 2. Figure 15.4 shows the results, comparing observed with simulated land use in 1997. Table 15.1 gives the observed and simulated percentage land use by category.

5.5. Results and implications of the modelling

The model outlined above essentially has five rules. As expected, Table 15.1 shows that the percentage of correctly classified pixels declines with each iteration of the model as we move further away from initialization. However, the fact that we still got a 65–70% success rate for our 1993 predictions implies that our simulations fit the general patterns observed in that year reasonably well. The forest boundary rule clearly played a large part in changing land use between 1973 and 1993. Farmers cultivated the area east of the Ghibe River much more heavily in 1993 than in 1973 and ceased cultivating the forest margin areas to the west, except for small pockets.

The resettlement policy also tended to clump cultivated areas together. Signs of these effects are seen in the simulated landscape for 1993. By 1997, with the relaxation of the resettlement policy (effective about 1992/93), farmers began to disperse back to the areas they lived in before 1986. Thus, some households moved back toward the forest. In general, large patches of crops around villages dispersed into smaller patches. Indications of this are apparent in the simulation for 1997, where smallholders are dispersing towards the lower elevations surrounding the main rivers.

In the absence of the strong elevation rule, proximity to roads, towns and villages becomes more important in determining the location of cultivated plots in the 1997 simulations. In view of the relative sizes of the weights apportioned, it is clear that, prior to 1997, elevation was the major physical

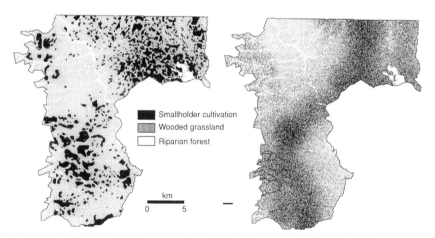

Fig. 15.4. Observed (left) and simulated (right) land use in Gullele, 1997.

determinant of where farmers located their cultivated areas. Because of the risk to animals and humans of trypanosomosis and malaria from the lower elevations and riparian forests, where fly densities are highest, up to 1993 smallholders preferred to live at the higher elevations in the landscape. By 1997, one can see the effect of the successful tsetse control in 1991. People began to move toward lower-elevation areas and to clear land for cultivation near the rivers.

6. Lessons Learned and Policy Implications

One major way in which trypanosomosis control can affect agricultural expansion is by encouraging use or first-time adoption of animal traction. African farmers using animal traction cultivate about twice as much land as farmers who use hand-hoes (Pingali et al., 1987). Nevertheless, across Africa, farmers have been slow to abandon the hand-hoe and to plough with oxen. The adoption of animal traction occurs principally in regions with good market access, little forest and high population densities. Markets provide profitable outlets for farm produce, allowing farmers to invest in animal traction. Farmers in forested areas adopt animal traction more slowly, because they have to destump to use a plough. High-population areas have higher agricultural intensity, so farmers have an incentive to invest in animal traction. In addition, farmers in areas with heavy soils or where land preparation takes a lot of time adopt animal traction earlier than farmers in areas with sandy soils or where land preparation is short.

The areas endemic with trypanosomosis often lack the conditions that encourage farmers to adopt animal traction. Regions with severe trypanosomosis problems tend to have low human populations, in part because limited human use leaves tsetse habitats intact (Jordan, 1986). Trypanosomosis occurs most in regions with more than 800 mm of rain per year, which can generally support dense woodlands or forests. Market access in low-human-population areas is often poor.

As mentioned earlier, we believe that six broad factors, which are largely linked to animal traction, increase the likelihood of agricultural expansion after trypanosomosis control: (i) agroecological conditions and the strength of the disease constraint; (ii) land availability; (iii) the type of technology and its likelihood of adoption (labour required, public/private-good benefits); (iv) the accessibility and functioning of markets (for labour, inputs, outputs and credit); (v) farmer characteristics (production goals); and (vi) culture.

6.1. Agroecological conditions and the strength of the disease constraint

In tsetse-infested areas of Zimbabwe and Burkina Faso, soils vary from sands to light clays, while, in south-western Ethiopia, many of the arable soils are heavy vertisols. The need for animal traction to substitute for human labour varies strongly across these sites. Similarly to Pingali *et al.* (1987), we hypothesize that disease control will increase the use of animal traction and the rate of agricultural expansion more in areas with heavy soils, which require more labour for land preparation.

If trypanosomosis is the most important constraint on agriculture, its control will probably lead to strong agricultural expansion. In south-western Ethiopia, the single most important constraint on cropping is the availability of oxen for ploughing (Omo–Ghibe River Project, unpublished data). In contrast, in south-western Burkina Faso, only 4% of migrants said that they moved to the area because of lower disease incidence. Almost half of the migrants came to the area in search of pastures and water (Kamuanga *et al.*, 1998). In this area, farmers cultivated about 30% of the land before tsetse control even began (R.S. Reid, personal observation).

6.2. Land availability

Land availability will influence how strongly disease control will affect agricultural expansion. In the studies we review here, disease control is occurring in areas with plenty of open land and thus control boosts agricultural expansion. In areas where arable land is more limited, we expect that disease control will be less likely to encourage further expansion of agriculture on to marginal lands (although increased grazing will probably occur on marginal lands). In land-scarce systems, farmers will receive the benefit of healthier animals, more livestock products and more efficient traction, but will be less likely to expand their cropping into uncropped areas. This supports the hypothesis that technological progress in land-extensive systems boosts agricultural expansion, while in land-intensive systems the same progress may not lead to significant agricultural expansion.

6.3. Type of technology and benefits

The amount of labour required to apply trypanosomosis control technologies and the type of benefits these technologies deliver to farmers varies. Pour-ons require little labour. They only need to be applied once every 1–2 months. They also provide both public and private benefits for farmers, which makes them particularly attractive. Another technology, therapeutic drugs, demands little labour and provides private benefits but is often only marginally effective, because of drug resistance. Targets and traps are a more labour-intensive way to control tsetse-flies than pour-ons and deliver principally public benefits. Government aerial and ground spraying programmes are often labour-free to farmers, provide public and private benefits, and are reasonably effective (but costly). The impact of control on agricultural expansion will differ from region to region, because different technologies will be available to farmers in each place.

Currently, the most successful control programmes use either pour-ons or a combination of pour-ons and other techniques. Such technology is most likely to control the disease and requires little labour and farmers derive most of the benefits, so farmers have the incentive to pay for treatment of their cattle. Use of this technology is thus most likely to lead to an expansion of agriculture, all other things being equal.

6.4. Accessibility and functioning of markets

As a labour-saving technology, which allows farmers to substitute animal traction for hand-hoes, trypanosomosis control can encourage agricultural expansion by relieving labour constraints. These constraints can be further relieved if there is a well-functioning labour market. In the Ghibe Valley, there are many opportunities to hire labour and off-farm economic opportunities are low. Farmers can quickly respond to increased availability of animal traction and plough more land. In addition, there is a well-developed exchange system for the use of animal traction, even for resource-poor farmers (Swallow *et al.*, 1995). Thus, many of the farmers in this area are particularly able to expand cropping in response to disease control. In other areas, where this is not the case, we expect farmers to expand the area that they cultivate in response to disease control more slowly.

In many areas with trypanosomosis control, in-migration is feasible and common. In Burkina Faso, Ethiopia and Zimbabwe, tsetse-infested areas are near high-population centres where land is scarce. Recent drought and resource degradation in these high-population regions have driven farmers to seek new land for agriculture in areas recently freed of the fly (Reid *et al.*, 2000). Much of the new land cultivated in the tsetse-free areas first comes under the plough of a recent migrant.

Access to input, output and credit markets gives farmers further incentives to expand the area of land that they cultivate. On the other hand, in many areas infested with the tsetse-fly, input use is low, credit availability is limited and markets for inputs and outputs are undeveloped. This militates against further expansion of agriculture in such areas.

6.5. Farmer characteristics

If farmers are poor and subsistence-orientated, they are less able to use new technologies and take advantage of market opportunities (and thus less likely to expand agriculture as a result of technology introduction). Indeed, most farmers in tsetse-infested areas are poor and subsistence-orientated compared with farmers in more highly developed areas with fewer constraints on agricultural production. This probably reduces the impact of disease control on agricultural expansion. However, as described above, current technologies to control tsetse often have public benefits, so use by wealthy farmers will allow poor farmers to benefit from the control at no cost.

6.6. Cultural practices

Lastly, cultural practices determine the extent to which people will use livestock to expand their cropping enterprises. In south-western Ethiopia, agropastoralists that we interviewed are more interested in increasing their herd sizes than in extending their cropping (R.S. Reid, unpublished data). In the same valley, the agriculturalists, who have a strong ox-plough culture (McCann, 1995), quickly expand cropping when they have access to additional oxen. In Zimbabwe, many of the people in the area with tsetse control were moved from the Lake Kariba lakeside in the 1950s. They are traditionally fisher people and thus may be slow to adopt mechanization.

A suite of factors can either strengthen or weaken the link between trypanosomosis control and agricultural expansion. These factors operate to different degrees in different locations across Africa. This explains why studies on the impacts of trypanosomosis control have apparently contradictory findings. Sometimes the link to agricultural expansion is strong, sometimes weak. Our case is a prime example where many of the 'right' conditions for disease control to drive agricultural expansion were in place. Of the studies available, it is also an exception.

Notes

1   We thank our colleagues, Woudyalew Mulatu, Brent Swallow, Joan Kagwanja and Stephen Leak, for stimulating discussions during the course of this work, and the editors and an anonymous referee for useful comments. This chapter is Land-Use Change Impacts and Dynamics (LUCID) Paper No. 4. A grant from the International Fund for Agricultural Development and funds from ILRI's 56 other donors supported this work.
2   The following is taken from Reid (1999).

References

Bourn, D. (1978) Cattle, rainfall and tsetse in Africa. *Journal of Arid Environments* 1, 49–61.
Bourn, D. (1983) Tsetse control, agricultural expansion and environmental change in Nigeria. PhD thesis, University of Oxford, UK.

Bourn, D. and Wilson, C.J. (1997) Galana Ranch, Coastal Province. In: Bourn, D.M. (ed.) *Draft Case Studies of Trypanosomosis Control in Kenya*. Technical Report of the KETRI and DFID Trypanosomosis Research Project, Muguga, Kenya.
CIRDES/ITC/ILRI (1997) *Joint Report of Accomplishments and Results*. Collaborative Research Programme on Trypanosomosis and Trypanotolerant Livestock in West Africa, International Livestock Research Institute, Nairobi, Kenya.
de Raadt, P. and Seed, J.R. (1977) Trypanosomes causing disease in man in Africa. In: Kreier, J.P. (ed.) *Parasitic Protozoa*. Academic Press, New York, pp. 175–237.
Dransfield, R.D., Williams, B.G. and Brightwell, R. (1991) Control of tsetse flies and trypanosomosis: myth or reality? *Parasitology Today* 7, 287–291.
EMA (Ethiopian Mapping Authority) (1988) *National Atlas of Ethiopia*. Ethiopian Mapping Authority, Addis Ababa.
Erdelen, W., Nagel, P. and Peveling, R. (1994) Tsetse control, land use dynamics and human impact on natural ecosystems. a conceptual framework and preliminary results of an interdisciplinary research project in Ivory Coast, West Africa. *Applied Geography and Development* 44, 17–31.
Ford, J. (1971) *The Role of the Trypanosomosis in African Ecology: a Study of the Tsetse-fly Problem*. Oxford University Press, Oxford.
Giblin, J. (1990) Trypanosomosis control in African history: an evaded issue? *Journal of African History* 31, 59–80.
Govereh, J. and Swallow, B.M. (1998) Impacts of tsetse and trypanosomosis control on migration into the Zambezi Valley of Zimbabwe. In: Reid, R.S. and Swallow, B.M. (eds) *An Integrated Approach to the Assessment of Trypanosomosis Control Technologies and their Impacts on Agricultural Production, Human Welfare and Natural Resources in Tsetse-affected Areas of Africa: Final Report for IFAD*. TAG Grant 284-ILRI, ILRI, Nairobi, Kenya, pp. 165–183.
Hanks, J. and Hogg, R. (1992) *The Impact of Tsetse Eradication: a Synthesis of the Findings from Livestock Production and Socio-economic Monitoring in Somalia Following the Eradication of Tsetse from the Middle Shabeelle Region*. Summary Report, Tsetse Monitoring Project, ODA.
Huss-Ashmore, R. (1992) *Nutritional Impacts of Intensified Dairy Production: an Assessment in Coast Province, Kenya*. Technical Report No. 1, International Laboratory for Research on Animal Diseases (ILRAD), Nairobi.
Jahnke, H.E., Tacher, G., Keil, P. and Rojat, D. (1988) Livestock production in tropical Africa, with special reference to the tsetse-affected areas. In: ILCA/ILRAD, *Livestock Production in Tsetse-affected Areas of Africa*. ILCA/ILRAD, Nairobi, Kenya, pp. 3–21.
Jordan, A.M. (1979) Trypanosomosis control and land use in Africa. *Outlook on Agriculture* 10, 123–129.
Jordan, A.M. (1986) *Trypanosomosis Control and African Rural Development*. Longman, London.
Jordan, A.M. (1992) Degradation of the environment: an inevitable consequence of trypanosomosis control? *World Animal Review* 70/71, 2–7.
Kaimowitz, D. and Angelsen, A. (1998) *Economic Models of Deforestation: a Review*. CIFOR, Bogor, Indonesia.
Kalu, A.U. (1991) An outbreak of trypanosomosis on the Jos Plateau, Nigeria. *Tropical Animal Health and Production* 23, 215–216.
Kamuanga, M., Antoine, C., Brasselle, A.-S. and Swallow, B.M. (1998) Impacts of trypanosomosis and tsetse control on migration and livestock production in the

Mouhoun Valley of southern Burkina Faso. In: Reid, R.S. and Swallow, B.M. (eds) *An Integrated Approach to the Assessment of Trypanosomosis Control Technologies and their Impacts on Agricultural Production, Human Welfare and Natural Resources in Tsetse-Affected Areas of Africa, Final report for IFAD.* TAG Grant 284-ILRI, ILRI, Nairobi, Kenya, pp. 222–228.

Kriesel, D.A. and Lemma, K. (1989) *Farming Systems Development Survey Report for the 'Chello' Service Cooperative.* FAO, Rome.

Lambin, E.F. (1994) *Modelling Deforestation Processes: a Review.* Research Report No. 1, TREES Series B, Ispra, Italy.

Leak, S.G.A., Mulatu, W., Authié, E., d'Ieteren, G.D.M., Peregrine A.S., Rowlands, G.J. and Trail, J.C.M. (1993) Epidemiology of bovine trypanosomosis in the Ghibe valley, southwest Ethiopia. 1. Tsetse challenge and its relationship to trypanosome prevalence in cattle. *Acta Tropica* 53, 121–134.

Leak, S.G.A., Mulatu, W., Rowlands, G.J. and d'Ieteren, G.D.M. (1995) A trial of a cypermethrin 'pour-on' insecticide to control *Glossina pallidipes, G. fuscipes fuscipes* and *G. morsitans submorsitans* (Diptera: Glossinidae) in southwest Ethiopia. *Bulletin of Entomological Research* 8520, 241–251.

McCann, J.C. (1995) *People of the Plow: an Agricultural History of Ethiopia, 1800–1990.* University of Wisconsin Press, Madison, Wisconsin.

Mills, A.P. (1995) Environmental impact of tsetse control in western Zimbabwe: historical study of land cover change, 1972–1993, using remote sensing and GIS. In: *Proceedings of Africa GIS '95 Congress: Integrated Geographic Information Systems Useful for a Sustainable Management of Natural Resources in Africa, 6–9 March, 1995, Abidjan, Côte d'Ivoire.*

Mugalla, C., Swallow, B.M. and Kamuanga, M. (1997) The effects of trypanosomosis risk on farmers' livestock portfolios: evidence from The Gambia. Poster presented at the 24th Meeting of the ISCTRC, 29 September–3 October 1997, Maputo, Mozambique.

Mulatu, W., Leak, S.G.A., Authie, E., d'Ieteren, G.D.M., Peregrine, A. and Rowlands, G.J. (1993) Preliminary results of a tsetse control campaign using deltamethrin impregnated targets to alleviate a drug-resistance problem in bovine trypanosomosis. In: *Proceedings of the 21st Meeting of the International Scientific Council for Trypanosomosis Research and Control, 1991, Yamoussoukro, Côte d'Ivoire,* pp. 203–204.

Muriuki, G. (1997) Olambwe Valley, western Kenya. In: Bourn, D.M. (ed.) *Draft Case Studies of Trypanosomosis Control in Kenya.* Technical Report of the KETRI and DFID Trypanosomosis Research Project, Muguga, Kenya.

Nagel, P. (1994) The effects of tsetse control on natural resources. In: *FAO Animal Production Health Paper 121.* FAO, Rome, pp. 104–119.

Nicholson, C.F., Thornton, P.K., Mohamed, L., Muinga, R.F., Elbasha, E.H., Staal, S.J. and Thorpe, W. (1999) Smallholder dairy technology in coastal Kenya: an adoption and impact study. In: *Proceedings of the 6th KARI Scientific Conference, Nairobi, 9–13 November 1998.*

Okoth, J.O. (1982) Further observations on the composition of *Glossina* population at Lugala, South Busoga, Uganda. *East African Medical Journal* 59, 582–584.

Oloo, G. (1997) Nguruman, southern Rift Valley. In: Bourn, D.M. (ed.) *Draft Case Studies of Trypanosomosis Control in Kenya.* Technical Report of the KETRI and DFID Trypanosomosis Research Project, Muguga, Kenya.

Ormerod, W.E. (1978) The relationship between economic development and ecological degradation: how degradation has occurred in West Africa and how its progress might be halted. *Journal of Arid Environments* 1, 357–379.

Ormerod, W.E. (1986) A critical study of the policy of tsetse eradication. *Land Use Policy* 3, 85–99.

Ormerod, W.E. (1990) Africa with and without tsetse. *Insect Science and Its Application* 11, 455–461.

Pender, J., Mills, A.P. and Rosenberg, L.J. (1997) Impact of tsetse control on land use in the semi-arid zone of Zimbabwe. Phase 2: Analysis of land use change by remote sensing imagery. *NRI Bulletin* 70, 40.

Pingali, P.L., Bigot, Y. and Binswanger, H. (1987) *Agricultural Mechanization and the Evolution of Farming Systems in sub-Saharan Africa*. Johns Hopkins University Press, Baltimore.

Reid, R.S. (1999) Impacts of controlling trypanosomosis on land-use and the environment: state of our knowledge and future directions. In: *Proceedings of the 24th Meeting of the International Scientific Council for Trypanosomosis Research and Control, Maputo, Mozambique, 29 September–3 October 1997*, pp. 500–514.

Reid, R.S., Wilson, C.J. and Kruska, R.L. (1996) The influence of human use on rangeland biodiversity in Ghibe Valley, Ethiopia, as affected by natural resource use changes and livestock disease control. In: West, N.E. (ed.) *Society for Range Management. Proceedings of the 5th International Rangeland Congress, 23–28 July 1995 in Salt Lake City, Utah, USA*. Denver, Colorado.

Reid, R.S., Wilson, C.J., Kruska, R.L. and Mulatu, W. (1997) Impacts of tsetse control and land-use on vegetative structure and tree species composition in southwestern Ethiopia. *Journal of Applied Ecology* 34, 731–747.

Reid, R.S., Kruska, R.L., Muthui, N., Wotton, S., Wilson, C.J., Taye, A. and Mulatu, W. (2000) Land-use and land-cover dynamics in response to changes in climatic, biological and socio-political forces: the case of southwestern Ethiopia. *Landscape Ecology* 15, 339–355.

Roderick, S., Stevenson, P. and Oloo, G. (1997) Land-use, grazing strategies and tsetse control by Maasai pastoralists. Paper presented at the 24th Meeting of the ISCTRC, Maputo, Mozambique, 29 September–3 October 1997.

Rogers, D.J., Packer, M.J. and Hay, S.I. (1996) *Identifying the Constraints on Livestock Productivity and Land-use in Africa Imposed by Trypanosomosis*. Final Technical Report for Natural Resources Institute (NRI) Extra Mural Contract X0239, Oxford University, Oxford.

Rutto, J. (1997) Busia District, on the border with Uganda. In: Bourn, D.M. (ed.) *Draft Case Studies of Trypanosomosis Control in Kenya*. Technical Report of the KETRI and DFID Trypanosomosis Research Project, Muguga, Kenya.

Slingenburgh, J. (1992) Tsetse control and agricultural development in Ethiopia. *World Animal Review* 70/71, 30–36.

Swallow, B.M., Mulatu, W. and Leak, S.G.A. (1995) Potential demand for a mixed public-private animal health input: evaluation of a pour-on insecticide for controlling tsetse-transmitted trypanosomosis in Ethiopia. *Preventive Veterinary Medicine* 24, 265–275.

Swallow, B.M., Kagwanja, J., Wangila, J., Onyango, O. and Tesfaemichael, N. (1998) Impacts of tsetse control on livestock production, migration and cultivation in the Ghibe Valley of southwestern Ethiopia. In: Reid, R.S. and Swallow, B.M. (eds)

An Integrated Approach to the Assessment of Trypanosomosis Control Technologies and their Impacts on Agricultural Production, Human Welfare and Natural Resources in Tsetse-Affected Areas of Africa: Final Report for IFAD. TAG Grant 284-ILRI, ILRI, Nairobi, Kenya, pp. 67–74.

Thornton, P.K. and Jones, P.G. (1997a) Towards a conceptual dynamic land use model. In: Teng, P.S., Kropff, M.J., ten Berge, H.F.M., Dent, J.B., Lansigan, F.P. and van Laar, H.H. (eds) *Applications of Systems Approaches at the Farm and Regional Levels,* Vol. 1. Kluwer Academic Publishers, Dordrecht, the Netherlands, pp. 341–356.

Thornton, P.K. and Jones, P.G. (1997b) A conceptual approach to dynamic agricultural land-use modelling. *Agricultural Systems* 57(4), 505–521.

Turton, D. (1988) Looking for a cool place: the Mursi, 1890s–1980s. In: Johnson, D.H. and Anderson, D.M. (eds) *Ecology of Survival.* Lester Crook Academic Publishing, London, pp. 261–282.

Wangui, E., Reid, R.S., Okello, O. and Gardiner, A.J. (1997) Changes in land-use and land-cover during tsetse control. In: Reid, R.S. and Gardiner, A.J. (eds) *Collaborative Research on the Environmental and Socio-economic Impacts of Tsetse and Trypanosomosis Control in Southern Africa, Final Technical Report for DFID.* Project Number R5874 (H), ILRI, Nairobi, Kenya, pp. 22–46.

Wilson, C.J., Reid, R.S., Stanton, N.L. and Perry, B.D. (1997) Ecological consequences of controlling the tsetse fly in southwestern Ethiopia: effects of land-use on bird species diversity. *Conservation Biology* 11, 435–447.

Tree Crops as Deforestation and Reforestation Agents: the Case of Cocoa in Côte d'Ivoire and Sulawesi

François Ruf

1. Introduction

World cocoa supply has grown steadily for four centuries. But behind this apparently sustainable supply of one of the world's main tree crops lie dramatic shifts in where cocoa is produced. The centre of world cocoa production shifted from Mexico to Central America in the 16th century. Then it went to the Caribbean in the 17th, Venezuela in the 18th, Ecuador and São Tomé in the 19th, Brazil, Ghana and Nigeria in the early 20th century and Côte d'Ivoire shortly after. Although Africa remained a major producer at the turn of the millennium, Asia – particularly Indonesia – has a chance to win first place in the 21st century.

These booms occurred in contexts with abundant and accessible forests, a large reservoir of potential migrants and rising cocoa prices (or at least expectations of rising prices). These are optimal conditions for massive migration to the forest frontier and deforestation. Under such conditions, technological progress will accelerate deforestation (Ruf, 1995a; Angelsen, 1999). The Côte d'Ivoire and Sulawesi (Indonesia) cases discussed in this chapter confirm that. Technological change in cocoa has mostly involved planting material and manual techniques. Fertilizers and chemicals came late. Almost no production is mechanized.

Our general conclusion is that technological progress in cocoa has accelerated deforestation, but the story is complex. Technological progress in tree crops, and cocoa in particular, may lead to different rates of deforestation, depending on the type of technology, the stage in the deforestation process in which the technology gets adopted, the ecological and institutional context and commodity market trends and price cycles.

Farmers may even seek out technological change in order to accelerate forest clearing. For migrants interested in getting fast returns, finding more efficient ways to clear forests is a high priority. More generally, some new agricultural technologies may be more a consequence than a cause of deforestation. Deforestation forces farmers to adapt their practices to new ecological conditions. This is a variant of Boserup's theory adapted to tree-crop cycles, in which deforestation leads to innovation (Ruf, 1991).

In any case, the issue may not be whether deforestation accelerates or slows down. In many areas, the forest is already gone. The key question is how technological change can encourage replanting and reforestation of fallow lands, especially grasslands.

The chapter is based on farm surveys and historical reviews of regional cocoa cycles. The surveys were done in 1980–1985 and 1997/98 in Côte d'Ivoire and in 1989–1999 in Sulawesi. Section 2 presents a qualitative model of cocoa cycles. Section 3 analyses the adoption of cocoa and new manual techniques of forest clearing and planting in Côte d'Ivoire. Section 4 discusses the adoption of chain-saws and herbicides and the impact of the Green Revolution in Sulawesi, Indonesia. We also briefly look at the impact of the unequal fertilizer adoptions in Sulawesi and Côte d'Ivoire. Tables 16.1 and 16.2 define all these technological changes and summarize their impact on deforestation.

2. A Model of Cocoa Cycles, Migration and Deforestation

2.1. Forest rent and the bioecological basis of cocoa cycles

Almost two centuries ago, David Ricardo (1815) introduced the concept of differential rent. As population and demand increased, farmers grew wheat on less and less suitable land. This led to a cost difference between varying ecological settings. As long as the price of wheat covered production costs in the least suitable areas, farmers cultivating the best land enjoyed extra profits, which Ricardo referred to as rents.

We can apply the same concept to cocoa. We define the differential forest rent as the difference in the costs of production of a tonne of cocoa from an area planted in recently cleared forest and a tonne of cocoa planted on fallow land or after felling the first plantation (Ruf, 1987, 1995a).

The cost difference between new and old cocoa plantations directly relates to the loss of a series of benefits provided by the forest. These benefits include low frequency of weeds, good topsoil fertility, moisture retention, due to high levels of organic matter in the soil, fewer problems with pests and diseases, protection against drying winds and the provision of food, timber and other forest products. When the cocoa trees grow older and most of the forest has been cleared, the forest rents vanish.

The older the trees, the higher the harvesting and maintenance costs. If the farmer waits too long before replanting, he/she will no longer be able to afford to replant. High tree mortality in replanted fields and the additional labour and/or other inputs needed to control tree mortality increase the risks and costs of replanting. Tree growth is slower in replanted fields and the trees require more labour and inputs. These are the biological factors underlying the replanting problem, which partially explains the constant regional shifts in cocoa supply.

In the Sulawesi uplands in Indonesia, planting cocoa on grassland was estimated to cost almost twice as much as planting cocoa in a recently cleared forest (Ruf and Zadi, 1998). As the smallholders observe: 'You have one hectare of cocoa after grassland or two hectares after forest'. Cocoa planted on former grassland also requires more maintenance and fertilizers. Farm budgets show that, in 1997, the cost of production was about 46 cents kg^{-1} on former grassland, compared with 36 cents kg^{-1} for plantations on land previously under forest, a difference of about 30%. This figure would approach 50% if all risks were included. Oswald (1997) obtained similar results for Côte d'Ivoire.

The notion of forest rent also applies to food crops intercropped with cocoa. The returns from food crops decrease when farmers grow them on fallows or grasslands instead of recently cleared forest, limiting the cash flow and the chances of buying inputs. This makes it harder for farmers to replant cocoa (Ruf, 1988; Temple and Fadani, 1997).

Cocoa farmers whose farms get old basically have three choices. They can face the much higher costs of replanting, move to a new area or give up cocoa production and possibly switch to off-farm activities. Jumping from one frontier to another is a classical pattern in cocoa history, especially in West Africa (Hill, 1964; Berry, 1976; Lena, 1979; Ruf, 1995a; Chauveau and Leonard, 1996). Typically, we see a decline in the production in the old cocoa regions, compensated by increasing production from new frontiers.

2.2. The economic, social and political basis of cocoa cycles

Cocoa cycles also have social, economic and political causes (Fig. 16.1). Family life cycles interact with tree life cycles. Farmers and their trees grow old together. By the time the plantations need to be replanted, the farmers have grown old and lack an available labour force, especially if they sent their children to school. Their cocoa yields decline at the very moment they need to invest in replanting. These different factors 'squeeze' them and drive them to look for new sources of credit and technology.

Land ownership evolves during the cocoa cycle. In most cases, when the boom begins, migrants find land cheap and can acquire it easily. Most booms can be interpreted as situations where local ethnic groups, who control land, or at least have a moral claim to it, meet up with migrants, who bring and control labour. In this meeting, migrants are often the winners, at least initially,

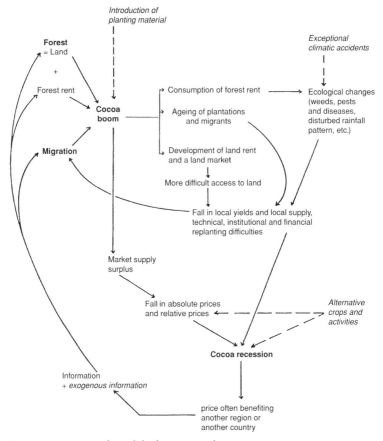

Fig. 16.1. A conceptual model of cocoa cycles.

when labour is scarce. Some 20–25 years later, land becomes scarce and the focus of increasing conflicts between local ethnic groups and migrants and possibly between generations within both groups as well. These land-tenure problems can even trigger local cocoa recessions.

These institutional changes and conflicts interact with the problem of technological change. Compared with the original inhabitants and the ageing migrants, new migrants usually have a higher labour-to-land ratio. Hence, they find it easier to adopt replanting techniques that demand additional labour. The prospects for replanting may therefore often hinge on a steady inflow of migrants, even after the forest has gone.

Massive expansions of cocoa supply affect world prices and can lead to price slumps. For example, when some 200,000 migrants poured into the south-west of Côte d'Ivoire, they put an additional half million tonnes of cocoa on the international market during the next 10 years, causing downward pressure on cocoa prices. The aggregate effect of the migrants' actions was ultimately to reinforce the processes that led to the abandonment of cocoa farms.

Policy decisions can also contribute to cocoa cycles. Governments are constantly tempted to maintain or even increase cocoa taxes when prices fall. They also face pressure to reduce the number of foreigners and thus the labour supply. These are two good ways to kill the goose that lays the golden egg.

All these factors typically interact to produce cycles lasting about 25 years. This is more or less the sequence of international price cycles (Ruf, 1995a, b). Cocoa production cycles and shifts of production between regions would occur even with stable prices. The rising production costs associated with the ageing of the trees and the reduced availability of forest are sufficient to trigger new migrations and new plantings rather than replanting (upper part of Fig. 16.1). Price factors only accelerate these processes (lower part of Fig. 16.1).

After most forest is gone in the region that has dominated world cocoa production, a new country with abundant forests and labour reserves takes over. This process will continue as long as there are countries where forests are cheap and available and other inputs are costly. Farmers' access to forests and the price ratios they face would have to change in order to modify their decisions and behaviour. Indeed, these are practically prerequisites for technological change to slow down deforestation.

To sum up, the basic equations of the forest rent and the cocoa boom are:

Forest rent = f (few weeds, few pests and diseases, good soil fertility, regular rainfall, timber, protein sources)

Cocoa boom = deforestation = f (forest rent, abundant land, abundant labour (migrants), accessibility, market outlets)

The history of cocoa shows that even partial accessibility is often enough. The bridge built over the Sassandra River in 1973 in Côte d'Ivoire made thousands of hectares accessible in the south-west and multiplied the number of migrants spectacularly. Logging trails have played similar roles. Once these primary connections are in place, farmers are great at starting plantations several kilometres away from even the worst roads. The forest rent in remote areas often more than compensates for the lower costs of transporting cocoa planted on fallows close to the road. Indeed, farmers may want to concentrate on their investments and not be bothered by relatives and neighbours! Many farmers assume that the transportation infrastructure will improve once their plantations start producing. Most cocoa pioneers are optimists.

2.3. The link between technological change, forest rent and the cocoa cycle

Based on the previous discussion, we distinguish between three types of technological changes according to where and when in the cocoa cycle they occur. First, technological progress may occur outside the cocoa region, in a neighbouring region or country (left side of Fig.16.1). In that case, if it includes labour-saving components and occurs before the cocoa boom, it helps trigger the boom by releasing labour. Technological change is then a push factor of cocoa booms and deforestation. Labour-saving technologies in paddy cultivation and the Green Revolution in Sulawesi have freed labour for the cocoa boom and thus deforestation (see section 5). Labour-intensive technologies will have the opposite effect.

Secondly, technological change may occur on the pioneer frontier at the beginning of the cocoa cycle (centre of Fig. 16.1). Even though financial capital has a limited role on the frontier, we shall see that changes in manual technologies can greatly affect deforestation and plantings, as can the introduction of chain-saws.

Thirdly, technological change may occur late in the cycle, when the cocoa region is already near the end of its cycle, on the brink of recession (right side of Fig. 16.1). The need to overcome the loss of the forest rent drives technology adoption, which normally includes a shift from manual techniques to herbicides and fertilizers. As weed invasion is a major problem in old cocoa areas, farmers pay particular attention to herbicides. These technologies may be agents of deforestation, since they save labour, but they may also be key reforestation agents as tools for replanting.

3. Cocoa and New Technologies in Côte d'Ivoire

The introduction of a new tree crop may itself be considered a technological change and can have contradictory effects on deforestation. A tree crop usually provides higher returns to labour and land than most annual food crops, which take up large areas of land. This might help to save forest. In most cases, however, profitable new tree crops attract migrants. Forest regions are suited to most commercial tropical tree crops and migrants that come to grow them significantly hasten deforestation. This is what happened in Côte d'Ivoire and in most of the earlier cocoa stories, from Brazil to Ghana (Monbeig, 1937; Hill, 1956; Lena, 1979; Ruf, 1991; Touzard, 1994; Clarence-Smith and Ruf, 1996). This section summarizes the evolution of cocoa production in central-western Côte d'Ivoire and the role of technological change in that process. It shows how technological changes have gone from promoting deforestation to encouraging reforestation, influenced in part by the context in which they have been introduced and adopted.

3.1. Local ethnic groups shift to clearing primary forest instead of secondary forest

In the early 20th century, central-western Côte d'Ivoire had a low population density. The local Bété population used a simple piece of iron attached to a wooden stick to clear forest (Oswald, 1997: 31). Most farmers grew food crops after clearing areas with 7–10-year-old fallows, which were easy to cut down and long enough to eliminate most weed seeds (Chabrolin, 1965; Boserup, 1970; Ruf, 1987; de Rouw, 1991; Levang et al., 1997; Oswald, 1997). Technical constraints and labour shortages protected primary forests.

In the 1920s, the use of the axe and the machete rapidly expanded (Oswald, 1997). Around that time, rice, coffee and cocoa were introduced to the region and farmers planted most of these crops on former 7–20-year-old secondary forests. Regular rainfall, a humid climate and few weeds made it easy to plant coffee after a 15-year fallow.

Then, thanks to the introduction of a more efficient axe, the percentage of upland rice planted in primary forest areas jumped from 5% to 40% in the early 1960s (Oswald, 1997). This was an early case of technological change promoting deforestation.

After Côte d'Ivoire got its independence in 1960, the first president, Houphouët-Boigny, was from a region inhabited by Baoulés. He encouraged Baoulé families to migrate to the forest zone and plant cocoa, and launched a new land policy with the motto 'the land belongs to those who develop it'. This triggered an interest in forest clearing as a way to establish land ownership. While the policy was intended to encourage migrants, local inhabitants understood that they also had to clear forest if they wanted to prevent migrants from taking over.

Data we collected on cocoa plantings in the region of Ouragahio in 1980 point to the dramatic impact this policy had on the forest. Seventy per cent of coffee and cocoa fields planted before 1965 were established on fallows and secondary forests of less than 25 years. In contrast, 80% of the cocoa that local ethnic groups planted between 1965 and 1980 was on cleared primary forest land. When we asked local farmers why they changed their pattern of forest clearing, 35% cited increased competition with Baoulé migrants. Fifteen per cent mentioned that the local inhabitants no longer feared the Forestry Service, since they saw that the Service did not bother migrants. Similar proportions attributed the change to the new axe introduced by logging companies, the fact that the logging companies had already cut the biggest trees and natural population growth.

The first migrants settled in remote forests far from the villages of the original ethnic groups, which presumably were of little value to the local inhabitants and were difficult for them to control. However, after a few years the locals realized that the migrants were moving rapidly towards their own villages. In response, they opened 'counter-pioneer fronts' and planted cocoa to keep the migrants from advancing. To a certain extent, the main factor driving the rapid growth in cocoa plantations planted by local inhabitants during the early 1970s was this desire to protect their territory, rather than an interest in increasing their incomes in the short run (Table 16.1).

Table 16.1. Features of the main technical changes studied in Côte d'Ivoire.

Type (and cost)	Description	Labour-saving?	Farmers' response	Deforestation impact
Adoption of cocoa (and coffee) during the 20th century (cheap)	Cocoa not only adopted by local ethnic groups, but also an opportunity for migrants to create new farms	Yes, compared with previous food-crop systems	Huge adoption over the 20th century with an exponential growth of migration since the mid-1960s	Enormous, mostly due to migration
New axe introduced in mid-1950s (considered expensive until the 1960s)	Introduction of strong axes, mostly by logging companies	Yes	General adoption in the late 1960s for establishing cocoa orchards	Moderate. The axe helped the original inhabitants clear forests to counter the advance of migrants
Forest clearing without felling giant trees since the mid-1960s (no monetary cost)	The method overcomes the difficulty of cutting down the big trees by burning them as they stand	Yes	Generally adopted by all migrants. The Burkinabé copied the Baoulé	Enormous. It greatly helped migrants to deal with primary and dense forests
New food crops, mainly yam instead of rice, since the mid-1960s (cheap)	A longer period of association of young cocoa and food crops made weed control more efficient and more productive	Labour-intensive. Implied regular labour inputs throughout the year	Adopted by all Baoulé migrants	Moderate. It improved the system's efficiency and made cocoa more attractive

Technique	Details	Labour	Adoption	Impact on deforestation
Planting with nurseries and plastic bags, mostly in the 1990s (moderate, but long considered an unnecessary expense)	Instead of direct sowing of beans at high density (around 5000 beans ha^{-1}), planting of 1200 seedlings grown in plastic bags in nurseries	Labour-intensive	Introduced by extension services in the 1970s, and adopted in the 1990s when most forest was gone	Limited since most forest had already disappeared. But it made farmers reconsider alternatives to forest clearing and might have discouraged them from moving to the last frontiers
Tree diversification in cocoa farms, mostly in the 1990s (no monetary cost)	Introduction of fruit-trees (avocado, orange-trees) and various other useful trees, either by favouring their natural regrowth (such as palm trees) or by planting	Labour-neutral and intensive	Increased adoption after farmers rediscovered the negative impact of monoculture, once plantations start ageing	No impact on deforestation, but important in a reafforestation strategy
Fertilizers, mainly in the 1990s (expensive, but the alternative may be to lose the trees and farms)	Application of fertilizers to ailing cocoa trees	Labour-intensive	Rediscovered in the 1990s by farmers in a specific region with poor soils. The objective was to prevent their cocoa trees from dying	No direct impact on deforestation, but it helps save cocoa trees and may favour replanting and reafforestation

3.2. Migrants swallow the primary forests

The local ethnic groups alone would not have harmed the primary forest much. But the migrant population increased by 10–20% each year, and that tremendously accelerated cocoa plantings and thus deforestation. More than 95% of cocoa farms that Baoulé migrants created between the mid-1960s and 1980 replaced primary or old secondary forests.

Most migrants were young and strongly motivated to make money quickly. This led them to clear forest and plant cocoa more aggressively than the local ethnic groups. Control by village elders or social pressure to devote time and cash to social ceremonies did not hamper the young migrants.

Technical change also helped the migrants succeed. The migrants introduced new techniques for clearing forests and new systems for associating food crops with young cocoa trees. The local ethnic groups tended to intercrop paddy with coffee and, to a lesser extent, cocoa during the first year of planting. Due to labour constraints, they only cleared forests or fallows for paddy and only planted coffee or cocoa every 5 years. The adoption of cocoa and more intensive techniques by local ethnic groups led to sharp competition for labour between paddy and tree crops. Baoulé migrants, who had much more available labour, began intercropping yams with cocoa. This had several advantages in terms of weed control and seasonal labour demand (Ruf, 1988, Vol. 2). Moreover, they planted cocoa every year, which resulted in much higher cocoa production, as well as deforestation.

Originally, when the local population cleared forest, the farmers left some large trees, in part because of the difficulty they had in felling them with the tools at their disposal. This led to cocoa farms under big trees. Then, the extension services began promoting total clearing, supposedly to maximize cocoa yields. Without chain-saws, this was a very labour-intensive practice. But the Baoulé migrants discovered a new labour-saving method for clear-felling, by burning the big trees as they stood, which did not require capital. They collected the dried undergrowth, cut a few weeks earlier, around each large tree and then set it on fire. This killed the trees and made them lose their leaves, resulting in a depressing landscape of huge dead upright trees. Nevertheless, it was a very efficient technique. It saved substantial labour. The cocoa trees grew as rapidly as they would have with any other type of total clearing. The falling branches and pieces of trunk from the dead giant trees provided free fertilizers. The system also seemed to suppress insects, at least for a while. Twenty years later, some migrants rediscovered some of the system's disadvantages, such as changes in microclimate and the lack of shade. Others had anticipated these problems and already moved on to other virgin forest areas.

The Baoulé method for forest clearing accelerated the already rapid deforestation in the 1970s and the massive clearings of protected forests between 1983 and 1988. The high and stable price of cocoa until 1988 and the 1983 El Niño drought reinforced this process. After 1988, deforestation slowed, due to the scarcity of remaining forest and increasing restrictions, which made access to the few remaining forests more difficult, risky and expensive.

3.3. Replanting cocoa on fallows in forest-scarce contexts

Chromolaena odorata is a perennial shrub native to South America. In the New World, it does not invade pasture or compete with plantation crops. Attacks by a large complex of insects and competition from related plants keep it from becoming overly aggressive (Cruttwell-Mcfadyen, 1991). In West Africa, however, where *C. odorata* was apparently introduced accidentally, the plant faces no major pests or diseases and produces spectacular quantities of biomass, which make it difficult to control. Thus, the shrub became a 'weed'. In 1979, farmers were already complaining about this problem. They nicknamed the shrub 'Sékou Touré', in reference to the fact that its effect on local agriculture had been just as revolutionary as the president of neighbouring Guinea had been. After the drought and fires related to El Niño in 1983, *C. odorata* took over the ecological niche created by forest and plantation fires, spread massively and changed the landscape.

Some smallholders who lost everything in the fires decided to migrate and plant cocoa elsewhere. Others stayed on and replanted, but many failed. The techniques they had developed over the years to clear and plant in forest areas did not work with *C. odorata* fallows. Many of the local planters and well-established migrants were getting old. However, a new generation of young Burkinabé migrants once again started arriving in Côte d'Ivoire, and they bought massive amounts of land covered with *C. odorata*.

The new Burkinabé farmers developed techniques for using the *C. odorata* land more efficiently. Most began to use nurseries to produce their cocoa plants, preferably with plastic bags. The extension services had pushed that technique in the 1970s as part of their technological package to promote more intensive (and thus less forest-consuming) cocoa farming, with little success. About a quarter of the farmers adopted the technique on 1 or 2 ha, mostly because it was part of a package that included highly coveted hybrid planting material and some \$50 ha^{-1} in cash subsidies. Back in the frontier days, very few farmers were really interested in investing labour in preparing planting material in nurseries using plastic bags. The 1983 El Niño and the disappearance of primary forest that farmers could clear to plant cocoa on changed that. Once most forests of the central-western region had gone and farmers had to take into account the higher climatic variability, the adoption of plastic bags took off after 1984. Ironically, that was about the same time that the extension services virtually died (Ruf, 1985).

Cocoa smallholders also introduced other innovations. They started to dig bigger holes and mix soil and grasses with the young seedling in the holes. This increased the seeds' chances of surviving droughts. Some sprayed the soil before planting cocoa to kill termites and other pests. Most even came to prefer replanting after a 5-year *C. odorata* fallow because the shrub helps them get rid of nematodes and other soil pests and diseases. The key to these farmers' success lies in their ability to control *C. odorata* regrowth. If they tried to slash it, the job would be endless. Instead they discovered that they had to uproot it by hand and destroy part of the roots with a hoe. Then they can slash the regrowth and use it as mulch. Up to 1999, few farmers used herbicides to replant, but that should come soon. Farmers also started growing maize (instead of paddy) with their young cocoa plants, using high planting densities, and increased the density of their plantain crops where soil fertility permitted it.

In the mid-1970s, one Burkinabé migrant started replanting cocoa below old coffee bushes and progressively cutting away the coffee. Once the neighbours verified that the migrant's method worked, many of them adopted it in the early 1980s.

The discovery of techniques for replanting cocoa on old *Amelonado* cocoa fields that had been seriously degraded constituted another important innovation. There are many old and heavily shaded *Amelonado* cocoa plantations that farmers from local ethnic groups planted between the 1940s and the 1970s and later abandoned. These cocoa fields are easy to cut down and replant. However, very few of the shadeless *Amelonado* fields that Baoulé migrants cleared by systematically burning all the large trees and then planted prior to the mid-1970s lasted more than 20 years or resisted severe droughts. Most were burnt or disappeared under pest attacks and became fallows with a few small surviving trees. To replant these fields with cocoa required a lot of innovation and technical change. Some Baoulé innovators introduced hoe ploughing and intercropping of yam and bananas to solve this problem in the mid-1980s and many others copied the technique in the 1990s (Ruf, 1991). This type of technological change has led to both replanting and reforestation.

One might ask whether it is reasonable to consider cocoa replanting a form of reforestation. If the replanting replaces secondary forest or old coffee plantations, it is not, since the result would be fewer trees and less biodiversity. But, if farmers plant areas that were previously in grassland and *C. odorata* fallows, we can rightfully refer to it as reforestation, since it increases both the number of trees and the amount of carbon sequestered.

Interplanting cocoa with timber trees provides particular benefits as a reforestation technique. Fortunately for the logging companies, local ethnic groups continue to control some old cocoa orchards in Côte d'Ivoire. These companies are now busy logging the trees the original local inhabitants left in their plantations. There is not much to take out in the migrant plantations, because of the forest-clearing techniques they used. By combining herbicides and leguminous trees, which make it easier to replant, with the intercropping of timber trees, farmers can maintain both their cocoa yields and a long-term source of increasingly scarce timber. The idea of intercropping timber trees dates back to the 1910s, but in a context of abundant forest resources generated little interest (Vuillet, 1925). Again, one finds that deforestation drives technological change.

3.4. Summary and partial conclusions

We can divide the history of cocoa, technological change and forest clearings in Côte d'Ivoire into three phases:

Phase 1: Tree-crop shifting cultivation, with moderate levels of cocoa production, mostly by local ethnic groups on secondary forest fallows and cocoa secondary fallows. This involved planting trees under forest cover and only increased deforestation slightly. Without the arrival of the migrants and the political support given to them, the introduction of cocoa production using 'primitive' technologies would not have harmed forests much. Competition for land and the introduction of a new axe made the local ethnic groups begin to clear more primary forest.

Phase 2: Cocoa boom, fuelled by massive inflow of, often foreign, migrants, at the expense of primary forests. Deforestation accelerated enormously. The introduction of apparently simple manual methods for clearing forest and planting cocoa constituted real technological changes, which farmers adopted to accelerate cocoa expansion, in a context of abundant and cheap forests. The new techniques spurred deforestation, although, under the circumstances, massive forest conversion would undoubtedly have happened anyway.

Phase 3: Replanting cocoa on grassland and degraded fallows, which is a kind of reforestation, mostly by foreign migrants. In a migrant village (Balikro), for example, after 1983 (El Niño) only 13% of cocoa farms were established by clearing forest. The rest involved various forms of replanting, mostly following the presence of *C. odorata*. Nurseries were widely adopted, but did not stimulate much deforestation. Other innovations, such as replacing paddy by maize, digging bigger holes and switching to mulch techniques, also formed part of an important technical transformation.

Land shortages and ecological changes related to deforestation encouraged these recent innovations. This demonstrates one of this chapter's main hypotheses: that deforestation also triggers technological change. These changes may even lower the rate of clearing Côte d'Ivoire's last surviving forests. Considering the high costs that regulations and remoteness put on the use of the country's remaining forests, replanting *C. odorata* fallows using the new technologies may well become more attractive. Indeed, a deforested environment and the new techniques may actually transform cocoa trees from being an agent of deforestation to becoming an agent of reforestation.

4. Cocoa in Sulawesi

4.1. Cocoa fever

The island of Sulawesi in Indonesia lies to the east of Borneo. Most of it is much more mountainous than Côte d'Ivoire. Its main ethnic group, the Bugis, is famous for its sea-sailing and trading skills. But the Bugis have proved to be excellent cocoa farmers too.

Sulawesi's cocoa boom is much more recent than West African's. Farmers planted the first trees in remote forested areas in the mid/late 1970s and real cocoa fever did not take hold until the mid-1980s. By then, 'pre-cocoa' migrations from the central south of the South Sulawesi province had already deforested a large part of the rich alluvial plains, where the cocoa fever first emerged. The opportunities to grow tobacco and soybeans in the early 1970s pulled migrants into the area, while drought and declining self-sufficiency in the villages of origin pushed them in that direction. Farmers planting cocoa preferred the already cleared alluvial plains, because of the area's favourable agroecological conditions and easy access to agricultural inputs. Thus, unlike most cocoa stories, the introduction of cocoa in Sulawesi was not initially associated with widespread deforestation.

Early in the cycle, cocoa even served as a tool to reafforest the plains. However, once most of the available deforested alluvial plains had been converted into cocoa farms, more and more migrants got the 'cocoa fever' and moved to the forested hillsides and the remaining forested plains in the provinces of South and South-east Sulawesi. By the late 1980s, a new wave of migrants, entirely motivated by the prospect of cocoa incomes, were looking for land everywhere, including the swampy plains of the Bone gulf, north of Noling. In the village of Pongo, for example, local cocoa planting boomed between 1990 and 1994. Around 60% of the cocoa orchards replaced forest. The real effect of cocoa planting on deforestation may have been even higher, as some migrants bought land that had just been cleared from the local inhabitants. In this way, they shared the social and political risk of buying forests, presumably owned by the state, with the sellers and escaped the physical risks of felling the big trees. The practice of buying recently cleared forest land instead of forests was even more widespread in the hilly areas, where the Forest Services supposedly exercised stricter control, due to the risks of erosion. Thus, in the village of Sambalameto, around 60% of the cocoa farmers established their farms, buying already cleared land, and 80–90% of the cocoa was planted on cleared forest land, including officially protected forests.

The Sulawesi case shows that adoption of a tree crop does not necessarily trigger deforestation in the short term. However, eventually, the high returns from cocoa led both established and new farmers to invest in clearing additional forest. It was therefore only a question of time – in this case, only a few years – before cocoa became a deforestation agent.

4.2. Introduction of the chain-saw (Table 16.2)

The Côte d'Ivoire case showed that technologies that reduce the labour requirements for clearing forests accelerate deforestation. Thus we should expect an efficient clearing tool, such as the chain-saw, to do the job it is made for, namely to cut trees. And that was the case. The chain-saw, now widely used in Sulawesi, speeded up deforestation.

Forests are not only a source of wealth and potential rent. They are also a source of problems. It is not easy to cut down the big trees of primary forests. It takes a lot of labour, is risky and requires experience and know-how. Thus, as Boserup noted in 1965, most shifting cultivators prefer to clear secondary forest, because it is easier.

Chain-saws and professional tree cutters changed all that. Most farmers in Sambalameto, for example, said they preferred to plant already cleared forest. That way they could avoid the problems involved in cutting down big trees and still get the advantages of planting on a cleared forest rather than a cleared fallow: high fertility, few weeds and rapid cocoa-tree growth.

Why did farmers in Sulawesi opt for chain-saws, rather than the labour-saving methods used to clear forest in Côte d'Ivoire? Several possible explanations exist. Sulawesi has less experience with cocoa. Further, the big trees in the Sulawesi forest are thinner but more numerous. Thus, the Baoulé method would take more time there. But the main reason might simply be the cost and availability of chain-saws in Sulawesi. A farmer in Sulawesi has to sell 1 t of cocoa to buy a chain-saw; his Ivorian counterpart must sell 2 t.

As often happens, technological change and various institutional and political factors interacted in this case. In Côte d'Ivoire, the Baoulé method for forest clearing using tree burning was perfectly adapted to bypass the control of the Forestry Service. It is a gradual and almost invisible method of clearing. The Forestry Services does not notice it until it is done – often months later, when the trees are dead. Chain-saws are much less discreet. Their wide use in the mountains of Sulawesi reflects the fact that local government representatives themselves got involved in the cocoa boom.

Most farmers cannot afford to buy a chain-saw during the first years after they migrate to an area. This and the political nature of forest clearing imply that only a few households tend to monopolize the use of chain-saws. Chain-saw owners usually have capital and are well connected with village and district authorities, which give them access to forest. This means that only a few people supply the cleared forest demanded by many migrants. In many cases, this leads to an oversupply of cleared forest. Several months and even years may pass before a cleared forest plot is sold. In the meantime, weeds invade and one component of the forest rent component is partially lost. The monopolization of chain-saw ownership has also encouraged the clearing of large blocks of forest and the total removal of shade.

Table 16.2. Features of the main technical changes studied in Sulawesi, Indonesia.

Type (and cost)	Description	Labour saving?	Farmers' response	Deforestation impact
Adoption of cocoa since the late 1970s (relatively cheap, though land is more expensive than in Côte d'Ivoire)	As in most countries, not only original dwellers adopted cocoa. It also offered an opportunity for migrants to create new farms	Yes, compared with previous food-crop systems	Massive adoption. Induced large-scale migration	Less pronounced during its early stage: the rich alluvial soils in plains were preferred. Forests had been already cleared by 'pre-cocoa' migrations driven by soybean, tobacco, coconut and clove. But massive migrations and high returns quickly turned the boom into a more typical deforestation pattern
Adoption of chain-saw in the 1980s (expensive for new migrants)	Piece-work tree cutting by a chain-saw team. Purchase of land already cleared by chain-saw	Yes, very much so	Monopolized by relatively rich and politically connected farmers, but wide impact through an active market for cleared forests	Strong. It accelerated deforestation, and made it easier for non-experts to establish cocoa farms by hiring chain-saw teams or buying already cleared forest
Hand tractor and other changes in paddy-fields in the 1980s and 1990s (expensive to buy, moderate to rent)	Purchase or renting of hand tractors	Yes	Important in the paddy sector	Significant effects. It freed labour and pushed farmers, their sons and ex-sharecroppers to cocoa frontiers. In some places it facilitated part-time paddy and cocoa farming, enhancing the transfer of capital and labour from paddy to cocoa

Herbicides, mostly in the 1990s	Regular purchase of herbicides, mostly on young plantations or after a drought on mature farms, which have fewer cocoa leaves, more light on the ground and more weeds	Yes, very much so	Massive and rapid adoption in the 1990s. Particularly important for new migrants during the investment stage (less important once cocoa trees form a canopy)	Helped reduce deforestation in the short run by making plains covered by grassland more attractive. But, as a labour-saving technology, it facilitated an 'accumulator' strategy by several larger farms. Cocoa revenues were invested at the expense of forests in the uplands
Fertilizers since the 1980s (relatively cheap. Much cheaper than in Côte d'Ivoire)	Regular fertilizer purchase, mostly devoted to mature farms	Labour-intensive	Massive and rapid adoption	Ambiguous. It helps farmers maintain their trees at a high level of productivity. But the increased return makes new plantings more attractive, both to established cocoa farmers and to newcomers

4.3. Labour-saving, Green-Revolution technologies

Many observers considered Indonesia to be a country that has undergone one of the most successful Green Revolutions. Within Indonesia, Sulawesi was one of the most successful regions. The introduction of new planting material, fertilizers, pesticides, herbicides and machinery, supported by subsidies and irrigation projects, significantly increased rice production there. Initially, the Green Revolution in Sulawesi probably reduced deforestation by increasing yields and supplies and thus reducing poor households' need to migrate to open new rice-fields. However, the rice self-sufficiency policy was combined with a transmigration policy, and the new irrigated rice-fields established by the transmigration programmes were among the major causes of deforestation in Sulawesi in the 1980s.

More importantly, most green-revolution technologies saved labour and freed it for employment on the cocoa frontier in the 1990s. Threshers and herbicides substantially reduced the demand for labour from the 1970s onwards (Naylor, 1992). Motor cultivators spread very rapidly after 1985/86, at least in regions where water management permits two or even three crops per year. They are accessible to a larger number of people on a rental basis. Compared with ploughing with buffaloes or cows, motor cultivators save $10-15$ man-days ha^{-1} $crop^{-1}$ and remove almost all labour constraints during soil preparation. In addition, mechanization makes tillage less laborious and improves social status. The work becomes more attractive and sought after by young people. This pushed fired paddy sharecroppers to the cocoa frontier.

Sharecroppers are not the only ones that move. Many Bugis migrants who own rice-fields establish cocoa plantations as quickly as possible and then return to the village to harvest the rice and prepare the land for the next cycle. Then they go back to the cocoa plantation. The rice surplus created by Green-Revolution technologies directly funds the migration and investment in cocoa (and deforestation). In some villages, as many as a quarter of the 'rice farmers' have become 'migrants' and 'cocoa cultivators' for parts of the year.

4.4. Herbicides – the key to deforestation/reforestation in cocoa

Low weed pressure after forest clearing is a key element of the forest rent. In principle, herbicides should help farmers overcome their weed problems without clearing forests and encourage cocoa pioneers to move to grasslands rather than forests. That should reduce deforestation. The counter-argument would be that, since herbicides are a labour-saving technology, they should stimulate deforestation. The Sulawesi case provides some evidence on this issue.

A cocoa planter told us:

> with 4 litres of Paracol, costing Rp 90,000, I can fill my sprayer tank 100 times. Allowing Rp 10,000 for the maintenance and cost of my sprayer, each tank costs me Rp 1000. A worker can spray 14 tankfuls per day. It therefore costs me Rp

14,000 and Rp 7000 in wages, that is to say Rp 21,000. If the job is done by hand, it will take 8 days and costs Rp 56,000. I therefore save Rp 35,000 and that pays for my fish!

This simple calculation explains why herbicides are rapidly gaining ground. They save time and money. If farmers hire labour, the cash saving can be substantial and add to the cocoa surplus, which can be reinvested in new cocoa farms. If they do the work themselves, they save time. Several ex-paddy farmers told us that they enjoyed cocoa planting with herbicides, because they could work in the morning and rest in the afternoon.

Herbicides are not only a labour-saving technology. They also constitute the best option for bringing land back into cultivation. Earlier, we noted that low weed pressure during the first 2 or 3 years of cultivation after forest clearing constitutes one of the major components of the forest rent. This fits perfectly with cocoa cultivation. After 3 years, the canopy is almost formed and weed control is no longer a constraint.

Our ideas about the differential forest rent were initially based on observations in Côte d'Ivoire. To see if they held up in a different environment we conducted a small survey in Sulawesi, where we asked farmers the following question: 'If I give you 1 ha of land, would you choose 1 ha of forest, 1 ha of *kabo kabo* (5–10-year-old fallow with some trees) or 1 ha of grassland, even *alang alang* (*Imperata cylindrica*)?' When we asked farmers in Côte d'Ivoire that question, 100% said they would prefer forest. In Sulawesi, we asked 14 farmers in the alluvial plains of Noling and Wotu and 51 farmers in the hills. Half of the farmers on the plains said they preferred grassland, but only two of the 51 farmers in the hills said that. This implies that the farmers in the rich alluvial plains have less fear of losing the forest rent. In addition to the plains' easy access and fertility, herbicides solve the weed problem there. This makes grassland fallows, even those with *Imperata*, much more attractive than before. A *kabo kabo* falls somewhere in between and farmers now appreciate that. They know that they are going to face weed problems and that their cocoa will grow more slowly, but it no longer worries them so much, since most of them can afford to buy herbicides.

In the hills and uplands, farmers still appreciate forests. Their soils are less fertile and more prone to erosion. The migrants who seek land there are generally poorer and fewer of them can afford herbicides, especially during the first years. The 40 Sambalameto farmers we asked about the relative advantages of forest, bush fallow and grassland expressed this explicitly. All the 22 farmers who opted spontaneously for forest stressed three basic factors: fertile soil, no weeds and fast growth. This confirms the importance of the fertility component of the forest rent on otherwise poor soils. The 18 farmers who opted for recently cleared forest and bush fallow lose the advantage of having 'no weeds', but easier land clearing compensates for that.

Herbicides appeared on village markets in the 1980s and their price has decreased at least since 1989. This substantially increased the value of grassland, especially land with *I. cylindrica*. The Bungku region in central Sulawesi, which was in the middle of a cocoa boom in 1997, illustrates this. On the plains, grassland near the river sold for an average of $US800 ha^{-1}, while *alang alang* land near the village went for up to $US1000. It is difficult to imagine a better demonstration of the value smallholders award to certain land under *I. cylindrica*. Admittedly, this price owes something to the fact that the land is close to the village and to roads and the alluvial plains are fertile. But, in 1985, it cost less than $180 ha^{-1} and the widespread use of herbicide since then has been a major factor behind the sky-rocketing grassland prices. The massive adoption of herbicides in the 1990s has totally modified transmigrants' perception of *alang alang* land.

In the foothills, forest is being 'sold' for about US$40 ha^{-1}, while 5–15-year-old fallows go for ten times that price. The low forest price reflects the lack of effective claimants to the land. Often the village chief or a person delegated by the head of the subdistrict takes the $40 ha^{-1}. Those who get $400 for 5–15-year-old fallow are mainly migrants who arrived in the 1950s and claimed land tenure rights by the virtue of clearing the forest.

What, then, has the net effect of herbicides been on forest clearing? 'The fish in the river always look thirsty', as one of the Noling leaders used to repeat, referring to his fellow farmers, who were not happy with the 4–6 ha of cocoa they already owned. Herbicides and the new attractiveness of *alang alang* land were not enough to quench migrants' thirst for forests, within a context of abundant forests and increasing cocoa prices (at least in nominal terms). Bungku provides a clear example of that. While herbicides were widely adopted in the plains and helped to rapidly increase the value of *alang alang* land, purchases of forests sky-rocketed at the same time. The acquisition of forest and cocoa planting started booming in 1995, 2 years before the economic crisis hit the country and a currency depreciation more than tripled the cocoa price. But, even at its 1994/95 level, the price of cocoa was attractive enough to trigger plenty of migration at the expense of forests. Moreover, the upward price trend from late 1993 to 1996 made farmers think that prices were going to increase further (and they were right up to mid-1998, although for reasons not foreseen). The cocoa fever had spread, and even a major breakthrough, such as the use of herbicides, was not sufficient to orientate this fever towards fallows instead of forests. Herbicides made grassland more attractive, but migrants also kept running to neighbouring forests. Farmers converted both grassland and forest to cocoa.

The experience of the hills of Tampumea, behind the plain of Noling, also points to the limits of the ability of herbicides to restrain deforestation in a context of abundant forests. Farmers there began to adopt cocoa in 1981. Three or four years later, the number of migrants increased dramatically in response to the new opportunities provided by cocoa. At the same time, they adopted herbicides, whose use grew exponentially. Herbicides helped farmers replant after fallows. Yet most cocoa farms were created on cleared forest lands. In 1998, when the cocoa price peaked, 22 out of 25 farmers that we surveyed who established cocoa farms did it in forests.

The increasing use of herbicides is almost an agricultural revolution in the humid tropics. It is bringing millions of hectares back into agricultural use, especially for tree crops. Herbicides may thus reduce deforestation by keeping migrants on their established farms, rather that having them looking for new forests to clear. However, like other technological changes, herbicides can only help to reduce deforestation when forest is already scarce or effectively protected. The positive potential of herbicides cannot outweigh the attractiveness of abundant and accessible forest land.

5. Fertilizer Use in Sulawesi and Côte d'Ivoire

In Sulawesi, farmers began adopting fertilizers from the beginning of the cocoa boom in the 1980s. They helped the farmers achieve impressive yields of close to 1500 kg ha^{-1} in the hills and often more than 2000 kg ha^{-1} in the plains. In Côte d'Ivoire, most cocoa farmers did not know about fertilizers until the 1990s and, for a long period, yields remained between 300 and 1000 kg ha^{-1}. The average farm in Côte d'Ivoire had around 6 ha, compared with between 2 and 4 ha in Sulawesi.

One could argue that fertilizers helped Sulawesi farmers keep the deforested area per household relatively low. However, the number of migrants was still high enough to clear forest at a spectacular rate, attracted in part by the high yields, also caused by fertilizers. Moreover, Sulawesi farmers invested the incomes their fertilizer helped make possible in new clearings and farms, often far from the first farm, and the average size of deforestation per family is increasing. Thus, the two regions seem to have ended up with a fairly similar rate of deforestation.

Nevertheless, our recent surveys in Côte d'Ivoire point to a slightly more positive role for fertilizers in reducing deforestation (Ruf and Zadi, 1998). Since the late 1980s, the small region of Soubré has experienced a rapid, and still unique by Côte d'Ivoire standards, process of fertilizer adoption on cocoa farms. Migrants faced unexpectedly early ageing of their cocoa trees and yields. Poor soils unsuitable for cocoa resulted in trees ageing and dying after 7–15 years, instead of the normal 20–25 years. This presented farmers with a problem. They did not have enough financial and psychological capital to move to a new frontier area. They had to innovate. The experimentation with fertilizers by innovative farmers took place during a period of low cocoa prices. Since then, a combination of price recovery (up to mid-1999) and spontaneous diffusion has substantially increased the use of fertilizers. Many cocoa farms recovered and were saved, and few migrants abandoned or sold their farms and went looking for new forests. The fact that forest land had become more difficult to find also contributed to the adoption of fertilizers. Thus this is yet another example of a technological change helping to reduce deforestation only when forest is already scarce.

6. Conclusions

Cocoa has been an active agent of deforestation. Its expansion in Côte d'Ivoire from the mid-1960s to the late 1980s has converted the few remaining Ivorian forests practically into 'prehistoric souvenirs'. In Sulawesi, cocoa adoption by migrants is still a major deforestation agent in the 1990s – possibly more than in the 1980s, when farmers chose mostly fallows and established coconut farms on rich alluvial soils to plant cocoa on. Despite – or perhaps because of – the extensive use of herbicides and fertilizers, cocoa expansion can be expected to continue to provoke widespread deforestation in the early 2000s.

The introduction of cocoa and, more generally, of tree crops represents a technological change that is more labour-intensive and provides higher yields than ranching, and requires much less forest per unit of profit. In the tiny spaces of southern Côte d'Ivoire and Sulawesi, where the migrant pressure per unit of forest land is higher than in the Amazon, this seems to be one reason why tree crops, rather than animals, drive frontier migration and deforestation.

Technological progress in cocoa cultivation has increased both yield and the opportunities for using fallow instead of forests. For instance, over a 20-year lifetime, 1 ha of cleared primary forest in Sulawesi may generate 20–40 t of cocoa, compared with 10–20 t in Côte d'Ivoire. However, this higher yield does not reduce farmers' interest in clearing forest in Sulawesi, since it makes cocoa cultivation more profitable.

In Sulawesi, labour-saving technologies, such as herbicides in cocoa and paddy and hand tractors in paddy, have freed labour. This has pushed migrants to the hilly areas and made it easier to manage larger cocoa farms.

Herbicides not only save labour, they also make replanting and the use of grasslands more attractive. This can reduce deforestation in the short run. In the long run, however, herbicides cannot prevent the clearing of neighbouring forests if they remain abundant. Herbicides can help to protect forests if they are already scarce, difficult to gain access to or protected by other means. Overall, herbicides can be expected to have a limited long-term impact on deforestation, but a large impact on reforestation. Agricultural policies that favour the adoption of non-remnant herbicides should be a priority.

Labour-intensive or labour-neutral technologies, such as yam and cocoa intercropping in Côte d'Ivoire in the late 1960s and fertilizer adoption in the 1980s in Sulawesi, can also promote deforestation agents, just as labour-saving technologies do. When labour is cheap (migrants) and forests are abundant, the higher labour demand they create is not sufficient to counterbalance the effect of the higher profits they provide.

Generally, the type of technology matters less than the stage in the cocoa cycle, the degree of deforestation and the availability of labour less with regard to the impact of technological change on deforestation. The studies of Côte d'Ivoire and Sulawesi show that technological progress helps reduce deforestation only after large areas of forest are gone. Then a few reserves and national parks can be protected at reasonable cost. It may look trivial, but technological change has a complementary role to play in conserving these last forest reserves of a country. Institutional rules and their enforcement have to keep access to these forest areas difficult and risky. Technological change will not save these forests alone, but can help divert farmers' interests to fallows and grasslands, rather than to the remaining forests.

The case-studies also demonstrate that deforestation can trigger technical progress. In particular, once massive deforestation has taken place, farmers seem to innovate and adopt technologies for weed control and replanting. In addition to looking for technologies that reduce deforestation, policies should pay more attention to technologies and institutional environments that can ease and encourage replanting on fallow land, and thus promote reforestation. More than deforestation, which is almost a historical issue in many areas, reforestation is what is really at stake in regard to technological change.

The original local population and migrants have different behavioural patterns and adopt distinct technologies, in line with their knowledge, time horizons and access to land and labour. In Côte d'Ivoire, the local ethnic groups' forest-clearing techniques were more environmentally friendly than the migrants' method. Migrants from other countries, in particular, never know how long they will be able to stay in an area, and therefore have less incentive to protect the environment. Even if they are not sent home, most of them are determined to retire in their home village. They came to make quick money. Building a patrimony to transfer to the children is only a secondary objective.

Policies should take into account the social and institutional dimensions of technological change. Who can adopt, adapt or create new technologies? Smallholders with access to labour and capital are in a better position to buy or clear new land, and then create or adopt new technologies. In Côte d'Ivoire in the 1990s, those farmers came from the neighbouring Burkina Faso. If one combines such a situation with a bad land policy and unclear property rights, it can trigger disastrous conflicts, where technological change plays a role. For example, if herbicides and fertilizers make fallows more profitable and foreign migrants can buy more of them, conflicts about property rights to fallows and old plantations will escalate.

In many places, it is almost too late to reflect on the interaction between technical change and land tenure during the deforestation phase. What matters now is not to ignore the interaction between technical change and more secure land tenure during the replanting and potential reforestation phase. The cases of Côte d'Ivoire and Sulawesi show that the lack of formal (statutory) property rights does not necessarily deter short- and medium-term investments in cocoa orchards. In an increasingly risky social environment, however, more secure land tenure may facilitate longer-term investments, such as replanting of cocoa with timber trees.

References

Angelsen, A. (1999) Agricultural expansion and deforestation: modelling the impact of population, market forces and property rights. *Journal of Development Economics* 58(1), 185–218.

Berry, S. (1976) Supply response reconsidered: cocoa in Western Nigeria, 1909–44. *Journal of Development Studies* 13(1), 4–17.

Boserup, E. (1970) *Evolution Agraire et Pression Démographique*. Flammarion, Paris.

Chabrolin, R. (1965) La riziculture de tavy Madagascar. *L'Agronomie Tropicale* 1, 9–23.

Chauveau, J.P. and Leonard, E. (1996) Côte d'Ivoire pioneer fronts: historical and political determinants of the spread of cocoa cultivation. In: Clarence-Smith, W.G. (ed.) *Cocoa Pioneer Fronts since 1800: the Role of Smallholders, Planters and Merchants*. Macmillan, London, pp. 176–194.

Clarence-Smith, W.G. and Ruf, F. (1996) Cocoa pioneer fronts: the historical determinants. In: Clarence-Smith, W.G. (ed.) *Cocoa Pioneer Fronts since 1800: the Role of Smallholders, Planters and Merchants*. Macmillan, London, pp. 1–22.

Cruttwell-McFayden, R.E. (1991) The ecology of *Chromolaena odorata* in the neotropics. In: OSTOM and SEAMEO BIOTROP (eds) *Ecology and Management of Chromolaena odorata*. Biotrop Special Publications No. 44, Bogor, Indonesia, pp. 1–9.

de Rouw, A. (1991) Rice, weeds and shifting cultivation in a tropical rain forest: a study of vegetation dynamics. PhD thesis, Wageningen Agricultural University, the Netherlands.

Hill, P. (1956) *The Gold Coast Farmer*. Oxford University Press, Oxford.

Hill, P. (1964) *Migrant Cocoa Farmers of Southern Ghana: a Study in Rural Capitalism*. Cambridge University Press, Cambridge.

Lena, P. (1979) Transformation de l'espace rural dans le front pionnier du sud-ouest ivoirien. Doctoral thesis in geography, University of Paris X.

Levang, P., Michon, G. and de Foresta, H. (1997) Agriculture forestière ou agroforestière? *Bois et Forêts des Tropiques* 251(1), 29–42.

Monbeig, P. (1937) Colonisation, peuplement et plantation de cacao dans le sud de l'Etat de Bahia. *Annales de Géographie* 46, 278–299.

Naylor, R. (1992) Labour-saving technologies in the Javanese rice economy: recent developments and a look into the 1990s. *Bulletin of Indonesian Economic Studies* 28(3), 71–89.

Oswald, M. (1997) Recomposition d'une société au travers de plusieurs crises: la société rurale Bété (Côte d'Ivoire). Doctoral thesis, INA-PG, Paris.

Ricardo, D. (1815) Essai sur l'influence des bas prix du blé. In: *Oeuvres complètes de D. Ricardo*. Guillaumin, Paris, 1847, pp. 543–570.

Ruf, F. (1985) Politiques et encadrement agricole: partage des tâches en Côte d'Ivoire. In: *CIRAD: Etats, Développement, Paysans*. CIRAD-MESRU, Montpellier, pp. 14–27.

Ruf, F. (1987) Eléments pour une théorie sur l'agriculture des régions tropicales humides. I – De la forêt, rente différentielle au cacaoyer, capital travail. *L'Agronomie Tropicale* 42(3), 218–232.

Ruf, F. (1988) Stratification sociale en économie de plantation ivoirienne. Doctoral thesis in geography, University of Paris X, 6 vols.

Ruf, F. (1991) Les crises cacaoyères: la malédiction des âges d'or. *Cahiers d'Etudes Africaines* 31(121–2), 83–134.

Ruf, F. (1995a) *Booms et Crises du Cacao. Les Vertiges de l'or Brun*. Karthala, Paris.

Ruf, F. (1995b) From forest rent to tree-capital: basic laws of cocoa supply. In: Ruf, F. and Siswoputanto, P.S. (eds) *Cocoa Cycles: the Economics of Cocoa Supply*. Woodhead Publishing, Cambridge, UK, pp. 1–53.

Ruf, F. and Zadi, H. (1998) Cocoa: from deforestation to reforestation. Paper presented at the First International Workshop on Sustainable Cocoa Growing, 29 March–3 April 1998, Smithsonian Institute, Panama. (www.si.edu/smbc)

Temple, L. and Fadani, A. (1997) Cultures d'exportation et cultures vivrières au Cameroun: l'éclairage d'une controverse par une analyse microéconomique. *Economie Rurale* 239, 40–48.

Touzard, J.M. (1994) *L'Economie Coloniale du Cacao en Amerique Centrale*. Collection Repères, CIRAD, Montpellier.

Vuillet, J. (1925). Notes sur la culture du cacaoyer en Côte d'Ivoire. *L'Agronomie Coloniale* 91, 1–10.

Agriculture and Deforestation in Tropical Asia: an Analytical Framework

Sisira Jayasuriya

1. Introduction[1]

Almost every tropical Asian country has lost significant amounts of forest in recent decades. In many countries, particularly in South-East Asia, this occurred in a context of major demographic and economic change. The population grew and became increasingly urbanized. The economies expanded rapidly. Manufacturing industries became more important and the share of agriculture fell. Agricultural productivity improved substantially. These wider changes conditioned the deforestation process. This chapter examines the interactions between productivity-enhancing technological change in agricultural sectors and deforestation in this broad economy-wide context.

Population growth and rising food demand led to considerable forest clearing in many parts of Asia, largely to plant upland food crops (coarse grains, rice, maize, vegetables) and to establish plantations of crops such as tea, rubber and oil-palm. For example, the area under rubber cultivation in Indonesia, Malaysia and Thailand expanded from 260,000 ha in 1910 to almost 7,000,000 ha in 1990 – mostly at the expense of forests (Barlow *et al.*, 1994). In Malaysia and Indonesia, state-sponsored settlement schemes were instrumental in clearing large areas of forests for plantation crops grown by smallholders and large estates. Commercial logging also facilitated the conversion of forests to agriculture, particularly where using land for agriculture conferred property rights over it (Repetto and Gillis, 1988; Deacon, 1994; Cropper *et al.*, 1999).

Rice, Asia's main staple, is grown most widely in wet lowlands. Converting forests to such wet rice systems is typically expensive and not always feasible. Nevertheless, in some places large state-initiated irrigation-cum-settlement projects have converted forests into wet rice (paddy) land. In Sri Lanka, for example, such projects expanded rice cultivation by almost 250,000 ha between 1956 and 1988 and contributed to substantial forest loss (Natural Resources, Energy and Science Authority of Sri Lanka, 1991).

All of the previously mentioned developments occurred in widely differing policy contexts and institutional settings and involved a wide range of actors. Given this great diversity of situations and the many factors that influence deforestation in complex and often location-specific ways, any sweeping generalization is bound to be misleading. This chapter addresses only one narrow aspect of the issue, the links between technological change in agriculture and deforestation, and abstracts this aspect from the multitude of other factors that affect forest clearing. It uses a simple trade-theoretic framework to analyse those links under various scenarios designed to reflect some of the main deforestation-relevant situations observed in tropical Asia. Its approach is 'macro', rather than 'micro'. Thus, for example, it generally disregards the complications introduced by the intricacies of decision-making in semi-subsistence farm households. Throughout, its emphasis is on highlighting the main linkages and mechanisms that tie developments in other sectors to forestry, rather than on formal rigour.

The chapter focuses on economic agents' responses to incentives stemming from market forces. Admittedly, this approach has strong limitations since non-market factors, including government policies, influence and at times drastically modify incentive structures. Non-market factors are also important in places where farmers who are only partially integrated into markets practise semi-subsistence farming. Nevertheless, this kind of analytical approach still has considerable value and relevance. Market factors dominate large areas of the economy and economic considerations temper government decisions, even if they do not entirely determine them.

Section 2 examines the links between technological change in agriculture and deforestation in a 'neoclassical' economy that produces two goods in two sectors, 'agriculture' and 'forestry', which compete with each other for land and labour. The first scenario in this section involves a small, open economy that faces exogenously determined output prices. The second scenario focuses on situations where technological change can affect output prices by influencing supply and demand.

Section 3 takes into account regional differences in agriculture and the fact that some factors are used to produce certain goods but not others. In the upland region agriculture competes with forestry for land, but in the lowlands it does not. Once again, we have one scenario with exogenous output prices and a second with endogenous output prices. In both cases, wages and land prices are endogenous. After that, we look at what happens when one introduces more labour-intensive forms of technological change. Then we consider situations, albeit in a simplistic fashion, when property rights on forested land are not well defined (or poorly enforced).

Section 4 examines the impact of technological changes in lowland agriculture, such as the Green Revolution. Here we examine not only the implications of different assumptions about output markets and property rights but also the income effects generated by technological change and the impact of technologies biased towards the use of capital.

By varying our assumptions regarding the structure of the economy, how output and factor markets behave and how the forest and agricultural sectors interact, we can obtain important insights about several of the most commonly observed situations involving deforestation in tropical Asia. These include situations where forests compete with internationally traded agricultural commodities, such as oil-palm, rice and rubber, as well as situations where they compete with subsistence crops (e.g. coarse grains) or products orientated to domestic markets (e.g. cool-climate vegetables). There are also important differences between upland agriculture, which directly competes with forests, and lowland agriculture, which for the most part does not. Distinguishing between these two types of agriculture provides richer, and sometimes distinct, insights than those gained from models that treat agriculture as a single undifferentiated sector. Table 17.1 summarizes the different scenarios covered in the chapter and their respective outcomes.

The chapter does not attempt to provide a rigorous welfare evaluation of the outcomes it analyses. How different outcomes affect welfare is not always clear. There are many equity considerations and forests provide important externalities whose welfare benefits are difficult to measure.

2. Model 1: One Region and Two Sectors

Our 'base' case draws insights from the standard neoclassical two goods–two factors model (the Heckscher–Ohlin model).[2] The economy allocates all its resources to producing two commodities, 'forestry products' (F) and 'agricultural products' (A), whose prices are P_F and P_A, respectively. Even though we call the non-forest sector 'agriculture', if we interpreted it more broadly as the 'rest of the economy' and assumed it produced a 'non-agricultural' good, that would not change the essence of our story.

We have two factors (resources): land (T) and labour (L). The production functions have the usual neoclassical properties. Factor proportions vary in response to factor prices. Both goods and factor markets are perfectly competitive. Production exhibits constant returns to scale. Factors can move freely and without cost between the forestry and agricultural sectors. Property rights are well defined and enforced. Agents have full information, so there are no risks or uncertainties. We abstract from time considerations by using a one-period model. Throughout our discussion we assume agriculture is more labour-intensive than forestry. While we have been unable to obtain reliable data on labour use in forestry, this assumption seems reasonable.

2.1. Case 1.1: the small open economy

Our first scenario assumes that the economy is a price-taker in world markets (i.e. 'a small country'). This implies that the world market exogenously determines all output prices. The production possibilities frontier (PPF_0) in Fig. 17.1 represents the maximum combinations of F and A the economy can produce with the initial set of resources and technology. It is concave, because the marginal cost of converting forest to other land uses rises as more area gets converted. Relative prices determine where on the PPF_0 the economy will end up. Given the initial PPF_0 and relative prices in Fig. 17.1, production will take place at point X_0 and land and labour will be allocated between the two sectors so as to produce F_0 and A_0.

Now consider the impact of a productivity-improving technological change in the agricultural sector. We assume for simplicity that technological change is factor-neutral (i.e. that the productivity of both factors increases by an equal amount). The PPF will move out, but not symmetrically: greater output with the same level of factor inputs is possible only in agriculture. The new production point is at a point such as X_1. At that point, agricultural output is higher than at point X_0 and forestry output is lower. Given that land and labour are the only two factors of production, this implies that both labour and land have moved out of forestry. In other words, deforestation has increased. Clearly, if agriculture becomes more productive, as long as commodity prices remain constant, it is rational to convert more forested land to agriculture. In this setting, a Green Revolution in agriculture will mean more deforestation. Similarly, if only the forestry sector experienced technological change, agriculture would contract.

This conclusion remains valid even though the higher national income that results from technological change increases demand for both agricultural

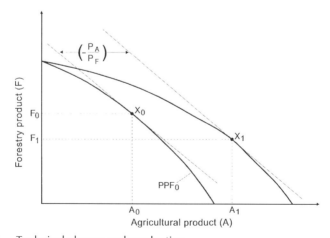

Fig. 17.1. Technical change and production.

and forest products. Given free trade, this higher demand does not translate into higher prices in a price-taking country. Imports can always meet excess demands at the prevailing international market price.

Technological progress also affects factor prices. Figure 17.2, which shows the so-called 'iso-profit' curves, illustrates the basic mechanism. These 'iso-profit' curves represent the combination of factor prices that are consistent with zero profits for a given technology and output price, and their shape reflects the elasticity of factor substitution.[3] Zero profits means that industries cannot earn anything more than the opportunity cost of the factors they invest in production. As long as markets are perfectly competitive, in equilibrium, profits must equal zero and factor price ratios must be consistent with that. Otherwise producers would expand production to take advantage of the available profits.

Technological progress allows each unit of a factor to produce more output. Because profits cannot rise above zero, this implies that producers must pay more for their factors. If technological change caused producers to make profits in the A sector, they would demand more labour and land and bid up factor prices until they reached the point where they could no longer make profits. At the initial production point X_0, the factor price ratio is $(w/r)_0$. However, if technological progress in agriculture shifts the iso-profit curve up and establishes a new equilibrium at a point such as X_1, factor prices will adjust to $(w/r)_1$. As long as agriculture always remains more labour-intensive than forestry and the economy continues to produce both A and F, w will increase relative to r.

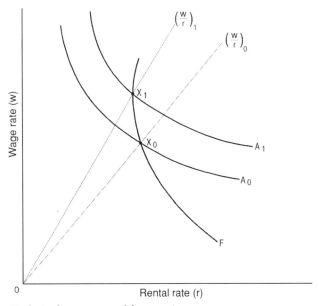

Fig. 17.2. Technical progress and factor prices.

Price increases and neutral technological changes have similar effects. They both lead to upward shifts in the iso-profit curve and similar changes in factor returns. If the technological change is not neutral, the new iso-profit curve will still shift upward but will not have the same shape. Its shape will reflect the factor bias incorporated in the new technology, which, in turn, will influence the resulting relative factor price configuration.

2.2. Case 1.2: endogenous output prices (Table 17.1)

Consider now the case where output prices are no longer exogenous. This implies that changes in the domestic supply of and demand for agricultural or forest products can affect prices. Consequently, the effects of higher supply and higher demand on agricultural prices will temper the impact of neutral productivity change in agriculture. Following technological change in agriculture, the greater physical productivity of factors devoted to agriculture attracts them to that sector, which expands at the expense of forestry. Technological change increases aggregate output and consequently real national income. This shifts the demand curve for both goods upwards. The magnitude of that shift depends on the income elasticity of demand for each product. At the same time, larger supplies depress prices, by an amount that is determined by the price elasticity of demand. The net impact on price depends on the relative weight of these two effects.

As long as an expansion in agricultural output makes its relative price go down, the relative price line will be flatter when prices are endogenous. Hence, the equilibrium will be at a point to the left of X_1. This implies that the new equilibrium level of agricultural output will be lower and the associated resource movement effects more muted. In the extreme case of very low demand elasticities for agricultural products, technological progress can reduce agricultural profitability so much via lower prices that ultimately agriculture may use fewer resources (even though its output will be higher than before). This may also be the result if the country is a net exporter of agricultural products and world demand is highly inelastic – the case of so-called 'immiserizing growth' (Bhagwati, 1968).

3. Model 2: Two Regions and Three Sectors: Technological Change in Upland Agriculture

Let us turn now to a more complex economy and to the more realistic case where land may not be able to easily move between sectors. In this section, we draw on land-use change models developed by Coxhead and Jayasuriya (1994, 1995) to analyse the interactions between agriculture and forestry when there are two separate types of agriculture. We use the Jones specific-factor model (the Ricardo–Viner model) to examine the question of factor mobility and factors that can only be used to produce certain goods (Jones, 1971).

In our new scenario, there are two regions – the uplands and the lowlands. Upland producers can use land either for forestry or 'upland' agriculture (UA). Hence, UA competes directly with forestry for land. In the lowland, land can be used only for 'lowland' agriculture (LA). Hence LA does not compete with forestry for land, although it still competes with forestry for labour.

Our analysis ignores the income effects of technological change in the uplands on the demand for agricultural and forestry products, both for simplicity and because in many real-world situations they are likely to be small. If such income effects were significant, they could alter the pattern of demand and have a particularly strong effect on the prices of non-tradable goods.

3.1. Case 2.1: output prices fixed, input prices endogenous

We start with the case where all forestry and agricultural products are internationally traded at prices determined by the world market and face perfectly elastic global demand. This implies that technological change will have no effect on output prices. We also assume that labour is fully mobile between all three sectors.[4]

Now consider the effect of a neutral technological change experienced by (and confined to) upland agriculture, UA.[5] Higher productivity in UA increases its profitability. As producers expand their output in response to the higher profitability, this generates a 'resource pull' effect, which attracts resources from other sectors in the economy. Agriculture will pull land (and labour) away from the forestry sector. In other words, there will be more deforestation and labour will migrate from the lowlands to the uplands. As a result, LA output and land values will fall, but national income will increase and wages will rise throughout the economy.

This model captures the kind of mechanisms that are at work in many South-East Asian countries when commercial crops (e.g. rubber, oil-palm) that compete with forests experience technological progress. Typically, these crops are exported and world prices are determined exogenously. Even when domestic supplies affect international prices, as in the case of rubber in Indonesia, coconuts in the Philippines and tea in Sri Lanka, the effect is small, since foreign demand elasticities are high. Hence the price-depressing effect of technological progress is minor. The dominant effect is to make these crops more profitable, which accelerates deforestation.

The extent to which wages increase as a result of technological change in UA depends partly on the elasticity of lowland labour supply. If the lowland sector is large relative to the uplands, UP producers should be able to attract the additional labour they need without increasing wages much. Thus, LA would probably not contract much but deforestation would advance rapidly,

Table 17.1. Summary of models, scenarios and effects on forest cover.

Case	Type of model	Assumptions	Technological change	Effects on forest cover
1.1	Two sectors (forestry and agriculture); two inputs (land and labour); agriculture more labour-intensive	Output prices fixed, factor prices endogenous	Technological change in agriculture	Reduced
1.2	As in case 1.1	Endogenous (agricultural) output price	As in case 1.1	Reduced (most probably) but less than in case 1.1
2.1	Upland: forestry and agriculture (UA); lowland: agriculture (LA)	All output prices fixed, factor prices endogenous	Technological change in UA	Reduced
2.2	As in case 2.1	UA price endogenous	As in case 2.1	Reduced (most probably)
2.3	As in case 2.1	All output prices fixed, factor prices endogenous	Labour-biased technological change in UA	Constrains short-term deforestation
2.4	As in case 2.1	Ill-defined property rights	As in case 2.3	Even larger reduction

3.1	As in case 2.1	Well-defined property rights, all output prices fixed	Technological change in LA	Increased
3.2	As in case 2.1	UA output price endogenous	As in case 3.1	Increased when income elasticity for UA products high, decreased when low
3.3	As in case 2.1	Well-defined property rights, all output prices fixed	Capital-biased technological change in LA	Increased (compared with neutral technological-change case)
3.4	As in case 2.1	Ill-defined property rights	As in case 3.3	Increased incentives for deforestation
3.5	As in case 2.1	Agricultural prices exogenous, forest-product prices endogenous	As in case 3.3	Increased when forest products are normal goods and property rights are well defined. Decreased if you have normal good but poorly defined property rights. If forest products are inferior good, contradictory labour pull (+) and land pull to UA (−)
3.6	As in case 2.1	Land mobile between UA and LA	As in case 3.3	Reduced
3.7	Upland: forestry and lowland: rest of the economy (ROE)	All output prices fixed, factor prices endogenous	High growth (technological progress) in ROE	Contradictory labour pull (+) and demand (−) effects

because wage increases would not dampen the stimulus for forest clearing generated by technological change. On the other hand, if the upland labour market is rather large compared with the lowland labour market, technological change may bid up wages, thus discouraging further deforestation.

3.2. Case 2.2: endogenous output and input prices

In this scenario, forestry and lowland agricultural products (e.g. rice) are tradable and the world market determines their price. However, UA either produces solely for the domestic market (e.g. vegetables) or produces for international markets in which the country has significant market power. In such a situation, changes in domestic supply and demand will influence the price of UA products. As long as the income effects are minor, increases in supply induced by technological progress will lower the price of UA products. The degree to which higher deforestation will occur will depend on how much higher supplies lower prices and how farmers respond to the price declines.

3.3. Case 2.3: labour-biased technological change

To get some insight into the impact of a non-neutral technological change, consider a situation where UA uses labour intensively and the upland labour supply is fixed (i.e. it cannot be 'imported' from the lowland region). One example might be a case where upland farmers have special skills required for implementing the new technology.

Now consider a technological change that is biased in favour of skilled labour. The adoption of the new technology increases the profitability of UA and the returns to labour. Unless the increase in output supply dramatically depresses prices, the sector will expand and pull in additional labour and land. Land, of course, comes from forests, implying deforestation. But, by definition, the supply of labour is fixed, and this constrains UA's ability to expand. Thus, limited labour availability reduces the scale of deforestation.

In a longer time frame, the new technology will enhance the incentives for acquisition of the upland skills. In such a situation, lowland labour will be encouraged to seek these skills and will demand services that can transfer these skills. Hence, the long-term supply elasticity of labour is likely to be higher and the technology is likely to generate more deforestation in the long term than in the short term.

3.4. Case 2.4: ill-defined property rights

So far we have maintained the assumption that forestry is like any other production sector. If forests were commercial plantations, with well-defined and enforced property rights, that might not be unreasonable. But most forests in tropical Asia are on state-owned land. Consider what may happen in the case where deforestation and subsequent conversion of land to agriculture may confer (or enhance the probability of obtaining) property rights to land. One observes such situations in many parts of tropical Asia (Angelsen, 1999).[6]

How technological progress in a given sector affects factor returns depends in general on the elasticity of factor substitution and on how commodity prices respond to changes in supply and demand. As illustrated earlier, in our simple two sector/two factor Heckcher–Ohlin-type economy, neutral technological progress in one sector has similar effects to an increase in its price. Among other things, it increases the price of both factors used in the expanding sector. Higher land values accelerate the 'race for property rights' (Anderson and Hill, 1990), since people now have a greater incentive to attempt to establish property rights by cutting down forests and 'squatting'. Hence any improvement in the productivity of agriculture in regions where it directly competes with forestry will accentuate deforestation even more than in situations where producers have well-defined property rights over forested land.

Using this same framework, we can also analyse the impact of higher timber prices in situations where property rights to the forest are not secure. Obviously, if timber prices went up permanently, logging would be more profitable. Thus, in a neoclassical economy with well-defined property rights, forestry would expand at the expense of agriculture. But, even if the higher prices were expected to be 'permanent', without secure property rights, the result would only be increased logging of current tree stocks.

Hence, if logging a forest makes it easier to convert forests to agricultural land, higher timber prices or technological change in forestry may accelerate deforestation rather than reducing it. Since such situations appear to occur frequently in practice, this explains why attractive timber prices may contribute to deforestation in much of tropical Asia.

4. Model 3: Technological Change in Lowland Agriculture

The 'Green Revolution' in tropical Asia is probably the best-known example of technological change in agriculture in recent years. It was associated with the use of high-yielding rice varieties in 'wet' lowlands and basically bypassed UA (Barker and Herdt with Rose, 1985). We can consider it an example of technological change in non-directly competing LA and analyse its impact on forests in that framework.

4.1. Case 3.1: the small open economy

Initially, we shall assume that LA produces tradable goods, with exogenously determined prices. Since rice has a well-established international market and most tropical Asian countries import or export it, this assumption is generally valid. (We would not want to push this assumption too far, however. The international rice market is rather 'thin' and major exporters, such as Thailand and Vietnam, and importers, such as Indonesia, have some market power.)

As long as higher output does not influence prices, the higher productivity in LA will always make the sector more profitable. This will raise the marginal product of labour in LA and induce lowland producers to offer higher wages to attract upland labourers. In other words, the Green Revolution will stimulate workers to migrate from the uplands to the lowlands. This process becomes all the more important as barriers to interregional labour mobility come down, a trend that has gathered momentum over time.[7] (If we allowed higher supply to depress the price of LA output, these effects would be dampened.)

What happens to the UA and forestry sectors? The upland economy as a whole will contract, because it loses labour to the 'booming' lowlands. But the impact on the two upland sectors differs. If we hold UA output prices constant, we can think of the uplands as a 'mini-Heckcher–Ohlin economy' whose labour force has contracted. Under these circumstances the Rybczynski theorem tells us that the more labour-intensive UA sector should contract. This should lead some marginal upland lands to shift from agriculture to forestry, which would be good for forests.

4.2. Case 3.2: endogenous output prices

However, if supply and demand changes influence UA output prices, the latter may not stay constant. The higher productivity of LA increases national income. This may raise or lower the demand for UA outputs and consequently influence their price. Such income effects may be quite large; note that the Green Revolution has been credited with a significant decline in real rice prices and rice is the main staple food in tropical Asia.

If UA outputs have low or negative income elasticities of demand, because they are 'less preferred' goods, such as coarse grains, it may take a substantial contraction in UA output to establish a new equilibrium. This would favour forestry. In that case, the negative effects of higher wages on agricultural production will outweigh the offsetting impact of higher demand (due to higher lowland incomes) and resources will flow out of the UA sector. If LA were to be thought of as irrigated rice, then the impact of a Green Revolution in irrigated rice would tend to contract the coarse-grains sector, thereby reducing deforestation.

If, on the other hand, demand for UA output is highly income-elastic (for example, if UA is 'temperate-climate fruits and vegetables'), the demand effect may dominate. Improvements in LA increase people's incomes. This leads them to demand more UA output, the price of which increases. In response, UA will expand and pull resources from forestry, thus aggravating deforestation.

Examples of such situations can be seen in some highland areas in Malaysia and Sri Lanka.

If higher LA output depresses the price of lowland products and LA and UA products are substitutes, technological change in LA will have two opposing effects on UA. First, the increase in profitability of LA will become more muted. This dampens the labour migration from uplands to lowlands and eases the labour-cost pressure on UA. On the other hand, the lower price of LA output pushes down the price of UA output and makes it less profitable to produce. The degree to which this adversely affects the UA sector depends on the relative strength of these forces. In many parts of Asia, the Green Revolution in rice has made it less attractive to grow substitute food crops in the uplands and this has undoubtedly had a positive effect on forests. However, government policies, such as granting protection to upland food crops, have at times undermined this effect (Coxhead, 2000).

4.3. Case 3.3: capital-biased technological change

Thanks to development strategies centred on promoting industries that substitute for imports, in the past new lowland technologies were often capital-intensive. The capital bias in technology reduces the upward pressure on labour demand and wages and thereby limits the pull of labour from the uplands. As a result, upland labour–land ratios remain high, which favours UA over forestry, since the former is more labour-intensive. This leads to greater deforestation than would occur in the neutral technological-change case.

Thus, productivity improvements in LA generate two opposing influences on UA. The cost effect tends to contract it, while the demand effect works to expand it. The net effect on UA, and hence on forestry products, cannot be predicted a priori. It will depend on the magnitudes of the relevant supply and demand parameters, including the demand for UA produce.

4.4. Case 3.4: ill-defined property rights

How the presence of poorly defined property rights influences the impact of the Green Revolution on forests depends crucially on whether the Green Revolution promotes or discourages UA. If it leads UA to contract, that effect will be less stronger when property rights are poorly defined or enforced, and so deforestation will decrease less than it would have otherwise. Arguably, though not quantified in empirical studies, this impact of the Green Revolution on deforestation has probably been quite important in many parts of Asia. On the other hand, in places where the Green Revolution encourages UA, that effect is also likely to be stronger in situations with poorly defined or enforced property rights.

4.5. Case 3.5: endogenous forest-product prices

Until now, we have assumed that forestry products have perfectly elastic demand. In many situations, this is not realistic. For example, nearby farm households consume a large part of forest produce in the form of fuel wood and timber. Consider a situation where domestic markets determine the price of forestry produce. To enable us to continue with a simple trade model and to focus on this aspect of the problem, let us now assume that UA produce is internationally traded and has perfectly elastic demand and that there are no property rights problems.

First consider the case where demand for forestry produce declines with income growth. In other words, these products are inferior goods. This may apply, for example, to fuel wood. In this situation, if property rights to forested land are well defined, lowland productivity improvements will have opposing effects on deforestation. Greater lowland employment opportunities will pull labour away from the uplands. As a result, UA, which is the relatively more labour-intensive upland sector, will contract and the forestry sector will tend to expand. However, reduced demand for forest produce will reduce the value of forests and encourage land to shift to UA. The outcome depends on the magnitudes of these forces.

If demand for forestry products (say, timber or amenity values) increases with higher lowland income, that is, they are normal goods, then both effects mentioned in the previous case are pro-forestry. The same basic insight carries through to the case where forested land can be converted to provide an intermediate good into lowland production. For example, upland forests can be cleared to provide irrigation and power, the demand for which increases with lowland growth, and this will tend to increase deforestation.

The lack of secure property rights would again modify these results. An increase in demand for forestry products, such as timber, may again lead to greater deforestation, as the incentives to log the current trees facilitates conversion of forests to agricultural land.

4.6. Case 3.6: land mobile between sectors

If it were technically possible to convert land in the upland region to 'lowlands' suitable for producing LA (e.g. convert 'dry lands' into wet rice lands), the incentives to do so would increase when technological progress raises the profitability of LA. Such changes will effectively make land mobile between sectors. This will tend to increase deforestation. Thus it cannot be assumed that more productive lowland technologies will invariably reduce deforestation. Since such technological changes encourage the conversion of uplands to lands suitable for LA, governments (and donors) may be encouraged to pursue schemes designed to produce these now more productive crops, as in the transmigration programme in

Indonesia and irrigation-cum-settlement schemes in Sri Lanka, with greater vigour.

4.7. Case 3.7: economic growth and interregional labour movements

While we have labelled the non-upland economy as the 'lowland' agriculture sector, it can be considered more generally as the 'rest of the economy' (ROE) and its land endowment can be thought of as composite, sector-specific capital stock. Many factors can stimulate growth in this sector, such as technological progress, an increase in its capital stock, due to, say, foreign investment, or an increase in the world price of its output. In all such cases, the impact on forestry will be mediated through the two main effects on the labour and commodity markets: the labour pull effect, which attracts labour away from the upland region, and the income growth-induced increase in demand for upland produce. In this framework, faster growth in the ROE will reduce deforestation, provided UA does not produce a highly income-elastic product, whose price might increase as per capita incomes go up.

If legal restrictions on regional labour mobility (such as in China) or socio-economic factors hindering labour mobility (such as costly transport) make it more difficult for workers to move from the highlands to the lowlands, the labour pull effect becomes weaker. Under such circumstances, lowland producers must pay wages high enough to overcome the 'transport cost' to attract upland labour. This reduces demand in the lowland region for labour originating in the upland. The reverse occurs when labour relocation costs are lowered.

A similar observation can be made with respect to commodity markets, if interregional transport is costly. Reductions in these barriers to mobility – due, for example, to better roads, communication facilities, etc. – all tend to increase the impact of ROE developments on the uplands. The pro-forestry effect of ROE growth increases with greater market integration.

The rapid economic growth in many parts of Asia has probably had its most important pro-forestry impact via the labour market. The pull of labour away from the uplands has probably strongly reduced the incentives to convert forests to food agriculture, though government policies that actively encouraged competing agriculture (e.g. agricultural settlement schemes such as the transmigration programme in Indonesia) have often counteracted this impact.

5. Concluding Remarks

In this chapter, we have used a number of situations observed in tropical Asia to motivate a simple trade-theoretical analysis of the implications of technological progress in agriculture. The models have recognized the factor and commodity market linkages between agriculture, forestry and other sectors in the economy, which serve as conduits through which technological progress and other changes in one sector transmit their influences throughout the economy.

We have ignored many aspects of the deforestation problem to focus on a few (in our opinion, quite important) issues. For example, our analysis does not consider the impacts of externalities and policy-induced distortions, and we have therefore refrained from making a welfare assessment of the outcomes. We also largely ignore the role of policy-induced distortions in both commodity and factor markets, which not only modify the impact of technological progress, but also influence the nature and pace of technology generation and adoption.[8] The analysis is static and does not explicitly address issues related to market imperfections (including missing markets). Thus we abstract from time-related issues as well as expectations, imperfect information and risk/uncertainty considerations. We treat property rights only in a very limited manner.

Despite these many limitations, even this simple analysis sheds light on some of the main mechanisms through which technological progress in agricultural sectors has an impact on deforestation, and helps to identify some important factors that condition the nature of that impact. In particular, it shows that the impact of technological progress in agriculture on forestry depends crucially on the degree to which agriculture that experiences such technological change directly competes with forestry for land. Thus, productivity improvements in crops such as rubber, tea, oil-palm or coffee, which are likely to compete for forested land, will aggravate deforestation, while the Green Revolution in wet rice agriculture, which reined in real food prices and increased agricultural employment, may have had a significant pro-forestry effect. However, the effect of low prices for food produced in the lowlands may not always be benign; lower food prices raise incomes and can stimulate demand for upland products, which may lead to increased deforestation.

Notes

1   With the usual caveat, I thank Ian Coxhead (much of the analysis in this chapter draws on collaborative work with him), Arild Angelsen, Mary Amiti, David Kaimowitz and participants at the Workshop on Technological Change in Agriculture and Deforestation in March 1999 at CATIE, in Costa Rica.
2   See any standard international economics text for the complete set of assumptions and properties of this model.
3   For an exposition in an agricultural setting, see Coxhead (1997).
4   Our 'upland' region is similar to Angelsen's (1999) model III, 'the small open economy with private property', but unlike in model III our wage rates are endogenous.
5   Here we draw on the so-called 'booming sector and Dutch disease economics' literature (see Corden, 1984).

6   For trade-theoretic models that analyse the impact of open access to forested land, see Brander and Scott Taylor (1994) and Deacon (1995).
7   For evidence from the Philippines, see David and Otsuka (1993).
8   The wider issue of technology generation in a distorted policy environment is addressed by Coxhead (1997).

References

Anderson, T.L. and Hill, P.J. (1990) The race for property rights. *Journal of Law and Economics* 33, 177–197.

Angelsen, A. (1999) Agricultural expansion and deforestation: modelling the impact of population, market forces and property rights. *Journal of Development Economics* 58(1), 185–218.

Barker, R. and Herdt, R.W. with Rose, B. (1985) *The Rice Economy of Asia.* Resources for the Future, Washington, DC.

Barlow, C., Jayasuriya, S. and Tan, S. (1994) *The World Rubber Industry.* Routledge, London.

Bhagwati, J.N. (1968) Distortions and immiserizing growth: a generalization. *Review of Economic Studies* 35, 481–485.

Brander, J.A. and Scott Taylor, M. (1994) *International Trade and Open Access Renewable Resources: the Small Open Economy Case.* NBER Working Paper Series No. 5021, National Bureau of Economic Research, Cambridge, Massachusetts.

Corden, W.M. (1984) Booming sector and Dutch disease economics: survey and consolidation. *Oxford Economic Papers* 36, 359–380.

Coxhead, I. (1997) Induced innovation and land degradation in developing country agriculture. *Australian Journal of Agricultural and Resource Economics* 41(3), 305–332.

Coxhead, I. (2000) Consequences of a food security strategy for economic welfare, income distribution and land degradation: the Philippine case. *World Development* 28(1), 111–128.

Coxhead, I. and Jayasuriya, S. (1994) Technical change in agriculture and land degradation in developing countries: a general equilibrium analysis. *Land Economics* 70(1), 20–37.

Coxhead, I and Jayasuriya, S. (1995) Trade and tax policy reform and the environment: the economics of soil erosion in developing countries. *American Journal of Agricultural Economics* 77(4), 631–634.

Cropper, M., Griffiths, C. and Mani, M. (1999) Roads, population pressures, and deforestation in Thailand, 1976–1989. *Land Economics* 75, 58–73.

David, C. and Otsuka, K. (1993) *Modern Rice Technology and Income Distribution in Asia.* L. Reinner Publishing, Boulder, Colorado.

Deacon, R.T. (1994) Deforestation and the rule of law in a cross-section of countries. *Land Economics* 70, 414–430.

Deacon, R.T. (1995) Assessing the relationship between government policy and deforestation. *Journal of Environmental Economics and Management* 28, 1–18.

Jones, R. (1971) A three factor model in theory, trade and history. In: Bhagwati, J., Mundell, R.A., Jones, R.W. and Vanek, J. (eds) *Trade, Balance of Payments and Growth: Essays in Honour of C.P. Kindleberger.* North Holland, Amsterdam.

Natural Resources, Energy and Science Authority of Sri Lanka (1991) *Natural Resources of Sri Lanka: Conditions and Trends*. Natural Resources, Energy and Science Authority of Sri Lanka, Colombo, Sri Lanka.

Repetto, R. and Gillis, M. (eds) (1988) *Public Policies and the Misuse of Forest Resources*. Cambridge University Press, Cambridge.

18 Deforestation, Irrigation, Employment and Cautious Optimism in Southern Palawan, the Philippines

Gerald Shively and Elmer Martinez

1. Introduction[1]

Rapid population growth in agricultural frontier regions contributes to forest depletion in developing countries. Thanks largely to in-migration from other parts of the country, population growth in the Philippine frontier province of Palawan has been particularly high (4.6% per annum) (Western, 1988). As a result, agriculture there has expanded into marginal and environmentally sensitive areas. Upland deforestation is acute (Sandalo, 1996) and the efforts of low-income individuals to earn incomes by establishing farms drive much of that. Finding ways to improve rural incomes without jeopardizing forest resources is important in Palawan, as elsewhere.

To intensify and raise agricultural production, the Philippine National Irrigation Administration (NIA) recently constructed or upgraded a number of small-scale communal irrigation systems in Palawan. These systems are in the lowlands, but most are adjacent to inhabited upland forest areas. A priori, the net impact of this new irrigation infrastructure on employment is ambiguous. Irrigation facilitates multiple cropping, thereby increasing the effective area under cultivation. This increases the demand for labour. But, at the same time, irrigation can induce farmers to adopt labour-saving production practices. For example, many researchers have observed that farmers who irrigate often adopt labour-saving methods, such as mechanization or chemical-based weed control (Castillo *et al.*, 1983; Kikuchi and Hayami, 1983; Coxhead and Jayasuriya, 1986; Boyce, 1993; Lingard, 1994).

This chapter examines how the introduction of lowland irrigation systems, a form of technical progress, has affected the local demand for labour

©CAB *International* 2001. *Agricultural Technologies and Tropical Deforestation* (eds A. Angelsen and D. Kaimowitz)

and, by extension, farmers' activities near the forest margins in Palawan. We compare agricultural outcomes in two adjacent and similar rice-farming communities, one newly irrigated and the other rain-fed. The central question we address is whether irrigation development has reduced pressure on upland forests through its impact on the labour market. To answer this question, first we measure how much irrigation has raised the demand for labour on lowland farms and local agricultural wages. Then we examine how upland farmers have responded to new off-farm employment opportunities. We show that by raising the opportunity cost of labour, the new job opportunities in the lowlands induce farmers to participate less in poorly remunerated activities, such as forest clearing and forest-product extraction. Employment on irrigated lowland farms acts as a magnet, drawing upland labour away from such activities.

Section 2 presents the basic framework of our analysis. Section 3 describes the data used, which come from a 1997 farm survey in southern Palawan. Section 4 reports the results. We found that irrigated farms demand less labour per hectare in each cropping season than rain-fed farms. Even so, since they grow more crops per year, it turns out that irrigated farms demand more total labour. Some of the additional workers hired came from the uplands, and upland households that obtained additional employment reduced their forest clearing by a small but statistically significant amount. Section 5 summarizes the key policy implications of our findings.

2. Lowland Technical Progress and Upland Labour Allocation

To analyse how introducing irrigation affects lowland farms, it is useful to think in the following terms. Imagine that, initially, lowland farms all use one pre-existing technology (rain-fed rice production) and local labour markets are in equilibrium. Farmers rely solely on family labour or combine family, shared and hired labour. They pay hired labour a fixed wage and keep hiring additional labour until the value of its marginal product equals the wage.[2] Now suppose an innovation, such as the construction of an irrigation system to store and deliver water, takes place. If this innovation raises the productivity of labour, farmers will hire more workers. This increase in employment may take the form of more labour being used during a single cropping season, a rise in the number of crop seasons per year, or both. We use the term effective labour demand to indicate the total amount of labour used on a hectare of land in a calendar year. The distinction between how much labour farmers require to farm 1 ha of land in a season and how much they demand in an entire year is important, since irrigation may induce farmers to use less labour in a given season but more labour over the course of the year. Any increase in the effective labour demand will raise the wage rate, since potential workers will require higher wages to be drawn away from alternative activities. Through this mechanism, technical innovation in the lowland sector may influence activities in the upland sector via labour demand and wage effects.

To fully understand how these mechanisms work, it helps to develop a formal framework for analysing how upland households allocate labour. For simplicity, we set aside several issues we cannot adequately explore here. Our framework is static and we assume that labour is the only resource households allocate. We further assume that households have a homogeneous pool of available labour, which they allocate to maximize their economic returns. This implies that how much labour households supply and how they allocate that labour do not depend on their levels of income.

We assume that upland households devote their labour to some combination of three income-generating activities: upland on-farm agricultural production (L_U); forest activity (L_F); and off-farm work in the lowland agricultural sector (L_O). The price of the outputs associated with upland farming and forest activities (P_U and P_F) determines the returns from those activities. The amount of output produced by the two activities depends exclusively on the amount of labour devoted to them. We assume that the production functions ($Y(L_U)$ for agriculture and $F(L_F)$ for forestry) exhibit decreasing returns to use of labour. When an upland resident works on a lowland farm, he receives an exogenously determined wage w. This lowland wage is set in a competitive market and depends on the technology of lowland production, which we denote θ. Upland households seek to maximize profit, defined as:

$$\pi = P_U(L_U) + P_F F(L_F) + w(\theta) L_O,$$

subject to a constraint on total available labour, namely $\bar{L} = L_U + L_F + L_O$. If a household engages in all three activities, then the optimal allocation of labour will occur where the value of the marginal product of labour is equal across activities. The amount of labour households allocate to a given activity will depend on labour productivity in the three activities and all prices, including the lowland wage. Admittedly, in many instances, not all households engage in each activity. In addition, the local economy may not demand as much hired labour as households wish to supply and, if wages do not fall enough to clear the market, employers may ration available jobs by non-price mechanisms. If no markets exist for certain products, households will allocate their labour based on implicit shadow prices, which may deviate from market prices due to transaction costs, risk aversion and the covariance of risks across activities (Sadoulet and de Janvry, 1995). None the less, the simple framework outlined above still provides a useful starting-point for analysing optimal labour allocation.

We are now in a position to develop our main hypothesis. Consider a change in lowland technology that leads to an increase in the agricultural wage. Households could now earn more if they shifted some of their labour from upland farming or forest clearing to working off-farm. In other words, the change in wage rates leads households to re-equate marginal returns to labour. If, as seems reasonable, all three activities exhibit diminishing returns to labour use, the only way they can do this is by reducing both L_U and L_F (so the values of their associated marginal products rise). Which falls more depends on the technical characteristics of production. Nevertheless, the underlying logic leads us to a testable hypothesis: irrigation development in the lowland agricultural sector reduces participation in forest-degrading activities.

3. The Data and the Survey Area

We collected the data used for this study on lowland and upland rice farms in two communities of southern Palawan in 1997. The lowland sample includes data from 56 farms in Marangas (municipality of Bataraza), of which 46 (82%) were irrigated, and data from 42 farms from Tamlang (municipality of Brooke's Point), of which 35 (83%) were rain-fed. This represents 38% and 34% of each community's population, respectively. The upland sample includes 104 farms adjacent to the lowland study areas (50 from Marangas and 54 from Tamlang), which are all on or near the forest margin. These represent approximately 30% of the underlying population. Figure 18.1 indicates the location of the study sites. Martinez and Shively (1998) discuss the sites and surveys in greater detail.

The study area has a distinct dry season from January to March, which makes it difficult for farmers to obtain multiple rice crops without irrigation. During the rest of the year, rainfall is generally adequate; annual rainfall typically exceeds 1600 mm. The region has slightly acidic clay loam soils with a pH of 5–6. The terrain in most upland farms in the sample had a slope of over 18%. Upland elevations extend to 1500 m above sea level. The area's main staple is rice and its main cash crop is maize (Garcia et al., 1995). Half of all lowland farmers reported receiving loans during the study period, but few upland farmers had access to credit.

Although the two lowland communities surveyed had similar demographic features and incomes, their average farm sizes differed significantly. Average farm size was 2.6 ha in Marangas and 5.1 ha in Tamlang. (The largest farm in the sample was 12 ha.)

4. Results

Table 18.1 illustrates differences observed between irrigated and rain-fed farms. Except for the amount of labour hired per hectare in each cropping season, the mean values for irrigated farms all differed significantly from those for rain-fed farms, at a 90% confidence level.

As expected, irrigated farmers had higher average physical yields (3639 kg ha^{-1} vs. 3200 kg ha^{-1}). The most important effect of irrigation, however, was to allow farmers to grow more crops on each parcel of land. Irrigated

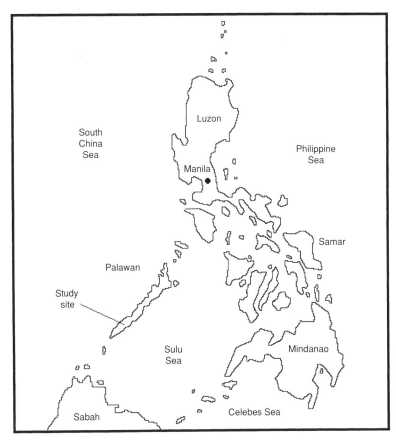

Fig. 18.1. Map of the Philippines indicating location of study site.

farms had an average cropping intensity of 1.9, whereas the cropping intensity of rain-fed farms was only 1.2.

Irrigated farms used both less family labour and less total labour per hectare during each cropping season (13 and 37 man-days ha^{-1} compared with 20 and 43 man-days ha^{-1}, respectively). However, they used more hired labour. This suggests that the introduction of irrigation led lowland farms to replace some of their family with hired labour.

On both irrigated and rain-fed farms, the amount of labour used per hectare decreased with farm size, especially family and shared labour. This reflects, in part, modest increases in use of tractors and chemicals (especially pesticides) on larger farms (Martinez and Shively, 1998).

Table 18.2 presents some mean values for the upland households surveyed. Eighty per cent of these households reported receiving earnings from off-farm work. While not all of that was on lowland irrigated farms, upland households in Marangas (the irrigated site) were far more likely to have one or more members working off-farm than those in Tamlang (the rain-fed site). On

Table 18.1. Characteristics of lowland rice farms, Palawan, 1996 (survey data).

	Irrigated	Rain-fed
Farm size (ha)	2.5	4.2
Household size (members)	5.8	4.8
Total income (pesos)	104,128	108,867
Income per capita (pesos)	22,604	25,364
Percentage of land owned	48%	78%
Rice yield (kg ha^{-1})	3,639	3,200
Number of crops year^{-1}	1.9	1.2
Fertilizer use (kg ha^{-1})	157	180
Pesticide use (pesos ha^{-1})	1,656	917
Total labour per crop (days ha^{-1})	37	43
Family and shared labour per crop (days ha^{-1})	13	20
Hired labour per crop (days ha^{-1})	25	23
Number of farms	53	45

At the time of the survey US$1 = 25 pesos.

Table 18.2. Characteristics of upland farms, Palawan, 1996 (from survey data).

	Without off-farm work	With off-farm work
Farm size (ha)	2.5	2.0
Household size (members)	4.8	4.9
Total income (pesos)	18,255	13,566
Income per capita (pesos)	4,586	3,224
Percentage with secure title	43%	42%
Rice yield (kg ha^{-1})	1,833	1,733
Fertilizer use (% > 0)	33%	29%
Pesticide use (% > 0)	10%	14%
Percentage reporting forest clearing	15%	16%
Average area of forest cleared (ha)	0.10	0.18
Percentage reporting charcoal or firewood sales	23%	27%
Number of farms	21	83

At time of the survey US$1 = 25 pesos.

average, households with members that worked off-farm had slightly smaller farms (2.0 ha vs. 2.5 ha) and lower incomes (P13,566 year^{-1} vs. P18,255 year^{-1}).

Even before irrigation was introduced, hired labour constituted 63% of all labour used on lowland farms. Lowland farmers used upland labour widely, in particular for land preparation and harvesting. Cruz *et al.* (1992) report similar links between upland and lowland communities elsewhere in the Philippines.

Upland households with off-farm workers participated more in activities with low returns, such as hunting, charcoal-making, resin collection and forest clearing. On average, upland households with off-farm employment cleared 0.18 ha year^{-1}, while households without off-farm work cleared only 0.10 ha. A separate study by Shively (1997) found similar strong links between poverty, farm size and forest pressure in the area.

Based on the deforestation rates reported by our survey respondents and the fact that we surveyed approximately 30% of the total population, we estimate that the total area of forest farmers cleared in the two communities was roughly 55 ha in 1996. This compares with an estimated total upland agricultural area of approximately 728 ha. Thus, recently cleared areas probably represented about 7% of the cropped area in the upland sample. Not all of the area cleared, however, involved destruction of primary forest. Other evidence from the study site suggests that about 30% of the area cleared in 1996 was virgin forest, 46% was degraded or secondary forest and shrubs and 24% was grassland.

Table 18.3 summarizes our main findings regarding irrigation's impact on lowland labour demand. As noted previously, irrigated farms used less labour per hectare during each cropping season than rain-fed farms, but had higher cropping intensities. As a result, irrigated farms had an effective labour demand of 70 days ha^{-1}, 27% higher than on rain-fed farms, which had an effective labour demand of only 55 days ha^{-1}. Both lowland and upland farmers reported that irrigation generated a 'boom' in the local labour market. As a result, wages rose from P45 day^{-1} in early 1996 to P75 day^{-1} by 1997.

Table 18.4 provides key data on the changes in lowland labour markets resulting from the irrigation projects and how upland households at the irrigated site responded to those changes. The proportion of households that cleared forest fell from 18% (19 households) to 12% (12 households) and the average area cleared (by those reporting clearing) declined from 2.5 ha to 1.9 ha. Taken together, these statistics suggest a 48% decline in the amount of forest upland households in the community cleared annually after the irrigation systems were installed (47.5 ha vs. 22.8 ha). The observed changes in the percentage of farmers clearing forest, in the average area cleared and in wages following the introduction of irrigation are all statistically significant.

Although the area planted to rice (the staple crop) remained the same following the introduction of the irrigation systems, the average area planted

Table 18.3. Labour use, cropping intensity and changes in effective labour demand.

	Labour per crop (days ha^{-1})	Effective cropping area (ha year^{-1})	Effective demand (days ha year^{-1})	% change from rain-fed case
Rain-fed	42.7	1.29	55.1	–
Irrigated (observed)	37.1	1.89	70.1	+27.2
Irrigated (predicted)	33.0	1.89	62.4	+13.2

Table 18.4. Forest conversion and upland indicators before and after irrigation (from survey data).

	Before irrigation	After irrigation
% households reporting forest clearing	18%	12%*
Average forest area cleared (ha year^{-1})	2.5	1.9*
Area in rice (ha)	0.95	0.94
Area in maize (ha)	1.20	1.05
Days of employment†	18	44
Average wage (P man-day^{-1})†	45	75*
Wage income (P year^{-1})†	1,150	3,226

*Means are significantly different at a 95% confidence level.
†The sample with off-farm employment only.

to maize (a cash crop) fell from 1.20 ha to 1.05 ha among households who planted maize. This suggests that farmers decided to obtain more of their cash income from off-farm employment after off-farm wages rose but continued to plant the same amount of crops for their own consumption.

Unfortunately, our data do not permit us to fully assess the changes in upland households' welfare following the introduction of irrigation. However, the considerable improvements in employment opportunities and off-farm wages suggest that lowland irrigation increased at least some upland households' welfare. Taken together, the data on employment and wages suggest that average wage income rose nearly threefold, following irrigation, among upland households that engaged in off-farm work. All upland households surveyed reported that lowland irrigation had either increased or at least not decreased their economic welfare.

Although our results are encouraging, we have been able to observe only the initial impact of irrigation on lowland labour demand. Irrigation arrived in the area recently and, while farmers are clearly enthusiastic about the new technology, many admitted facing difficulties in managing their farms using irrigation. Thus, it seems prudent to analyse whether current farmer practices are likely to continue. What if lowland farms are in a process of adjustment? What if the logic of profit maximization eventually leads lowland farmers to use less labour-intensive technologies? Can farmers sustain the observed rates of labour use? Could the beneficial impacts of irrigation on rates of deforestation disappear?

To get at these questions, we derived an estimate of 'optimal' labour use on irrigated farms from a production function based on plot-level data from the lowland sample. We report results in the final row of Table 18.3. This estimate is designed to provide insights into the possible long-run impact of irrigation on lowland labour demand. To compute the estimate we followed a standard – if somewhat simplistic – approach to forecasting labour demand. We first formulated a Cobb–Douglas production function and estimated it econometrically.

Table 18.5. Production function results.

Independent variables	Estimated coefficient (standard error)
Constant	6.7796**
	(0.3988)
Log of labour (man-days ha^{-1})	0.1106*
	(0.0603)
Log of fertilizer (kg ha^{-1})	0.1335**
	(0.0499)
Log of pesticide (pesos ha^{-1})	0.0397
	(0.0292)
Season {0 = dry season,	0.2366**
1 = wet (rainy) season}	(0.0515)
Farm size (ha)	−0.0200**
	(0.0010)
R^2	0.32
Number of observations	105

Regressions were conducted at the plot level and were corrected for heteroskedasticity. The symbols * and ** denote significance at 10% and 5% test levels, respectively.

We then used the results to derive profit-maximizing input levels, given observed input and output prices. By estimating optimal labour use under irrigated conditions and comparing these results with input levels observed among representative rain-fed and irrigated farms, we can draw inferences about possible future changes in labour demand.[3]

Table 18.5 contains results from the production function used to derive our labour estimates. All the parameter estimates have the expected signs and, except for the estimate for pesticide use, are significant at the 95% confidence level. As expected, the results suggest strongly diminishing returns to input use. The negative sign on the farm-size variable implies that smaller irrigated farms in the sample were either more efficient in their production or occupied more productive land.

To think about how irrigation might eventually affect labour demand, it helps to imagine a stylized two-stage process. In stage I, rain-fed farms become irrigated and employ (possibly suboptimal) input levels such as those observed among the irrigated sample. In stage II, these newly irrigated farms adjust to employing profit-maximizing factor proportions and levels. Our results suggest that whether farmers use more or less fertilizers and pesticides following the adoption of irrigated systems depends on whether they are in stage I or stage II. As we saw previously, since adopting irrigated systems (stage I), the amount of labour each farmer demanded per hectare during each cropping season fell from 42.7 days to 37.1 days. Based on our regression results, in stage II we would expect farmers to further reduce labour demand per hectare to 33 days. This suggests that some of the observed gains in employment arising from irrigation – and associated reductions in forest clearing – could evaporate if lowland farmers reallocate inputs to profit-maximizing levels. However, as the final columns of Table 18.3 indicate, effective labour demand would probably still remain higher than in the rain-fed situation. Further simulations with our model show that suboptimizing behaviour in the dry season could lead to an overall reduction in annual labour use of up to 2% ha^{-1} compared with the amount used by rain-fed farms. The story that emerges, therefore, is that, while the short-run impact of irrigation on forests may be beneficial, the long-run impact will depend on whether irrigated farms seek and achieve profit-maximizing factor intensities and, if so, whether irrigation in the delivery area is fully utilized during the dry season.

5. Conclusions and Policy Implications

In Palawan, natural population growth and migration rates influence the size of the agricultural labour force. As in other frontier areas, agricultural expansion, timber and fuel-wood collection and charcoal-making threaten the province's forests. The persistence of activities that degrade forests reflects lack of economic opportunity and low economic returns from current agricultural options. This study examined the pathway though which investments in lowland irrigation development increase agricultural productivity and wages, and how these, in turn, generate employment opportunities for households that rely on forests for agricultural land and timber. Where upland and lowland communities are close to one another, increased employment resulting from irrigation development can draw pressure away from the forest margin.

Our results suggest that lowland irrigation projects can raise employment among upland residents and improve their welfare. In the example studied here, this change led households to allocate less time to upland forest clearing and hillside farming – especially of cash crops. This implies that lowland agricultural intensification can have beneficial impacts on adjacent upland forests. However, we must qualify these conclusions in four regards. First, the area described here is unusual, in that the upland area is physically adjacent to the lowland area. For most upland households, working on lowland farms required only a 1 h trip on foot. If larger distances separated the lowland and upland areas, the opportunity cost of travel for upland households would be much higher and could discourage upland households from seeking employment on lowland farms. Secondly, the adoption of mechanization, direct seeding, chemical-based weed control and other labour-saving technologies could lead irrigated lowland farms to shed additional labour in the future and hence partially reverse the employment gains we observe. Thirdly, our study has not addressed the role of input pricing policies. The relative costs of labour and inputs that can substitute for labour partially determine how much labour farmers demand. Some sets of relative prices could undermine the labour absorption we observed in this case. Thus policy-makers should take into account the environmental gains associated with labour-intensive production in frontier areas, when considering economy-wide policies that discourage labour use, by reducing the relative prices of fertilizer, pesticides and machinery, such as tractors. Finally, since irrigation may significantly increase farmers' incomes, policy-makers should pay attention to how these higher incomes translate into new patterns of consumption and investment. While many potential investments by upland farmers do not necessarily pose a threat to adjacent forests, others – such as purchases of livestock or chain-saws – clearly do.

Our analysis demonstrates that irrigation reduces labour demand per hectare per cropping but raises total labour use per hectare in a calendar year and that encouraging labour use in lowland agriculture can reduce upland deforestation. To the extent that off-farm labour displaces environment- and forest-degrading activities with lower rates of return, shifts in time allocation may increase incomes at the same time as they reduce environmental pressure. The more important policy lesson, however, is that the labour market plays a key role in facilitating environmental improvements. For this reason, policy-makers should embrace opportunities to expand employment and labour-market participation, especially in areas where upland deforestation is a continuing problem.

Notes

1 We gratefully acknowledge comments from the editors, Ed Barbier, Ian Coxhead, John Lee, Will Masters and an anonymous reviewer. We also thank Richard Yao and the staff of the Natural Resources Management Programme at the South-East Asian Regional Centre for Graduate Study and Research in Agriculture (SEARCA), in Los Baños, for their field assistance. A grant from the Ford Foundation funded part of this research.
2 Hired labour and family labour are often imperfect substitutes, especially if supervision is difficult or costly. This can make farmers reluctant to replace family labour with hired labour.
3 One might reasonably question whether a simple Cobb–Douglas production function is the most appropriate functional form and whether it accurately represents the technology available to farmers. One might also question whether farmers' only goal is to maximize profits. Nevertheless, this exercise allows us to roughly compare observed levels of labour use on rain-fed farms and those that might be expected on irrigated farms under a plausible set of conditions. Elsewhere (Martinez and Shively, 1998) we show that, on average, irrigated farms are operating below profit-maximizing levels. We also argue that the observed increase in labour use associated with irrigation may partially reflect other farmer concerns such as risk aversion.

References

Boyce, J.K. (1993) *The Philippines: the Political Economy of Growth and Impoverishment in the Marcos Era*. University of Hawaii Press, Honolulu.

Castillo, L., Gascon, F. and Jayasuriya, S. (1983) Off-farm employment of farm households in Laguna, Philippines. In: Shand, R.T. (ed.) *Off-farm Employment in the Development of Rural Asia*. Vol. 2. *Papers Presented at a Conference Held in Chiang Mai, Thailand, 23 to 26 August 1983*. National Centre for Development Studies, Australian National University, Canberra, pp. 133–146.

Coxhead, I. and Jayasuriya, S. (1986) Labour-shedding with falling real wages in Philippine agriculture. *Asian Survey* 26(10), 1056–1066.

Cruz, M.C., Meyer, C.A., Repetto, R. and Woodward, R. (1992) *Population Growth, Poverty, and Environmental Stress: Frontier Migration in the Philippines and Costa Rica*. World Resources Institute, Washington, DC.

Garcia, H.N.M., Gerrits, R.V., Cramb, R.A., Saguiguit, G.C., Jr, Perez, A.S., Conchada, J.J., Yao, R.T., Mamicpic, M. and Bernardo, R.G. (1995) *Soil Conservation in an Upland Farming System in Palawan: a Socio-economic Survey*. SEARCA-UQ Upland Research Project Survey Report 4, South-East Asian Regional Centre for Graduate Study and Research in Agriculture, Los Baños.

Kikuchi, M. and Hayami, Y. (1983) New rice technology, intrarural migration, and institutional innovation in the Philippines. *Population and Development Review* 9(2), 247–257.

Lingard, J. (1994) Farm mechanization and rural development in the Philippines. In: Lloyd, T. and Morrisey, O. (eds) *Case Studies in Economic Development*. St Martin's Press, New York, and Macmillan Press, London.

Martinez, E. and Shively, G. (1998) Irrigation, employment, and the environment in southern Palawan. *Journal of Agricultural Economics and Development* 26(1–2), 112–135.

Sadoulet, E. and de Janvry, A. (1995) *Quantitative Development Policy Analysis*. Johns Hopkins University Press, Baltimore.

Sandalo, R.M. (1996) Sustainable development and the environmental plan for Palawan. In: Eder, J.F. and Fernandez, J.O. (eds) *Palawan at the Crossroads*. Ateneo de Manila University Press, Manila, pp. 127–135.

Shively, G.E. (1997) Poverty, technology, and wildlife hunting in Palawan. *Environmental Conservation* 24(1), 57–63.

Western, S. (1988) Carrying capacity, population growth and sustainable development: a case study from the Philippines. *Journal of Environmental Management* 27, 347–367.

Agricultural Development Policies and Land Expansion in a Southern Philippine Watershed

Ian Coxhead, Gerald Shively and Xiaobing Shuai

1. Introduction[1]

In spite of mounting evidence of economic and environmental costs associated with upland agricultural growth, Philippine agricultural policy – broadly defined to include both price and technology policies – continues to focus primarily on increasing production and yields. These price and technology policies interact. In addition to their usual effects on supply, price supports increase farm profitability, and this spurs both the demand for innovations and the investments in R&D intended to increase their supply. In this setting, policy-makers and farmers give only secondary attention to long-run environmental concerns and thus fail to anticipate many of the environmental effects of technical progress.

Philippine government policies reinforce both expansion and intensification in marginal agricultural lands. Agricultural expansion, when it occurs, may have severe environmental consequences. For example, much of the Philippines' biodiversity is currently threatened, in large part as a result of rapid population growth, which stimulates forest clearing, compresses fallow cycles and degrades habitat (IUCN, 1988; Myers, 1988; Cox, 1991; Goodland, 1992). Between 1960 and 1987, the upland area devoted to agriculture in the Philippines increased sixfold and much of this increase coincided with a decline in forest cover (Cruz et al., 1992). Total forest losses for the country are estimated to have been as much as 2000–3000 km^2 year^{-1} in recent decades (Bee, 1987; Kummer, 1992), and this rapid deforestation greatly threatens Philippine wildlife. As an example, studies indicate that, in areas where natural forest cover has been removed, only one in ten endemic bird species has successfully adapted to habitat changes (Rabor, 1977).

To better understand how agricultural policies affect the incentives for agricultural expansion, this chapter provides an *ex ante* evaluation of factors affecting farmers' land use. We use survey data gathered from low-income maize and vegetable farmers in a southern Philippine watershed at the forest margin to evaluate the roles expected prices and yields and their variances play in agricultural land allocation. We find that land allocation within farms responds to relative crop prices and yields. However, each crop elicits a different response. Some crop expansion takes place primarily through the substitution of one crop for another (and, to a lesser extent, through intensified input use). Changes in prices or yields of other crops provoke an expansion of total farm area. Land and family labour constraints bind at different points, depending on the crop involved. These results suggest that, just as multiple agricultural development policies interact, environmental policies must also have multiple strands if they are to eliminate the incentives for further land expansion.

In our study area, as in many other upland areas of developing market economies, commercial agriculture is the norm but farmers are poor, and therefore concerned about risks. Thus, our study site shares many other characteristics with similar sites elsewhere.

Next we briefly review recent land-use trends and the policies that have influenced resource allocation at the agricultural margin. Section 3 presents a model of land allocation among crops by risk-averse farmers and the equations we used for econometric estimation. In section 4, we discuss data and our econometric approach. We present our main results in section 5, and in section 6 we discuss how our findings might contribute to answering the question, 'Do investments in agricultural productivity for upland farms promote deforestation?'.

2. Historical Background and Context

2.1. Recent agricultural development trends in Lantapan, Bukidnon

Our study site is Lantapan, a municipality in central Bukidnon province, northern Mindanao. The municipality covers one side of the Upper Manupali River watershed. The river's left bank bounds it to the south and the Mt Kitanglad Range National Park defines the limit to the north. From east to west, the landscape rises from irrigated lowland rice-fields at about 500 m a.s.l., through rolling sugar-cane and maize areas and a strip of maize and coffee, and into a mid-to-high-altitude maize–vegetable system that extends from 800 m a.s.l. into the buffer zone of the national park. Much of the park remains heavily forested. Immigration has driven the growth in farmed area. In the decade beginning in 1970, Lantapan's population increased at an average annual rate of 4.6%, from 14,500 to 22,700 (NSO, 1990). By 1994, it had 39,500 inhabitants (Municipality of Lantapan, Municipal Agricultural and Demographic Database). Annual population growth since 1980 has thus averaged 4%, much higher than the Philippine average of 2.4%.

Agriculture dominates the Lantapan economy. The area in temporary crops totalled 14,350 ha in 1973 – about 28% of total land area. By 1994, this figure had risen to over 25,000 ha, or half of the land area. Neither of these figures includes large areas of coffee, rubber, abaca and other tree and shrub crops. In 1988, agriculture provided 71% of provincial employment, compared with 5% in industry and 23% in services, and was the primary source of income for 68% of Bukidnon households (NSO, 1990). As is typical of a recently settled area, in 1980 – the last year for which agricultural census data are available – most Lantapan farms (about 70%, covering 80% of total farm area) were managed by their owners or by people who had 'owner-like possession'. Farms are small by upland standards. The modal farm size class (1–3 ha) contained 46% of farms in Lantapan in 1980 and 75% of all farms were smaller than 5 ha. Most households live close to the poverty line. In 1988, food, fuel and clothing accounted for 59%, 4% and 5% of household expenditures, respectively (NSO, 1990).

Since the 1950s, agricultural land has expanded substantially in Lantapan, as just noted, and farmers have substituted certain crops for others in response to new commercial opportunities. At the end of the Second World War, forest covered most sloping and high-altitude land. Farmers in the mid- and high-altitude villages primarily produced maize, cassava and coffee, using various forms of long-fallow shifting cultivation. Presumably, they also harvested logs and non-timber forest products. In the 1950s, migrants from northern Luzon introduced commercial cultivation of potatoes, cabbages and other temperate-climate vegetables. The success of these crops, as well as the introduction of new maize varieties and the replacement of coffee and shrub crops by annual crops, all indicate steady land-use intensification.

Since the late 1970s, commercial agriculture in Bukidnon has thrived, thanks to infrastructure improvements, greater integration of the province's economy in national agricultural markets and increasing national demand for maize and temperate-climate vegetables. Maize production has flourished, becoming a major commercial crop, where formerly it had been traded very little outside northern Mindanao. Vegetable cultivation has also continued to increase in area and economic importance. Now people sometimes describe the upper watershed of Lantapan as a 'second Benguet', in reference to the Philippines' primary temperate-climate vegetable production area in northern Luzon.

Annual crops have replaced large areas of forest and perennials. This can be seen clearly in data constructed from satellite imagery (Fig. 19.1). Over a 20-year period ending in 1994, the permanent forest area shrank from about half to a little over one-quarter of the total area. Part of that land went into shrubs or secondary forest, but farmers converted a much larger part to

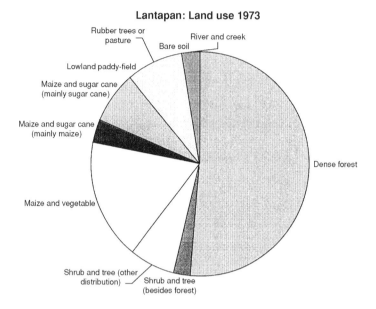

Fig. 19.1. Land-use changes, municipality of Lantapan, 1973–1994 (from Li Bin, 1994: Tables 5.9 and 6.12).

annual crops, especially maize–vegetable systems, which expanded from 17% to 33% of total land area. As the data in Table 19.1 indicate, annual crops have also moved steadily into areas with higher slopes.

A number of phenomena have influenced agricultural expansion and intensification. Relative crop prices have changed over time, but so too have input prices and, since the major crops differ widely in the factor intensity of production, this probably influenced the product mix. After five decades of economic growth with rapidly increasing population in the Philippines, agriculture remains the largest employment sector and, until recently at least, most industrial production was highly capital-intensive. The relative abundance of labour favoured agriculture and, within the sector, relatively labour-intensive crops, such as annual crops. For a long time, the frontier served as the employer of last resort for underemployed, unskilled labour. Over time, land scarcity promoted intensification, which further increased labour demand and raised the returns to land used for intensive production. Only within the last decade has non-agricultural growth shown signs of absorbing labour at rates significantly faster than labour-force growth, foreshadowing a slow-down in the net growth of upland populations. Lantapan, whose population grew rapidly in past decades, is just now beginning to display signs of labour shortage.

2.2. Agricultural development policy in the Philippines

Although soil quality, moisture, temperature and (for some vegetable crops) the presence of soil-borne pathogens all condition agricultural land use, farmers in Lantapan usually explain their land-use decisions in terms of the relative economic benefits of different crops. Over time, a number of Philippine government policies have directly and indirectly affected the profitability of cultivating maize and vegetables. These consist mainly of market interventions directed at stabilizing farm prices; trade interventions designed to

Table 19.1. Land use by slope (10% and greater), 1973 and 1994 (from Li Bin, 1994: Tables 5.5 and 5.11).

Land-use class	10–20%		20–40%		40–90%	
	1973	1994	1973	1994	1973	1994
Dense forest	69.5	38.9	88.3	59.9	91.7	57.3
Shrub and tree (besides forest)	3.0	11.1	6.2	22.7	3.9	32.5
Shrub and tree (other distribution)	4.0	5.2	1.2	1.7	1.4	0.9
Agriculture	17.6	41.8	3.4	13.1	1.9	7.0
Grass	4.1	..	0.17	..	0.85	..
Bare soil	0.1	1.3	0.2	2.0	0.1	2.3
River and creek	1.7	1.7	0.5	0.5	0.1	0.1

.. indicates data not available.

reduce dependence on imports and defend the livelihood of upland farmers; and public support for research and extension aimed at raising yields and reducing the prevalence of pests and diseases.

Maize and temperate-climate vegetables are import substitutes in the Philippines, and import restrictions and domestic price supports have considerably encouraged producers – mainly upland farmers – to expand their production (Coxhead, 1997, 2000). Quantitative restrictions on maize, cabbage and potato imports (recently converted to tariffs at the maximum allowable rate under the World Trade Organization) have raised their domestic prices relative to world prices. For these crops, nominal protection has been so high that it more than offset the prevailing bias against agriculture introduced through industrial promotion and exchange rate policies (Bautista et al., 1979; Intal and Power, 1990). Even in the recent era of declining protectionism, protection of vegetable producers has been stable and that of maize producers has risen: the implicit tariff on maize rose from near zero in the early 1970s to close to 100% by the early 1990s (Intal and Power, 1990; Pagulayan, 1998). Conversely, direct and indirect export taxes on coffee, an important commercial crop in the watershed in earlier years and one in which the Asian Development Bank (ADB) has identified Mindanao as having a comparative advantage (ADB, 1993), have discouraged its cultivation. As a result, regional coffee production has deteriorated in both quantity and quality, and processing and marketing infrastructure, extension support and other assistance to the industry have all but disappeared.

Technology policies have likewise promoted maize and vegetable production. The Philippine government designated Bukidnon province as a 'key production area (KPA)' for maize in its Grain Production Enhancement Programme (GPEP). Farmers in KPA zones are eligible for subsidies and supports directed at increasing maize production, and are the first beneficiaries of research and development directed at increasing maize yields (Department of Agriculture, 1994). As a result, the area planted to maize has risen steadily in Bukidnon, even as it has declined nationally.[2]

Vegetable producers have also been the beneficiaries of disproportionate amounts of research funding and effort (Librero and Rola, 1994; Coxhead, 1997). The Philippine Department of Agriculture recently identified potato, a cool-climate crop that is widely grown in Lantapan in some years, as a 'high-valued crop', placing it in a category with high priority for research and extension allocations. Foreign agencies also support potato research, which is regionally concentrated in Department of Agriculture facilities in northern Luzon and in Bukidnon and strongly promoted by industrialists in the potato-processing industry. Bacterial wilt, cyst nematodes, late blight and various insect pests threaten potato production. Research concentrates on developing and disseminating planting materials, such as true potato seed (TPS), which, under suitable management regimes, greatly reduce the risk of crop losses through disease. Studies of the Philippine potato industry indicate that, if TPS or similar improvements became widely available, production costs would fall, yields increase and the variability of yields decline (Brons, 1996). A similar story applies in cabbage and other temperate vegetable crops, where pests and disease pose the greatest threats to yields and maintaining crop health is a large component of production costs. Philippine cabbage research appears largely to focus on reducing yield variability and input costs by addressing pest and disease problems.

Although maize yields have risen over time with the development and spread of improved varieties, vegetables have not progressed to the same degree. However, technological breakthroughs, if they emerge, will be at least as important for dampening the volatility of vegetable yields as for increasing expected profits. If the main effect of vegetable research is to reduce variability of returns, then technical progress could have a substantial impact on the land-use decisions of risk-averse farmers. Other things being equal, existing vegetable farmers will opt to increase production, and other farmers not currently growing vegetables may switch existing land or expand planted area to begin. However, the magnitude of the land-area response will depend on product prices and their volatility and the availability of inputs. For vegetable farmers, credit for inputs and the managerial skills required by technologically advanced vegetable production are both likely to significantly constrain land-area expansion. With this in mind, we conducted an *ex ante* analysis of the probable land-use effects of technological improvements in Philippine vegetable production.

3. Determinants of Land Allocation under Uncertainty

This section highlights factors influencing farmers' land-use responses to economic and technological stimuli. The discussion here is necessarily brief. We encourage interested readers to consult the appendix to this chapter and the more formal exposition in Coxhead *et al.* (1999).

Our main goal in this study is to measure how land and labour allocations for various crops respond to changes in expected output prices, expected yields or price or yield volatility. The model we base our analysis on assumes that farmers are endowed with land and family labour, which they use to produce a combination of maize and vegetable crops. They can either use all the land at their disposal or leave some fallow. They purchase other inputs, whose farm-gate prices (as well as those of outputs) are determined by distance from a central market. Given family labour availability and the prices of inputs (including hired labour), each farmer decides at the beginning of a season: (i) how much land to plant; and (ii) what fraction of the land to allocate to each crop.

Since prices and yields are stochastic, we assume that farmers make choices to maximize expected utility. Uncertainty has two sources: prices and production. Production or yield risk arises both from the characteristics of the land (its slope and quality, for example) and family labour endowments and from external events, such as weather, disease and pest infestations. Price risk arises because, at the time farmers decide how to allocate their land, they do not know with certainty what crop prices will be at harvest time. From our survey, we observe that, in this kind of uncertain environment, farmers have three basic responses to external shocks. On the extensive margin, they can increase or decrease the total cultivated area by bringing new plots into production or by leaving part of their land fallow. On the intensive margin, they can adjust labour and input use by crop, using more or less of each to attain a desired production target. In between, farmers can also adjust land allocation among different crops.

This reasoning suggests a series of equations describing land allocation to crops, labour use and changes in total crop area. Focusing on the most widely planted crops in Lantapan, maize and vegetables, we use four equations. These are:

$$T_c^* = T_c(\theta_i, \phi_i^2, \mu_i, \sigma_i^2, W, A_{-1}, AD, \text{Others}), \quad i = \text{maize, vegetables} \quad (1)$$
$$T_v^* = T_v(\theta_i, \phi_i^2, \mu_i, \sigma_i^2, W, A_{-1}, AD, \text{Others}), \quad i = \text{maize, vegetables} \quad (2)$$
$$FL^* = FL_c^* + FL_v^* = L(\theta_i, \phi_i^2, \mu_i, \sigma_i^2, W, A_{-1}, AD, \text{Others}),$$
$$\quad i = \text{maize, vegetables} \quad (3)$$
$$\Delta A^* = \Delta A(\theta_i, \phi_i^2, \mu_i, \sigma_i^2, W, A_{-1}, AD, \text{Others}), \quad i = \text{maize, vegetables} \quad (4)$$

where variables are defined as follows:

T_c^*	Total area planted to corn (maize)
T_v^*	Total area planted to vegetables
FL^*	Total family labour used in agriculture[3]
ΔA^*	Net change in cultivated area
θ_i	Expected price for crop i
ϕ_i^2	Price variance for crop i
μ_i	Expected yield for crop i
σ_i^2	Yield variance for crop i
W	A vector of variable input prices
AD	Number of adult family members
A_{-1}	Total area cultivated in the previous crop season

In each of these equations, we include several variables intended to control for farm characteristics that might serve as additional constraints on land-use behaviour. For all equations, we add a variable representing tenure security, which can take several values, ranging from low (most secure) to high (least secure). We also include a 'credit constraint' variable, which takes a value of 1 for farms reporting that they did not plant a crop or that they altered total land area, because they were unable to obtain credit (or reported being credit-constrained in some other similar way). In the total land equation, we also include dummy variables representing other possible reasons for changes in land area, notably contractual reasons such as the expiry of a 3-year lease. A dummy variable for 1995 is also added to each regression equation. On the basis of our conceptual model, we observe the following.

First, we expect that the area planted to a crop will respond positively to increases in its price or yield and negatively to increases in input prices. For risk-averse farmers, increases in price or yield variances will have an unambiguously negative impact. When maize prices or yields rise, risk-neutral farmers will expand their maize area more than risk-averse farmers, since an increase in maize production also implies an increase in the associated variance in income from maize.

The reasoning holds for vegetables, although empirically, since vegetable prices and production are more volatile than those of maize, we expect that small increases in expected price or expected yield may elicit very small (or even zero) responses among risk-averse farmers. Exogenous changes in variances may have more measurable effects.

A land constraint implies that maize and vegetables are substitutes. Thus we expect an increase in price or yield variability for one crop to encourage production of the other. Once again, responses of risk-averse farmers should not be as strong as those of risk-neutral farmers. Similarly, an increase in the expected yield of one crop should reduce the land planted to the other.

In a single-crop, risk-neutral production model, a rise in the price of some input would have a negative effect on land use. In our model, however, we have two crops, so the response of land use in each crop to a given input price shock will depend on relative input intensities of the crops. Since vegetable production is more intensive in fertilizer and chemical use, we expect input prices to have a strong negative effect on vegetable land area. For maize land, the positive substitution effect may dominate the direct negative effect; thus the same input price shock might have a positive effect on area planted. As before, risk aversion also plays a role here by reducing the magnitude of responses.

Quite a few farmers in our sample grow no vegetables, only maize. Though the risk-aversion model does not explain why they grow only maize in the first place, it can shed light on why they might feel reluctant to change to vegetables. For example, in some cases, only a sizeable jump in expected vegetable price or a fall in maize price (or equivalent shifts in relative yield) will provide the farmer sufficient incentives to diversify. Once again, if exogenous shocks, such as price policies or technological innovations, change the variances, then a risk-averse farmer might find it profitable to make non-marginal changes in his/her land use.

Finally, we note the role of land and labour constraints. Our conceptual model permits farmers to add new land to the farm at the beginning of each period, but at a cost. This cost might represent the cost of preparing fallow land for cultivation or the cost of establishing a claim to cultivate new land, whether through colonization of forest or fallow land, negotiation of a tenancy contract or other means. The nature of these costs implies that family labour availability is likely to constrain land acquisition. Family labour constraints also operate differently between crops, since vegetables are generally more management-intensive. Whereas farmers can expand their maize production by hiring more labour (assuming they have available land), the same may not hold, or at least not to the same extent, for vegetables. The fact that we have land and labour constraints in our model implies that it is a short-run model, since in the long run the constraints are less likely to be binding.

4. Data and Econometric Method

We used data drawn from three annual surveys of production, prices and household, plot and farm characteristics of a sample of farmers in the maize–vegetable zone of Lantapan to estimate equations (1)–(4). Table 19.2 summarizes the major features of the sample. The data provide direct observations of land use, technology, input use, production and plot/farm/household characteristics. We constructed variables representing expected prices and their variances from independent data.[4] Variables representing expected yields and their variances were constructed from the predicted values and residuals of production functions fitted to the data. Coxhead *et al.* (1999) outline these calculations in detail.

The equation system (1)–(4) is a reduced form, in which individual equations explain the land-area decision, the allocation of land between crops and total labour use for all crops. The equations can be estimated independently. Because the equations contain lagged values, we use only the data from the second and third years (1995 and 1996) in our estimation. We construct farm-level crop-area, labour-use and land-characteristics variables by aggregating plot-level data using area weights. Since there was no variation in wages in our data, we were forced to exclude wages from the set of explanatory

Table 19.2. Summary of farm-level data.

Variable	Units	Mean	SD
No. adults resident in households		3.416	2.055
Total farm area	ha	2.769	2.772
No. plots per farm		1.682	0.822
Average area added year^{-1}	ha	0.064	0.326
Average area reduced year^{-1}	ha	0.458	2.359
Maize			
Expected price	pesos kg^{-1}	6.336	0.814
Variance of price		0.637	0.156
Expected yield	kg ha^{-1}	362.93	557.48
Yield variability		2.8225	0.6983
Vegetables			
Expected price	pesos kg^{-1}	8.936	1.686
Variance of price		4.499	2.825
Expected yield	kg ha^{-1}	2787.6	3278.6
Yield variability		7.267	4.171

variables we used in our estimation. For chemicals, the difficulty of imputing a price per unit of active ingredient and of aggregating these across different chemicals also kept us from including them in the estimation.

5. Results

Table 19.3 reports ordinary least squares (OLS) estimates of equations (1)–(4) in elasticity form. (Table A-1 in the Appendix provides coefficient estimates from which the elasticities were computed.) Most estimates exhibit the expected signs but, overall, Table 19.3 shows that the efficiency of the estimates is low. This may be due to genuinely weak economic relationships or to the fact that data are measured with error, as is typical in studies of this kind. Moreover, we find that the expected yield variables are highly correlated ($r = 0.96$), as are expected yields and the dummy variable for 1995 (average $r = -0.95$).[5]

In the regressions in which planted area serves as dependent variable, estimated responses to own prices are positive and estimated responses to cross-prices are negative. Input prices also exhibit the expected signs. The maize area declines when the price of nitrogen rises. A rise in the price of manure, which is used most intensively on vegetable plots, reduces vegetable area. However, none of the crop prices and only the two input prices just mentioned have statistically significant relationships with the dependent variables.

Table 19.3. Estimated elasticities of crop- and farm-area response functions.

Variables	Maize area	Vegetable area	Area change
Expected maize price (peso kg^{-1})	0.3769	−0.7607	0.0089
Expected vegetable price (peso kg^{-1})	−0.6600	0.9789	0.1124
Expected maize yield	−0.1382	0.3016	0.6817
Expected vegetable yield	0.2320	0.2826	−0.8489
Variance of maize price	−1.3120	0.8564	−1.3173
Variance of vegetable price	0.6983[c]	−0.7432	0.5005
Maize yield variability	−1.4896[a]	2.6042[a]	1.1114
Vegetable yield variability	0.5321[a]	−0.5766[b]	−0.0657
Price of nitrogen (peso kg^{-1})	−0.9027[a]	−0.1407	0.5240
Price of manure (peso kg^{-1})	0.4898	−3.7306[a]	−1.3240[c]
Total farm area last year (ha)	0.9921[a]	−0.0937	−1.1171[a]
Number of adults in the household	0.0010	0.7002[a]	1.2998[a]
Average tenure of the farm	−0.0370	−0.2297	−0.6616[a]
Credit constraint	−0.5564[a]	−0.0931[a]	−0.3407[a]
Contractual reason for dropping plot	–	–	−0.1678[a]
Other reason for dropping plot	–	–	0.1096

Superscript letters a, b and c indicate significance at 1%, 5% and 10%, respectively.

More explanatory power resides with the variables indicating risk aversion. Area changes are negatively correlated with increases in own-price variances and positively correlated with increases in cross-price variances. Area changes are also negatively correlated with increases in the variability of own yields and positively correlated with increases in cross-yield variability. These results, which are statistically robust, indicate that farmers are risk-averse. The elasticity measures in Table 19.3 show that changes in the riskiness of maize are more important than changes in the riskiness of vegetables – for both maize- and vegetable-area decisions.

Land and labour constraints are clearly important and the pattern of statistical significance of coefficient estimates reveals the expected differences between crops. As we expected, the land-area constraint (lagged farm area) binds for maize but not for vegetables. If new land were to be added to the farm, it would go mainly into maize production. Conversely, the number of adults in the household limits the area planted to vegetables, but not that planted to maize. These findings accord with our hypothesis that vegetable production is more intensive in use of the managerial and supervisory skills best provided by family members. Finally, lack of credit constrains the area of both crops.

The third equation captures change in total farm area. As in the crop equations, prices have no measurable effect on the year-on-year farm-area change. Nor do price and yield variability significantly affect farm area, although we note that increases in the variability of maize yield are positively associated with the growth of farmed area, while instability of vegetable yields has the opposite sign. In any case, farmers apparently reduce risk mostly through their crop portfolios rather than by planting larger areas. The fact that expected prices, yields and input prices have low explanatory power is perhaps not surprising, given that we are estimating a short-run model.

As expected, increases in family labour and greater access to credit are both correlated with the addition of new land to the farm. The empirical link between credit availability and farm area expansion accords with predictions from a formal intertemporal model of a credit-constrained farm household presented by Barbier and López (1999). These authors have argued that, while the effects of credit constraints on incentives for indebted households to invest in natural resources are ambiguous, it may be rational for severely indebted households to degrade resources at a greater rate when liquidity is increased.

6. Conclusions and Implications for Policy and Environmental Outcomes

The econometric results presented allow us to speculate about the effects of economic policies on agricultural intensification and extensification. This section seeks to assess how policy-driven exogenous changes in prices, yields and variances influence land use and land expansion in Lantapan and similar sites – bearing in mind that some of our results have a rather low degree of statistical confidence.

From a policy perspective, the pronounced pattern of risk-averting behaviour observed among the sample farmers is of great importance. In the short run, it appears that farmers alter their crop shares more or less predictably, in line with changes in expected prices and yields. But, more significantly, we find that farmers will switch land among crops to avoid the uncertainty associated with income volatility, especially as driven by yield variability. Yield risk, rather than price risk, appears to best express risk aversion in our sample. Furthermore, our estimates of changes in total farm area indicate a safety-first motive among farmers: increases in the volatility of maize yields induce farmers to expand farm size, while higher vegetable yield volatility, if it has any effect at all, reduces incentives to expand farm area. These results accord with findings from other frontier areas of the Philippines, where farmers appear to take into account risk both when choosing between annual and perennial crops (Shively, 1998) and when investing in soil conservation (Shively, 1997). Taken together, the main policy message behind these findings is that policies that reduce economic risks are likely to be environmentally favourable: farmers overuse resources, in part, as insurance against loss.

We now return to our earlier discussion of price and technology, in light of these results. Recall that the most important policies, from the perspective of upland or frontier farming areas, either encourage production of staple grains or seek to reduce pest- and disease-induced yield variability in commercial vegetables, such as cabbage and potato. For maize, our results suggest that policies to support and stabilize prices (e.g. through import restrictions) do not affect land use much in the short run. Technical progress aimed at reducing the variability in maize yields, in contrast, will raise the share of area planted with maize, but may actually reduce total area planted. In other words, improving the stability of maize income may be sufficient to discourage area expansion, even if expected incomes do not rise.

For vegetables, price supports and price stabilization will also increase allocation of existing land to these crops. Technical progress that reduces the volatility of vegetable yields will result in land-use substitution towards vegetables, but we expect little impact at the extensive margin. This is because, in the short run, access to credit and the availability of the special skills and attention that family members bring to land and crop care, as opposed to hired labour, constrain the expansion of total farm area.

These latter findings draw attention to potentially relevant interactions among economic and technology policies as they affect upland land use. First, the perception that maize and vegetables generate potentially high incomes for farmers drives much of the Philippine investment in improving these crops' productivity. We have seen, however, that these high incomes come largely from price supports, particularly those involving trade-policy interventions. For potato, which the Philippine government classifies as a 'high-value crop' and has targeted for additional research and development expenditures, domestic production might not even exist if it were not for past barriers to imports (Coxhead, 1997). However, now that economic policies have brought it into existence, large shifts in the production function (including reductions in yield volatility) could make the vegetable industry economically viable even at free-trade prices. Similarly, the widespread replacement of coffee by maize in Lantapan – a pronounced shift from permanent to annual crops – can be attributed both to policy distortions and to the effects of yield-increasing research and development investments in maize, but not in coffee.[6]

Finally, in the broader policy context of Philippine economic development, past policies that failed to set the country on a path of stable aggregate growth and labour-intensive industrialization greatly favoured continuing migration to the agricultural frontier. Policy reforms in the 1990s have addressed these failings through sweeping macroeconomic, trade, finance and banking reforms, which have raised the growth rate of the gross national product (GNP). Over time, the reorientation of the Philippine economy should raise the opportunity cost of farm labour. This is likely to diminish incentives to expand agricultural area in spite of technical progress in agriculture. Of course, growth outside agriculture, especially in the manufacturing sector, will generate other environmental concerns. Nevertheless, a realignment of economic incentives could reduce demand for innovations in upland farming and might also reduce the number of households seeking a livelihood at the forest margin, with the long-run result that upland agricultural area ceases to expand.

Notes

1 The SANREM CRSP (USAID Contract No. CPE A-00-98-00019-00) and the Graduate School of the University of Wisconsin supported research for this chapter. We especially thank Agnes Rola (University of the Philippines, Los Baños) for her role in gathering data and the research upon which the chapter is built. We also acknowledge helpful comments from the editors and an anonymous reviewer.
2 Experiments with an economy-wide model of the Philippines indicate that, at constant prices, technical progress in maize production, which has the same effects on farm profitability as a price rise, would substantially increase the area planted to maize (Coxhead and Shively, 1998).
3 In our survey, it is difficult to allocate labour to different crops on the same plots. Thus we only have the total days of labour used.
4 Coxhead and Rola (1998) report these data and provide the results of Granger causality tests, which demonstrate that commodity prices are exogenous to producers in Lantapan, i.e. that an expansion of production in the watershed will not affect market prices of crops.
5 This multicollinearity arises because we cannot directly observe expected yields and therefore must use a sample-wide mean adjusted by plot-level characteristics and other variables to impute expected yields. As a result of this procedure, many observations have similar values.
6 Coffee is indicative. Policy distortions have affected other perennials similarly. Evidence from other areas of the Philippines suggests that appropriate price incentives

can induce smallholders to plant large amounts of commercially valuable trees. See Shively (1998).

References

ADB (Asian Development Bank) (1993) *Industrial Crops in Asia*. ADB, Manila.
Anderson, J.R., Dillon, J.L. and Hardaker, J.B. (1976) *Agricultural Decision Analysis*. Iowa State University Press, Ames.
Barbier, E.B. and López, R. (1999) Debt, poverty and resource management in a rural smallholder economy. Paper prepared for Royal Economic Society Conference, University of Nottingham, 29 March–1 April 1999.
Bautista, R.M., Power, J. and Associates (1979) *Industrial Promotion Policies in the Philippines*. Philippine Institute for Development Studies, Manila.
Bee, O.J. (1987) *Depletion of Forest Reserves in the Philippines*. Field Report Series No. 18, Institute of South-East Asian Studies, Singapore.
Brons, J. (1996) True potato stories: costs and benefits of planting materials. CIP/UPWARD, Los Baños (manuscript).
Cox, R. (1991) Philippines. In: Collins, N.M., Sayer, J.A. and Whitmore, T.C. (eds) *The Conservation Atlas of Tropical Forests, Asia and the Pacific*. Simon & Shuster for the World Conservation Union (IUCN), New York, pp. 192–200.
Coxhead, I. (1997) Induced innovation and land degradation in developing countries. *Australian Journal of Agricultural and Resource Economics* 41(3), 305–332.
Coxhead, I. (2000) The consequences of Philippine food self-sufficiency policies for economic welfare and agricultural land degradation. *World Development* 28(1), 111–128.
Coxhead, I. and Rola, A.C. (1998) Economic development, agricultural growth and environmental management: what are the linkages in Lantapan? Paper presented at a conference on Economic Growth and Sustainable Natural Resource Management: Are They Compatible? Malaybalay, Bukidnon, Philippines, 18–21 May 1998.
Coxhead, I. and Shively, G.E. (1998) Some economic and environmental implications of technical progress in Philippine corn agriculture: an economy-wide perspective. *Journal of Agricultural Economics and Development* 26(1–2), 60–90.
Coxhead, I., Shively, G.E. and Shuai, X. (1999) *Development Policies, Resource Constraints, and Agricultural Expansion on the Philippine Land Frontier*. Department of Agricultural and Applied Economics Staff Paper Series No. 425 (October), University of Wisconsin, Madison.
Cruz, M.C., Meyer, C.A., Repetto, R. and Woodward, R. (1992) *Population Growth, Poverty, and Environmental Stress: Frontier Migration in the Philippines and Costa Rica*. World Resources Institute, Washington, DC.
Department of Agriculture (1994) *Grain Production Enhancement Program*. Philippine Department of Agriculture, Quezon City.
Goodland, R.J.A. (1992) Neotropical moist forests: priorities for the next two decades. In: Redford, K.H. and Padoch, C. (eds) *Conservation of Neotropical Forests*. Columbia University Press, New York, pp. 416–433.
Intal, P. and Power, J. (1990) *Trade, Exchange Rate and Agricultural Pricing Policies in the Philippines*. World Bank, Washington, DC.

IUCN (1988) *The Conservation Status of Biological Resources in the Philippines.* IUCN Conservation Monitoring Centre, Cambridge.

Kummer, D.M. (1992) *Deforestation in the Postwar Philippines.* Ateneo de Manila University Press, Manila.

Li Bin (1994) Impact assessment of land use changes in a watershed area using remote sensing and GIS: a case study of Manupali watershed, the Philippines. MS thesis, Asian Institute of Technology, Bangkok.

Librero, A.R. and Rola, A.C. (1994) Vegetable economics in the Philippines. Paper presented at the AVRDC/GTZ Workshop on Agricultural Economics Research on Vegetable Production Systems and Consumption Patterns in Asia, Bangkok, Thailand, 11–13 October 1994.

Myers, N. (1988) Environmental degradation and some economic consequences in the Philippines. *Environmental Conservation* 15, 205–214.

NSO (National Statistics Office) (1990) *Bukidnon Provincial Profile.* NSO, Manila.

Pagulayan, A.C., Jr (1998) Philippines. In: *Agricultural Price Policy in Asia and the Pacific.* Asian Productivity Organization, Tokyo, pp. 265–278.

Rabor, D.S. (1977) *Philippine Birds and Mammals.* University of the Philippines Press, Quezon City.

Sandmo, A. (1971) On the theory of the competitive firm under price uncertainty. *American Economic Review* 61, 65–73.

Shively, G.E. (1997) Consumption risk, farm characteristics, and soil conservation adoption among low-income farmers in the Philippines. *Agricultural Economics* 17(2), 165–177.

Shively, G.E. (1998) Economic policies and the environment: the case of tree planting on low-income farms in the Philippines. *Environment and Development Economics* 3(1), 15–27.

Appendix

This appendix briefly outlines the conceptual model used to derive equations (1)–(4) in the text and contains econometric results for these equations. For complete details regarding the model and its estimation, see Coxhead et al. (1999).

Following Sandmo (1971) and Anderson et al. (1976), we begin by constructing an expected utility function EU in terms of expected profit and its variance:

$$EU = U(E(\pi), \text{var}(\pi)) \tag{A1}$$

where we adopt the conventional assumptions, i.e. $\partial U/\partial E(\pi) > 0$, $\partial U/\partial \text{var}(\pi) \leq 0$.

Assuming no joint production, the production function for each crop is:

$$Y_i = \varepsilon_i F_i(T_i, FL_i, X_i) \quad i = c, v \tag{A2}$$
$$E(\varepsilon_i) = \mu_i; \text{var}(\varepsilon_i) = \sigma_i^2, i = c, v;$$
$$\partial F_i/\partial T_i > 0, \partial F_i/\partial L_i > 0, \partial F_i/\partial X_{ik} > 0, \forall \text{ variable inputs } k$$

where T_i is area planted to the ith crop, FL_i is family labour, X_i is a vector of other inputs such as hired labour, fertilizer, manure and chemicals and ε_i is a random variable representing multiplicative production risk. We assume that σ_i^2 captures production (or yield) risk from all sources.

The land constraint can be written as:

$$\sum_{i=c,v} T_i \leq A_{-1} + \Delta A \tag{A3}$$

where A_{-1} is the total area cultivated in the previous crop season and ΔA is the interseasonal area change. There is a cost associated with bringing new land into cultivation, which we write by $M(\Delta A)$, $M' > 0$.

In general, family labour and hired labour are not perfect substitutes in the sense that family labour usually embodies supervisory capacity as well as farm-specific land and crop management skills. In the short run, it seems reasonable to assume family labour is fixed in supply. We assume that each unit of land cultivated requires s units of family labour for management and supervision, in addition to labour used in usual farming tasks. So we can write the constraint for family labour as:

$$\sum_{i=c,v} FL_i + s(A_{-1} + \Delta A) \leq AD \tag{A4}$$

where AD is the number of adult family members.

Defining a vector W_i of variable input prices, the current period profit function is:

$$\pi = \sum_{i=c,v}\left[P_i\varepsilon_i F_i(\bullet) - X_i W_i\right] - \delta M(\Delta A) \tag{A5}$$

where $\delta = \begin{cases} 1 & \text{when } \Delta A > 0 \\ 0 & \text{otherwise} \end{cases}$

For simplicity, we assume price risk and yield risk are independent of each other. Define expected prices and the variances of prices as: $E(P_i) = \theta_i$, $\text{var}(P_i) = \sigma_i^2$. Then expected profit is:

$$E(\pi) = \sum_{i=c,v}\left[\theta_i \mu_i F_i(\bullet) - X_i W_i\right] - \delta M(\Delta A) \tag{A6}$$

and its variance is:

$$\text{var}(\pi) = \sum_{i=c,v} F_i^2(\bullet)(\phi_i^2\sigma_i^2 + \phi_i^2\mu_i^2 + \theta_i^2\sigma_i^2) \tag{A7}$$

Given these definitions, maximization of expected utility subject to the land and family labour constraints gives a system of equations that can be represented in a reduced form of $2k + 5$ equations. Since each endogenous variable depends only on the set of exogenous variables, we can estimate each equation independently. For a two-crop portfolio corresponding to maize and vegetables, this model implies equations (1)–(4) presented in the text.

Table A-1. Estimated crop area and land area response functions.

Variable	Area planted: maize (corn) (T_c)	Area planted: vegetable (T_v)	Total labour use (L)	Net area added (ΔA)
Expected maize price	0.0613	−0.064	2.5820	0.0006
	(0.428)	(−0.479)	(0.455)	(0.006)
Expect vegetable price	−0.0761	0.0581	0.1302	0.0049
	(−1.575)	(1.329)	(0.067)	(0.161)
Variance of maize price	−2.1229	0.7136	−41.094	−0.8154
	(−1.406)	(0.500)	(−0.654)	(−0.812)
Variance of vegetable price	0.1599	−0.0877	−1.3270	0.0439
	(1.936)[c]	(−1.126)	(−0.388)	(0.803)
Expected maize yield	−0.1425	0.1601	26.332	0.2688
	(−0.452)	(0.5434)	(2.058)[b]	(1.312)
Expected vegetable yield	0.2391	0.1500	−29.455	−0.3347
	(0.516)	(0.3524)	(−1.641)	(−1.171)
Maize yield variability	−1.5352	1.3821	−9.168	0.4382
	(−2.736)[a]	(2.664)[a]	(−0.394)	(1.183)
Vegetable yield variability	0.5484	−0.3060	−10.718	−0.0259
	(3.475)[a]	(−2.09)[b]	(−1.655)[c]	(−0.248)
Price of nitrogen from fertilizer	−0.0752	−0.0060	−2.8150	0.0167
	(−3.407)[a]	(−0.350)	(−3.696)[a]	(1.371)
Price of manure	0.0473	−0.1856	3.8211	−0.4894
	(1.127)	(−4.923)[a]	(2.292)[a]	(−1.774)[c]
Lagged farm area	0.3233	−0.1572	6.8516	−0.1392
	(11.661)[a]	(−0.560)	(4.342)[a]	(−5.166)[a]
Adults in household	−0.0003	0.1088	2.3443	0.1500
	(0.007)	(2.649)[a]	(1.273)	(5.090)[a]
Tenure	−0.0110	−0.0353	−2.2707	−0.0754
	(−0.314)	(−1.101)	(−1.576)	(−3.261)[a]
Credit constraint	−1.2401	−1.0684	−22.941	−2.9051
	−2.997)[a]	(−2.736)[a]	(−1.299)	(−10.12)[a]
Contractual constraint	−	−	−	−1.9076
				(−6.516)[a]
Other constraint	−	−	−	0.3740
				(0.788)
Year 1995 = 1	0.3162	1.2537	15.282	0.3112
	(0.413)	(1.756)[c]	(0.460)	(0.5699)
Constant	0.6514	−1.2614	166.62	1.3975
	(0.202)	(−0.421)	(1.270)	(0.659)
R^2 adjusted	0.612	0.304	0.224	0.645
Observations	158	162	169	170

T-statistics in parentheses. Superscript letters a, b and c indicate significance at 1%, 5% and 10%, respectively.

The Impact of Rubber on the Forest Landscape in Borneo

20

Wil de Jong

1. Introduction[1]

Rubber is the most widespread smallholder tree crop in South-East Asia. Although initially large estates planted the bulk of the region's rubber, smallholders soon captured most of the production. Currently, Indonesia's rubber plantations cover 3.4 million ha, of which smallholders account for more than 75% (BPS, 1999). In peninsular Malaysia, the area in rubber has declined since the 1970s, but rubber remains the second most common tree crop in terms of area, with 1.5 million ha in 1990 (Vincent and Ali, 1997). Large estates produce much of Malaysia's rubber, but smallholders dominate rubber production in the State of Sarawak (Cramb, 1988).

Some analysts blame the expansion of rubber for greatly contributing to the conversion of mature tropical forest in both Indonesia and Malaysia (Vincent and Hadi, 1993). This chapter critically examines to what extent smallholder rubber production actually led to forest conversion in West Kalimantan (Indonesia) and neighbouring Sarawak (Malaysia). Although we shall not discuss either Sumatra or mainland Malaysia in detail, it appears that parts of these regions went through similar processes (Vincent and Hadi, 1993; Angelsen, 1995).

This chapter presents two main arguments. First, as long as there was low pressure for land, swidden-fallow farmers who grew rice could easily incorporate rubber into the fallow component of their production systems. The introduction of rubber did not lead to encroachment into primary forest, nor did it greatly affect the broader forest landscape, comprised of primary forest, secondary forest and forest gardens. We use evidence related to the adoption of

©CAB International 2001. *Agricultural Technologies and Tropical Deforestation* (eds A. Angelsen and D. Kaimowitz)

rubber by Iban Dayak in the Second Division of Sarawak and by the Kantu Dayak in the eastern part of West Kalimantan to support this argument. The Iban case and two other cases presented below suggest that, in areas where land pressure became important long after rubber was introduced, local respect for forest remnants and authorities constrained the expansion of agricultural land into unclaimed forests. As a result, rubber production did affect the amount of fallow (secondary) forest in the landscape, but not the remaining primary forest. Rubber gardens basically replaced swidden fallows.

Secondly, we argue that the introduction of rubber by swidden agriculturalists actually had a positive effect on reforestation and therefore on the total forest landscape. Many farmers combine conversion of tropical forests for agriculture with the active creation of forests, such as structurally complex and floristically diverse forest gardens (Padoch and Peters, 1991; de Jong, 1995). We develop this argument using evidence on rubber's impact in three Bidayuh Dayak villages in West Kalimantan. In particular, we look at the expansion of managed forests in the subdistrict of Noyan (de Jong, 1995) and forest management in the subdistrict of Batang Tarah (Padoch and Peters, 1991) and in Sinkawang (Peluso, 1990).

The chapter first summarizes how rubber arrived in Malaysia and Indonesia. Section 3 discusses why swidden farmers easily adopted rubber and the effect this had on the forest landscape. Section 4 demonstrates rubber's contribution to traditional reforestation practices. Section 5 draws general conclusions from the cases discussed.

2. Rubber in Indonesia and Malaysia

2.1. The arrival of rubber

The island of Borneo is geographically divided into two Malaysian States, Sarawak and Sabah, and four Indonesian provinces, West, East, Central and South Kalimantan. The local indigenous population includes many linguistic and cultural groups, commonly referred to as Dayaks. In the past, these groups all subsisted – and many still do – by growing upland rice in swiddens cleared yearly and by hunting and collecting forest products.

Rubber was first introduced in Borneo at the beginning of the 20th century and expanded rapidly. Table 20.1 gives data on the expansion of rubber in the region. By 1921, the area grown in South-East Asia had reached 1.6 million ha and smallholders already accounted for one-third of that (van Hall and van de Koppel, 1950). The crop expanded in a parallel fashion in Sarawak and West Kalimantan. Of the 86,000 ha produced in Sarawak in 1930, smallholders grew 90%. In 1924, exports from West Kalimantan (then Dutch Borneo) reached 15,247 t, implying an area of between 40,000 and 100,000 ha (Uljée, 1925).

Table 20.1. Historical development of rubber in Borneo (from Uljée, 1925; van Hall and van de Koppel, 1950; Cramb, 1988).

Year	Region	Area (ha)	Production (tonnes)
1910	All of South-East Asia	500,000	
1921	All of South-East Asia	1,600,000	
1911	West Kalimantan	500–1,000	128
1924	West Kalimantan	40,000–100,000	15,247
1930	Sarawak	30,000	
1940	Sarawak	97,000	
1961	Sarawak	148,000	
1960	Second Division	25,000	50,000
1971	Second Division	36,000	19,000

The rate of rubber's expansion fluctuated over time. In 1912, the territory now called Indonesia (then Dutch East Indies) had the world's second largest area of rubber plantations after Malaysia (then referred to as British Malaya) (van Hall and van de Koppel, 1950). High rubber prices in the mid-1920s, resulting largely from restrictions on the international rubber trade associated with the 'Stevenson Reduction Scheme', led to rapid expansion in production (McHale, 1967; Ishikawa, 1998). However, by the end of the decade, expanding rubber production in the Dutch East Indies had depressed world prices, which remained low during the early 1930s (McHale, 1967). In 1934, both the Dutch East Indies and Sarawak joined the International Rubber Regulation Agreement (IRRA), which severely limited the expansion of rubber. The agreement established a coupon system, which restricted how much rubber producers could sell and traders could buy. This especially affected smallholders (van Hall and van de Koppel, 1950; McHale, 1967; Barlow, 1978). Prices boomed again in 1950/51, leading to a new surge in rubber planting and tapping in Sarawak (Cramb, 1988) and probably in West Kalimantan as well.

Between 1960 and 1971, rubber exports from Sarawak gradually declined from 50,000 t to 19,000 t. Interest in replanting among small farmers declined, but, thanks to a government rubber planting scheme, total area increased from 25,000 ha to 36,000 ha in 1971. The scheme provided cash advances to farmers who established new rubber gardens. Between 1971 and 1977, when the scheme was temporarily halted, no new planting took place. During this period, pepper also became a prominent cash crop. In subsequent years, farmers have shifted their primary focus back and forth between pepper, rubber and off-farm work (Cramb, 1988).

Coastal Chinese and Malay farmers initially planted most of the rubber in West Kalimantan (Dove, 1993). In the 1930s, inland Dayak swidden agriculturalists widely adopted the crop. This may seem surprising given the colonial restrictions on rubber expansion at the time. But apparently many traders from Sarawak were able to obtain extra coupons, despite the restrictions, and used them to buy cheaper rubber from Dutch Borneo. Smuggling rubber from Dutch Borneo to Sarawak was common (Ishikawa, 1998) and remains so to this day. All rubber officially exported from West Kalimantan came out of Pontianak, the provincial capital. Since no good roads connected the province's remote Dayak villages to the capital, the only way villages' inhabitants could sell their rubber was to send it to Sarawak. This situation continued until better roads finally connected most of West Kalimantan to Pontianak during the early 1980s. Since that time, the Dayaks have sold all of the rubber they produce in the provincial capital. This has made rubber production more attractive and led farmers to plant more rubber.

2.2. Adoption of rubber in swidden agricultural systems

Several authors (Cramb, 1988; Dove, 1993; Gouyon et al., 1993) point out that rubber production fitted the Dayak farmers' traditional swidden-agriculture systems well. In the prevailing swidden systems in Borneo, each year farmers slash-and-burn a field and plant rice. They may also plant small amounts of other crops or tree species just prior to, together with or shortly after planting rice. Once they harvest the rice at the end of the year, they devote less labour to the field. If they planted manioc there, they will still come back the following year to harvest. They also harvest fruit species and may continue to plant additional fruit-trees during the following years. However, after the third year or so, the field gradually reverts into secondary forest, with or without any planted trees. If the field contains many planted or tended trees, farmers will gradually start to clear around them. Otherwise, they will convert the field into a swidden again, once the fallow vegetation has developed sufficiently.

Rubber fits nicely into the swidden system. Farmers can plant it during the swidden stage, often before rice is planted, and then leave it virtually unattended until the trees are large enough to tap, about 10 years later. Cramb (1988) portrays rubber gardens as simply managed fallows that make the swidden-fallow cycle more productive. Farmers were already familiar with the low labour-input technique required to establish tree crops in fallow areas, as they had used them to cultivate indigenous tree crops, such as fruit, illipe nut and gutta-percha (Cramb, 1988; Padoch and Peters, 1991; de Jong, 1995). Rubber's seasonal labour demands complement those of rice cultivation. Farmers cultivate rice during the rainy season, while rubber is fairly flexible and provides work and income during the dry season. Farmers can easily dispose of the output of rubber, which provides a regular source of cash. Although rubber has quite a low ratio of value to weight, it can be stored for long periods and marketed when convenient. For many swidden-fallow farmers, rubber constitutes their main source of cash. Moreover, rubber provides a convenient bank account that can be tapped – literally – as the need arises, for example, in periods of natural and economic shocks.[2]

3. Incorporating Rubber in the Swidden-fallow Cycle

3.1. Rubber among the Iban in Sarawak's Second Division

This section discusses two cases where upland rice farmers in Borneo incorporated rubber into their swidden-fallow cycle: the case of the Iban Dayak farmers in Sarawak's Second Division (Cramb, 1988) and that of the Kantu Dayak farmers in the village of Tikul Batu, in eastern West Kalimantan (Dove, 1993).

The Iban Dayak arrived in Sarawak's Second Division in the 16th century. During the next 200 years, they converted most of the original primary forest into secondary forest, leaving only remnants of primary forest (Cramb, 1988). They farmed their swidden fields for 1 year and then left them in fallow for an average of 15–20 years. This was well beyond the minimum fallow period required to restore the nutrient content in the vegetation and avoid excessive weed invasion after slashing, which was about 7 years. Before they started growing rubber around 1910, the Iban had been growing coffee and pepper commercially for around a decade. The Sarawak government heavily promoted smallholder rubber production and the Iban took up the activity with enthusiasm. Initially, only wealthy communities could afford the plantation costs, at that time equal to about 750 kg of rice ha^{-1}. Once rubber gardens were more widely established, however, seeds and seedlings became cheaper and just about any interested household could plant the new crop. After rubber's initial expansion, planting continued more or less progressively, even during periods of low prices or trade restrictions, such as the 1920s (Cramb, 1988; Ishikawa, 1998).

The introduction of rubber led farmers to reduce their fallow periods and begin planting three to four consecutive rice crops, after which they would plant rubber and leave it there for many decades. This led to higher pressure on the remaining fallow land, but did not transform the forest landscape much. Similar areas of land remained under tree cover. As traditional rubber gardens are rich in plant species, there may have been little impact on species diversity (de Foresta, 1992; Rosnani, 1996). Farmers converted some of their previous fallow land into rubber gardens, but these contained a large amount of secondary vegetation, which developed together with the rubber. The age distribution of fields with secondary forest or rubber gardens that included secondary vegetation may have shifted, but the total forest landscape probably did not change much, nor did encroachment into primary forest accelerate.

During the 1930s, reports emerged that excessive rubber planting had caused shortages of land for rice. While some areas did experience shortages, they were isolated cases where households had only 1–2 ha of rubber, mainly planted on land that was not suited for rice in any case. The cultural importance of rice kept people from planting rubber on fallow land where they could produce rice again.

Cramb (1988) suggests that, by the time rice land became scarce, governments were able to monitor the expansion of agricultural land and to keep farmers from clearing primary forest without permission. The government widely announced that farmers could not expand their agricultural holdings into uncleared forest. Visits of government officials to villages were probably at least partially effective in enforcing these measures.

In the decades following the Second World War, population growth increased the pressure on remaining fallow land. Farmers were forced to rely more on cash crops and devote less land to growing rice. Price booms boosted rubber planting, but farmers did not respond to periods of low prices by reducing their rubber gardens. By 1960, rubber covered half of the entire territory in some Iban villages in the Second Division, and the villages had ceased to be self-sufficient in rice. Some people preferred rubber and only produced rice when they felt they had enough land to do both. Others looked for off-farm income or migrated to remote areas. By the 1980s, many Iban rubber gardens in the Second Division had gone through at least two rice–rubber cycles and hill rice farming had become only a supplementary activity for most farmers (Cramb, 1988). The province of Riau in Sumatra went through a similar process (Angelsen, 1995).

In the Iban case, by the time rubber became the dominant crop, farmers had stopped expanding into primary forest and swidden-fallow land had already expanded a great deal. Additionally, government prohibitions on converting primary forest limited further encroachment into primary forest areas. Had this not been the case, swidden cultivation might have expanded more and rubber could have played a role in that. The government in Sarawak did not consider secondary forest off limits and did not restrict rubber from replacing it.

3.2. Rubber among the Kantu in eastern West Kalimantan

The Kantu in eastern West Kalimantan underwent a process similar to the one just described. The Kantu received their first rubber seeds from their Iban neighbours, living in Sarawak's Second Division. By the Second World War, the majority of farmers reportedly had rubber, but few had full-grown rubber gardens. In the mid-1980s, an average Kantu household had two dozen plots on 52 ha of land, of which two or three plots were used each year to produce rice and an average of five plots or 4.6 ha was in rubber. Although this was mainly on land that farmers had once used for swiddens, the land was of poor quality and therefore had little value within the swidden system. These sites are, for instance, located along the river-banks or on poor heath soils. Today rubber provides the principal cash income among the Kantu and complements non-monetary incomes from agriculture and forest collection.

At least until the late 1980s, rubber gardens had no significant effect on agricultural expansion into the forests (Dove, 1993). Apparently, as in the Iban case, the Kantu have enough fallow land where they could plant rubber for them not to need to convert primary forest. Some of that land is of poor quality and farmers are willing to take it out of the rice production cycle. Given the abundance of fallow land, they could put some land aside to grow rubber without drastically reducing the length of the fallow and thereby rice yields.

The Kantu swidden agricultural labour system was apparently flexible enough to allow farmers to allocate some of their time to rubber tapping and occasional weeding of rubber gardens without significantly affecting their other main economic activities. They do not devote labour they would otherwise allocate to cultivating rice to producing rubber. Most of the time they spend on rubber production would probably otherwise go into activities such as hunting, forest-product collecting, house maintenance or leisure.

4. Rubber as an Agent of Forest Reconstruction

The introduction of rubber has not only affected the clearing of forest by Dayak farmers but also their reforestation activities. Elsewhere, we have argued that, while Dayak farmers throughout Borneo convert some forested land into agricultural land, they also transform other non-forested land back into forest (de Jong, 1997). Many of these human-made forests are similar in structure and diversity to the original primary forest (de Jong, 1995, 1999). This section discusses three cases to show the impact of rubber in this process, all of which involve the Bidayuh Dayak, who live in central and western West Kalimantan. It goes into greatest detail in the case of Maté-maté farmers in the village of Ngira, central West Kalimantan (de Jong, 1995, 1997, 1999). It also makes reference to the village of Tae, 150 km south-west of Ngira, occupied by Jangkang Dayak, and to Bagak, a village located much closer to the coast, near the border between West Kalimantan and Sarawak (Peluso, 1990; Padoch and Peters, 1991; Padoch, 1998).

4.1. Rubber in Ngira

Maté-maté Dayak, a linguistically separate group of what are identified as Bidayuh or Land Dayak, inhabit the village of Ngira (King, 1993). In 1994, the village had a population density of 14 km^{-2}. Farmers in Ngira first adopted rubber production during the mid-1930s. Much rubber was exported to Malaysia, since the road to Pontianak was very poor. As late as 1980, even though many farmers had rubber fields, rubber still occupied only a small portion of the land.

Road improvements, which made the region more accessible, changed this situation. Farmers became more integrated into the cash economy and began to consume more goods from outside the region. Rubber is now the main cash crop and many of the current rubber gardens were planted during the last 20 years. Hence, to a certain extent, the village's rubber expansion remains fairly recent compared with the coastal regions in West Kalimantan or the Second Division in Sarawak.

Villagers grow rubber both in rubber gardens and in *tembawang*, or forest gardens. *Tembawang* are forests that farmers have actively created on previous agricultural land to produce tree products and mark the sites they have occupied (de Jong, 1999). When the owners plant rubber in it, one can consider *tembawang* as mixed rubber–forest gardens.

The evolution of Ngira's forest landscape between 1984 and 1993 clearly indicates that introducing rubber can have a positive effect on land use. In 1994, the village had 1688 ha that had been slashed for swidden production at one point or another (Fig. 20.1). A small part of this was currently under swidden production (125 ha). More than half of it was in fallow (954 ha). There was 95 ha of full-grown *tembawang* forest, much of it with rubber inside. Rubber gardens covered 344 ha, of which 121 had been planted within the last 10 years. An additional 251 ha of fallow land had rubber planted on it but, since the trees were still small, the land was classified as fallow rather than rubber garden. In total, 692 ha had planted tree vegetation – 40% of the total cultivated land.

Of the 692 ha planted with trees, 280 ha had either a mixture of rubber, fruit- and other trees or fruit-trees and other species. Most of these areas were adjacent to *tembawang* areas, since villagers prefer to keep their rubber gardens close to the village. This is because they tap the rubber in the morning and collect the latex just before noon. If the villagers ultimately decide to allow abundant secondary regrowth in those fields, which appears likely, they will end up creating an additional 280 ha of *tembawang*.

In total, we calculated that 512 ha of land was replanted with trees during the last 10 years to create *tembawang* or mixed rubber gardens, which combine rubber, other planted species and spontaneous vegetation (de Foresta, 1992; Rosnani, 1996). In that same period, farmers converted only 360 ha from natural forest to agricultural land. Moreover, this increase in effective forest cover took place at the same time as population grew annually by 2.9%.

Our data suggest that introducing rubber into Ngira greatly contributed to the reforestation just described. It encouraged the expansion of forest

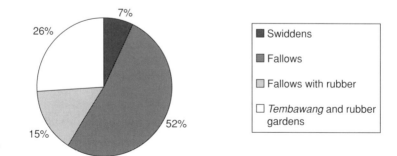

Fig. 20.1. Land use in Ngira, 1994.

gardens and the transformation of swidden fallows into rubber forests. The next two cases provide some indication of how this type of process might play out in the long run.

4.2. Rubber in Tae and Bagak

In both Tae and Bagak, the adoption of rubber has resulted in the sustained presence of forests. Tae is a village located in the subdistrict of Batang Tara, about 150 km south-west of Ngira, and has a population density of about 80 people km^{-2}. People use motor cycles on a well-kept dirt road to take valuable durian fruits (*Durio zibethinus*) to traders, who come from Sarawak (Padoch and Peters, 1991). Many farmers have turned to wet rice cultivation in permanent paddy-fields, while maintaining some upland swiddens and rubber gardens. Only the peaks of the highest mountains still have unclaimed primary forest, although significant areas of communal primary forest protected by the communities and *tembawang* remain.

Bagak, a Dayak village located along the northern coast of West Kalimantan, near Singkawang, represents another example. It has a population density of 120 people km^{-2}. It is strictly forbidden to open new fields in the Gunung Raya Pasir nature reserve, which borders the village territory, even though the village has no other remaining areas of natural forest it could convert to agriculture (Peluso, 1990). In 1990, 11% of the 1800 ha of cultivated land in the village was under paddy rice and 19% under swiddens and swidden fallows. Another 16% of the area consisted of improved rubber plantation, established in 1981 and 1982, while 39% of the land was under mixed tree cover, similar to *tembawang*. Secondary forests preserved by the community accounted for another 3% (Fig. 20.2).

These last two cases indicate that the presence of the forests tends to stabilize when swidden fallow and forest management reach an advanced state of land use. Respect for individual ownership of forest gardens, communal protection of forest remnants and agreements between communities and

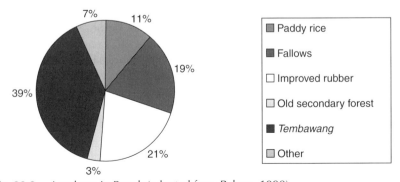

Fig. 20.2. Land use in Bagak (adapted from Peluso, 1990).

governments to preserve protected areas largely explain this tendency of forest area to stabilize. Table 20.2 summarizes the main characteristics of the five cases discussed.

5. The Effect of Rubber-like Technologies on Forest Landscapes

In several of the cases discussed above, introducing rubber had the following effects on forest clearing. At the time rubber was first introduced, farmers already had substantial areas of fallow and only planted a small portion of it with rubber. They mostly planted rubber on land that was not vital for rice production, either because they had enough other land to grow their rice or because the rubber land was of poor quality. Subsequently, further planting of rubber in fallows coincided with population growth and increased pressure on land. Farmers no longer had enough land to sustain swidden rice production and remain self-sufficient in rice. This led them to seek alternative sources of income, either cash-cropping or off-farm employment. While the expansion of rubber appears to have accelerated the abandonment of rice self-sufficiency, it probably did not result in forest encroachment. Rather than clearing additional forests to plant rice, most farmers chose to take up off-farm employment or obtain income through other means. Partly this happened because the government has, in some cases, been able to persuade communities to stop expansion of agricultural land into the remaining primary forest areas.

In some areas of West Kalimantan, the introduction of rubber seems to have actually increased forest cover. In Ngira, the expansion of *tembawang* and rubber gardens appears to have offset forest encroachment for agriculture between 1984 and 1993. The existing forest management practices can easily incorporate rubber, and rubber actually appears to have stimulated the expansion of these human-made forests, which have a diverse structure and floristic composition. On balance, rubber appears to have increased total forest cover in this area. The cases of Tae and Bagak suggest that eventually the process of forest transformation reported in Ngira will stabilize and lead to a mixture of agricultural land, mixed rubber gardens and forest gardens and primary forest, which villagers and the government do not allow farmers to convert to other uses.

One can draw several general conclusions from these cases. At the time rubber was introduced into the pre-existing extensive land-use system, significant areas of primary forest had already been converted for agricultural use. This made it possible to incorporate rubber without creating a significant demand for new land from primary forest. Rubber did not require much additional labour and the labour it did require was largely during periods when it was not needed by other agricultural activities. Farmers had little need for cash and adjusted their level of effort to what was required to meet that need (Dove, 1993). They maintained yearly swiddens as long as that was politically

Table 20.2. General characteristics of the five cases presented in this chapter.

Attribute	Second Division, Sarawak	West Kalimantan	Ngira	Tae	Bagak
Ethnic group	Iban	Kantu	Bidayuh (Maté-maté)	Bidayuh (Jangkang)	Bidayuh
Year rubber introduced	1910s	1940s	1930s	1930s	1930s
Population density	Not available	Not available	11 person km^{-2}	80 person km^{-2}	110 person km^{-2}
Accessibility	Good	Poor	Poor	Regular	Good
Stage of development	Rice cultivation being abandoned because of land pressure. Rubber replaced fallow land. Much rubber also abandoned	Rubber incorporated into swidden fallow cycle. Still little impact of rubber on general land use	Rubber fully incorporated into swidden agricultural cycle. Rubber boosts expansions of forest gardens	Rubber stable part of land use. Upland agricultural fields increasingly converted to mixed rubber fields. Rice production converted to irrigated fields. No further encroachment into forest area	All stages of forests and tree vegetation, including rubber, have stabilized. No further encroachment into forest area
Key factors that influence impact of rubber technology on forest landscape	Population pressure; government control; abandoning of rubber gardens	Incorporation of technology in extensive land cultivation	Increased cash production; existing forest management technology	Local customs related to forest ownership; communal management of forest reserves	Local customs related to forest ownership; communal management of forest reserves; control of protected area

feasible and there was sufficient land for new swiddens. The fact that they continued to grow rice also reflected their cultural preference to produce their principal staple in their own private fields.

Once population density increased, the pressure for land limited farmers' flexibility. They maintained their rubber gardens, but gradually stopped cultivating rice. In three of the five cases discussed, local and/or national authorities increasingly circumscribed encroachment on additional forest areas. Pressure from within the communities to preserve the remaining forests increased and governments persuaded farmers to stop further encroachment. Simultaneously, the state and its local representatives increased their presence. In West Kalimantan, as in many other places in the world, the laws largely prohibit farmers from encroaching on forests. However, such rules only became relevant once the government had sufficient presence to enforce them.

The introduction of a new cash-based production system coincided with and was a catalyst for a number of cultural and socio-political changes, including the increased presence of the state. The rising importance of cash-based economic transactions and improved infrastructure allowed for better communication between state officials and communities. Officials at the regional level adopted national concerns about forest encroachment, and that facilitated enforcement of forest regulations.

Lastly, the cases discussed above demonstrate that new technologies involving some kind of tree or forest production may contribute to reforestation. The presence of tree crops also influences what happens to the forest landscape when other changes occur in local agricultural systems and demographic patterns. For example, land already under tree vegetation is much more likely to revert to forest when farmers shift from upland rice to wet rice cultivation or migrate to the cities. This is taking place in West Kalimantan in areas with out-migration from rural areas. One observes many old *tembawang* and rubber gardens that have developed into closed dense forests.

Tree-planting technologies, like rubber production, have a low impact on the forest landscape when they are incorporated into long-existing extensive agricultural systems. However, when population pressure and market integration increase alongside each other, these effects change. When these technologies are introduced at an early stage in a resource-use continuum from extensive to more intensive land use, socio-economic progress allows for a consensus on land use that preserves forests. This may offset negative effects that might otherwise have been caused by the impact of the technology under changing conditions, such as increased land pressure caused by higher population densities. Tree-planting technologies may be incorporated in local forest management practices and subsequently have a positive effect on the forest landscape.

These findings suggest important policy recommendations. In general, tree technologies have significant advantages when trying to improve local agriculture. Before promoting new technologies, policy-makers should take into account the degree of government presence and negotiations with communities over preservation of certain areas. The promotion of new technologies should always be considered in the light of local resource (forest) management practices, to obtain positive synergies and achieve an outcome acceptable to local farmers and national authorities, as well as limiting negative environmental impacts.

6. Conclusions

The introduction of rubber in West Kalimantan contributed little to encroachment into primary forest. On the other hand, it apparently favoured the restoration of forests in areas where land use became less intensive. It needs to be emphasized, however, that specific conditions in the local context allowed this to take place. If, for example, adoption of rubber had been accompanied by substantial migration into rural areas, that would probably have resulted in encroachment into forest areas. This has happened in places in Sumatra (see also Chapter 16 in this volume by Ruf). The impact of a new agricultural technology on forest conversion depends on the technology itself, but also on the economic and socio-political circumstances in which it happens. In addition, the impact changes over time, in part as a result of parallel economic and socio-political changes.

Tree technologies should be preferred when trying to improve local agriculture. Policy-makers should consider the degree of government presence and negotiated agreements concerning forest conservation before promoting new technologies in forested regions. Incorporation of local resource management technologies, especially tree-planting or forest-management technologies, may enhance positive outcomes in terms of increased income and forest preservation.

Notes

1   The results presented here stem partly from research conducted between 1992 and 1996 on Dayak forest management in the subdistrict of Noyan, West Kalimantan, Indonesia. The New York Botanical Garden, the Tropenbos Foundation and the Rainforest Alliance through their Kleinhans Fellowship funded the research and the Indonesian Academy of Sciences and Tanjungpura University sponsored it. I thank Noboru Ishikawa, Patrice Levang, the editors and an external reviewer for their comments.

2   A recent Center for International Forestry Research (CIFOR) study documents a sharp increase in rubber planting during the recent economic crisis in Indonesia, including West Kalimantan. The future income security and flexibility rubber provides are probably among the main reasons why farmers planted rubber in the midst of the crisis (Sunderlin *et al.*, 2000).

References

Angelsen, A. (1995) Shifting cultivation and 'deforestation': a study from Indonesia. *World Development* 23(10), 1713–1729.

Barlow, C. (1978) *The Natural Rubber Industry: Its Development, Technology and Economy in Malaysia.* Oxford University Press, Kuala Lumpur.

BPS (1999) *Statistical Data on Food and Tree Crops.* Kantor Statistik Nasional, Jakarta.

Cramb, R.A. (1988) The commercialisation of Iban agriculture. In: Cramb, R.A. and Reece, R.H.W. (eds) *Development in Sarawak: History and Contemporary Perspectives.* Monash Paper on Southeast Asia, No. 17, Center for Southeast Asian Studies, Monash University.

de Foresta, H. (1992) Botany contribution to the understanding of smallholder rubber plantation in Indonesia: an example from South Sumatra. Paper presented at the workshop Sumatra: Lingkungan den Pembangunan, BIOTROP, Bogor, 16–18 September 1992.

de Jong, W. (1995) Recreating the forest: successful examples of ethno-conservation among Dayak groups in West Kalimantan. In: Sandbukt, O. (ed.) *Management of Tropical Forests: Towards an Integrated Perspective.* Centre for Development and the Environment, University of Oslo, Norway.

de Jong, W. (1997) Developing swidden agriculture and the threat of biodiversity loss. *Agriculture, Ecosystems and Environment* 62, 187–197.

de Jong, W. (1999) Taking NTFP out of the forest: management, production and biodiversity conservation. In: *NTFP Research in the Tropenbos Programme: Results and Perspectives.* Tropenbos Foundation, Wageningen, the Netherlands, pp. 145–167.

Dove, M.R. (1993) Smallholder rubber and swidden agriculture in Borneo: sustainable adaptation to the ecology and economy of the tropical forest. *Economic Botany* 17(2), 136–147.

Gouyon, A., de Foresta, H. and Levang, P. (1993) Does 'jungle rubber' deserve its name? An analysis of rubber agroforestry systems in southeast Sumatra. *Agroforestry Systems* 22, 181–206.

Ishikawa, N. (1998) Between frontiers: the formation and marginalization of a borderland Malay community in southwestern Sarawak, Malaysia, 1870s–1990s. Dissertation, Faculty of Anthropology, City University, New York.

King, V. (1993) *The Peoples of Borneo.* Blackwell Scientific Publications, Oxford.

McHale, T.R. (1967) *Rubber and the Malaysian Economy.* University Handbook Series on Malaysia and Singapore 3, Singapore.

Padoch, C. (1998) Swidden, sawah, and in-between: agricultural transformation in Borneo. *Human Ecology* 26(1), 3–20.

Padoch, C. and Peters, C.M. (1991) Managed forest gardens in West Kalimantan, Indonesia. In: Potter, C.S., Cohen, J.I. and Janczewski, D. (eds) *Perspectives on Biodiversity: Case Studies of Genetic Resource Conservation and Development.* AAAS, Washington, pp. 167–176.

Peluso, N. (1990) *The Impact of Social and Environmental Change on Forest Management: a Case Study from West Kalimantan.* Food and Agriculture Organization of the United Nations, Rome.

Rosnani, Y. (1996) *Studi Tentang Biodiversitas Tumbuhan Pada Kebun Karet Dengan Tiga Tingkat Umur Yang Berbeda di Dusun Ngira Kecamatan Noyan Kabupaten Dati II,*

Sanggau. Skripsi Fakultas Pertanian, Universitas Tanjungpura, Pontianak, West Kalimantan, Indonesia.

Sunderlin, W.D., Resosudarmo, I.A.P., Rianto, E. and Angelsen, A. (2000) *The Effect of Indonesia's Economic Crisis on Small Farmers and Natural Forest Cover in the Outer Islands.* Report, Center for International Forestry Research (CIFOR), Bogor, Indonesia.

Uljée, G.L. (1925) *Handboek voor de residentie Westerafdeeling van Borneo.* N.V. Boekhandel Visser, Weltevreden, the Netherlands.

van Hall, C.J.J. and van de Koppel, C. (1950) *De landbouw in de indische archipel.* N.V. Uitgeverij W. van Hoeve, 's-Gravenhage, the Netherlands.

Vincent J.R. and Ali, R.M. (1997) *Environment and Development in a Resource-rich Economy: Malaysia under the New Economic Policy.* Harvard University Press, Cambridge, Massachusetts.

Vincent, J.R. and Hadi, Y. (1993) Malaysia. In: National Research Council (ed.) *Sustainable Agriculture and the Environment in the Humid Tropics.* National Academy Press, Washington, DC.

Agricultural Technology and Forests: a Recapitulation

21

Arild Angelsen and David Kaimowitz

1. Viewing the Link in the Larger Context

The opportunity for farmers or companies to capture a forest rent by converting forest to pasture or cropland largely drives deforestation. A number of factors help create such opportunities beside agricultural technologies. These include high output prices, road construction and maintenance in forested areas and cheap and abundant labour and capital, among others.

To understand the link between agricultural technology and tropical deforestation, one must view it within this larger context. The contributors to this book have sought to keep a clear focus on the link between technology and deforestation, without losing sight of the context in which that link occurs. This chapter summarizes their main findings and draws general lessons.

Section 1 presents six representative situations with regard to the technology–deforestation link in different agricultural systems or contexts and recapitulates the key points from each chapter. Section 2 looks at the main conditioning factors that determine how technological change affects forests at a more general level.

2. Six Typical Technology–Deforestation Stories

We divided the cases presented in this book into six categories: developed countries, commodity booms, shifting cultivation, permanent upland (rain-fed) agriculture, irrigated (lowland) agriculture and cattle production. Several of the book's chapters deal with more than one category (Table 21.1).

Table 21.1. The six typical technology–deforestation stories discussed in the book.

Context	Chapters	Key lessons
1. Developed-country history	Mather (3) Rudel (4)	Reforestation and agricultural yield increase can be part of a development that also includes technologies less suited for marginal land, new off-farm jobs and forest regulations
2. Commodity booms	Kaimowitz and Smith (11) Ruf (16) Wunder (10)	When the right (wrong?) conditions exist (technology, export market, cheap labour and capital, abundant forest, favourable policies), large-scale deforestation difficult to stop by economic incentives. Direct regulations (e.g. protected areas) needed
3. Shifting-cultivation systems	Holden (14) de Jong (20) Ruf (16) Yanggen and Reardon (12)	New technologies can, in principle, reduce the need for land, but farmers often choose to expand land area. If migration is also attractive, the innovations easily become deforestation agents
4. Permanent upland cultivation	Cattaneo (5) Coxhead et al. (19) Holden (14) Jayasuriya (17) Pichon et al. (9) Reardon and Barrett (13) Reid et al. (15) (Roebeling and Ruben (8))	Outcome depends on factor intensities and market conditions. A potential for 'win–win' if new technologies can shift resources from more extensive systems. But long-term effects on migration can increase pressure on forests, and higher farm surplus can be invested in forest clearing
5. Irrigated, intensive agriculture	Jayasuriya (17) Ruf (16) Shively and Martinez (18) (Rudel (4))	Probably good due to supply effects and possibly also labour-market effects, but several caveats: might be labour-saving, relax capital constraints and increase demand for upland crops, all of which stimulate deforestation
6. Cattle in Latin America	Cattaneo (5) Roebeling and Ruben (8) Vosti et al. (7) White et al. (6) (Pichon et al. (9))	Improved pasture technologies, if adopted, tend to increase farm income and deforestation, presenting a win–lose situation. Policy packages have the potential for 'win–win'

2.1. Developed countries

The concept of a forest transition plays a central role in the historical reviews by Rudel and Mather. This implies that forest cover declines before it levels out and slowly increases again. Since the first half of the 19th century, forest cover has risen in several European countries, including the three studied by Mather: Denmark, France and Switzerland. At the same time, agricultural yields have steadily increased. This might suggest that growth in yields helped reverse the decline in forest cover. But, as both authors note, these processes occurred in the context of radical social changes, which undoubtedly had their own large impact on forest cover, which makes it difficult to assess the marginal effect of technology on forest cover.

What were these other changes? In both Europe and the USA, better transport networks made agriculture more commercially orientated, weakened the connection between local population growth and agricultural expansion and allowed for more specialized production based on local conditions. These factors, combined with new agricultural methods that tended to be more suited to fertile lands, helped shift agricultural production from marginal to fertile regions. Some abandoned marginal agricultural lands reverted to forest, either through natural regrowth or through tree planting. Deforestation continued in some more favourable agricultural areas, but reforestation of marginal lands more than offset it.

The rural exodus, largely driven by new industrial jobs in the cities, reduced the labour available for agriculture, grazing and fuel-wood collection, thus raising its cost. Even though the mechanization of the American South allowed farmers to cultivate the same area with less labour, a steadily expanding urban labour demand absorbed the labour that mechanization expelled, so deforestation did not rise as a result. In many European countries, migration to 'the New World' relieved the pressure from otherwise growing rural populations. Shifts in energy supply from wood to coal and later other fossil fuels reduced demand for fuel wood. This also contributed significantly to the forest transition.

Political and cultural changes played a central role as well. The state emerged during the 19th century as a legislative and technical agent of environmental management. The enclosure movement and special laws separated woodland from farmland in Europe. As Mather notes, society began to view forests differently; in particular, forests became 'more than timber'.

The history of the present high-income countries offers several relevant lessons for today's low-income countries. The European and North American experiences demonstrate that new agricultural technologies and yield improvements can go together with increasing forest cover. But other elements of development, such as the growth in urban employment, policies that clearly separated forest from agricultural land and an active state, willing and able to enforce environmental regulations, were at least as important as agricultural technologies.

While the history of the developed world and present-day developing countries have certain similarities, one must take care not to go too far in drawing the parallels. The world economy has changed over the years. Developing countries are increasingly integrated into a global economy, which is very different from the one that existed 100 years ago. The policy environment also differs. Rudel notes that 'The American state launched more programmes that affected forests than the contemporary neoliberal states of the developing world will ever do.'

2.2. Commodity booms

Commodity booms offer the most sensational stories of how new technologies – combined with other factors – can convert millions of hectares of forest to cropland in short periods of time. Wunder's chapter on bananas in Ecuador, Ruf's work on cocoa in Côte d'Ivoire and Sulawesi (Indonesia) and the piece by Kaimowitz and Smith on soybean in Brazil and Bolivia provide examples of such booms. Note, however, that the factor intensities of the new technologies differed sharply in the three cases. Soybean is highly capital-intensive, while cocoa is labour-intensive. During the early stages, banana production was labour-intensive but it became increasingly capital-intensive.

Commodity booms and deforestation generally occur when five conditions coincide:

1. International markets can absorb the additional supply without significantly depressing the price.
2. Policies stimulate forest conversion to the new crop.
3. Production can expand into abundant forest areas.
4. Cheap labour is available to plant the new crop.
5. Someone provides the capital to finance the expansion.

The history of cocoa over the last four centuries resembles a cyclone that moves from country to country, wreaking destruction on large tracts of tropical forest. The key to understanding this process is the concept of forest rent, defined as the extra surplus (reduced costs) farmers get by producing a crop on recently cleared forest, rather than in an area that has had crops for some time. Pests, diseases and weeds make it much more expensive to produce cocoa in old cocoa plantations. In addition, the rural labour pool ages and becomes less productive after the first few decades of agricultural colonization. To a lesser extent, this applies to other crops as well. Thus, once farmers have exhausted the forest frontier in a given region or country, some other region with large forest areas will attain a cost advantage and take over. It is Adam Smith at work. The invisible hand guides production to regions with the lowest production costs, sometimes assisted by both visible and invisible political lobbying aimed at facilitating the move.

Commodity booms are linked to labour migration or large pools of underemployed labour, at least when they involve labour-intensive production systems. Again, migration often responds not only to market signals pointing to new economic opportunities, but also to policies that encourage people to move. The first president of Côte d'Ivoire actively promoted migration to the forest frontier, with the slogan 'the land belongs to those that develop it'. Similarly, Indonesia's transmigration programme moved several million people from Java and Bali to the outer islands, including Sulawesi. In both cases, not only did the new migrants provide the necessary labour for the cocoa booms, but immigration also stimulated local inhabitants to 'race for land' with the new migrants by planting cocoa.

As has occurred with agriculture in general, production in the three cases has become more input-intensive over time. In the banana case, the new 'Cavendish' variety required more infrastructure and inputs than the old 'Gros Michel' variety, besides being more perishable. This made production less mobile, which benefited the forest, but excluded smallholders, who lacked sufficient capital to grow the new variety. Similarly, Ruf provides examples where farmers have begun to apply herbicides in their cocoa plantations and notes that this could help prevent the weed problem associated with replanting old plantations.

In both the banana and cocoa cases, greater use of capital inputs had the potential to reduce pressure on forests, but farmers expressed little interest in this type of technology until they ran out of forest where they could expand their operations. In Ecuador, farmers adopted the 'Cavendish' variety after extensive expansion had reached its limit. Ruf also notes that forest scarcity was a major force driving herbicide adoption. Moreover, even in cases when farmers adopted herbicides while a forest frontier still existed, herbicides apparently did not reduce cocoa expansion into forest. As a village leader in Sulawesi noted, 'the fish in the river always look thirsty'.

The dynamic investment effect Wunder and Ruf observe partly explains why technological change did not reduce deforestation in these cases. Ruf describes how the Sulawesi cocoa farmers invest their surplus in expanding their cocoa farms. Wunder points to macro-level investment effects. The banana boom generated additional income for the government, which the latter used in part to invest in infrastructure and credit for further agricultural expansion.

The soybean story differs significantly from cocoa and, to a lesser extent, bananas. It involved heavy capital costs for transportation, storage, processing and marketing. This permitted the industry to achieve regional economies of scale. Once technology and favourable policies helped production reach a critical level, economies of scale and cheap land on the frontier combined to greatly stimulate forest conversion. High capital requirements could have constrained soybean expansion, but they did not. Subsidized credit was plentiful in Brazil until the 1990s and Bolivian farmers had access to private credit.

Not all the soybean expansion took place at the expense of natural forest. Some of it replaced other types of natural vegetation or other crops. But, even where soybean replaced other crops, as in southern Brazil, it had wide-ranging effects on forests. Soybean cultivation there displaced labour and many small farmers could not afford the machinery and chemicals that growing soybean requires. Many sold their land and moved to the Amazon frontier, where they cleared forest for crops and pasture. Cattaneo's simulation model of Brazil backs up this story. He found that capital-intensive innovations outside the Amazon that were labour-saving led to substantial increases in deforestation within the Amazon.

2.3. Shifting cultivation

Some 200–300 million farmers in the tropics practise shifting cultivation. After one or a few years of crops, they leave their land fallow for some longer period to allow the soil to recuperate. For a long time, Boserup (1965) and others have argued that this system is a rational response on the part of farmers to situations of land abundance. Boserup further argued that, as long as land remains abundant, expanding the area cultivated will generally provide higher returns per day of labour than cultivating existing agricultural land more intensively. As a result, farmers tend to exploit the extensive margin before they exploit the intensive one. This is bad news for those interested in forest conservation, since it implies that farmers will not intensify until the forest has already disappeared. Mounting evidence supports Boserup's main claim, pointing to a key dilemma.

Still, in some instances, farmers intensify even when forest is still available. This volume discusses three important forms of intensification of shifting-cultivation systems. Holden examines the introduction of a new principal crop – cassava – in the *chitemene* shifting-cultivation system in Zambia. De Jong and Ruf look at the introduction of commercial tree crops (rubber and cocoa, respectively) following annual crops. This can be seen as enriching the fallows, but often the new cash crop becomes the dominant reason for expansion. Yanggen and Reardon explore the impact of kudzu fallows in Peru, which farmers use to help fallow soil to recuperate more rapidly, enabling shorter rotations.

Intensification of shifting-cultivation systems can increase the output of total agricultural land (cropping and fallow) several-fold. The adoption of cassava in northern Zambia multiplied the carrying capacity of the shifting-cultivation system two to six times. It also allowed farmers to convert to a short-rotation fallow system with a carrying capacity about ten times higher. It was labour-saving and coincided with a major out-migration of males, who went to work in the copper industry. Thanks to the exodus of workers, the new technology did not result in freeing additional labour to clear more land. Thus, in the short term, the replacement of finger millet by cassava reduced deforestation. However, Holden argues that, in the long run, the introduction of cassava will probably induce more deforestation and soil degradation, since higher agricultural productivity has paved the way for higher population densities and allowed families to settle in new marginal areas.

In the past, many analysts have blamed the introduction of rubber into shifting-cultivation systems in South-East Asia for provoking large-scale forest conversion. De Jong critically examines this claim. In the areas of West Kalimantan (Indonesia) and Sarawak (Malaysia) that he studied, rubber contributed little to encroachment into primary forest. On the contrary, it favoured the incorporation of additional tree cover in lands previously used for agriculture, and rubber gardens (or rubber-enriched fallows) produce various economic and ecological benefits. This result contrasts with both Ruf's description of the cocoa story in this volume and the experience with rubber elsewhere in Indonesia, where tree crops have promoted deforestation.

What makes de Jong's cases different? First, farmers had a reserve of already cleared land where they could plant rubber. Secondly, many of these areas were quite isolated and had low in-migration. Thirdly, government enforcement of forest regulations constrained forest encroachment. In the case of cocoa in the neighbouring island of Sulawesi and of rubber in other regions, these conditions did not hold. In these cases, a large labour reserve, reasonable accessibility and new economic opportunities induced migration, while government forest regulation provided few checks. In fact, the state apparatus actively encouraged forest conversion. Under these circumstances, tree crops can rapidly reduce the primary forest cover.

Improved fallows provide a third way to intensify shifting-cultivation systems. Kudzu is a leguminous vine that fixes nitrogen and makes more nutrients available to the soil, which speeds up soil recuperation. It also suppresses weeds, reducing the demand for labour for clearing and weeding. Kudzu therefore permits shorter fallow periods. This should reduce the stock of fallow land, allowing for a larger forest area. It is a low-cost, labour-saving technology that increases yields (of total agricultural land, including fallow) and could potentially save forests. What more could you wish for? But, as Yanggen and Reardon point out, no one can guarantee the forest-saving part. Indeed, higher productivity and labour saving pull in the opposite direction. The two authors' econometric analysis shows that kudzu reduces primary-forest clearing, but boosts secondary-forest clearing, with the net effect being a modest rise in total forest clearing.

In all cases examined, intensification greatly increased yields at a low cost. Under such circumstances, farmers will be prone to intensify their shifting-cultivation systems, even where forest remains abundant. However, it does not necessarily follow that intensification will reduce deforestation. One may well get intensification *and* expansion.

Finally, shifting-cultivation systems beg us to clarify what we mean by 'forest'. Does it include secondary forest and fallow and tree crops, such as rubber? We return to this issue later, in the next and final chapter.

2.4. Permanent upland (rain-fed) cultivation

Permanent upland cultivation (PUC) is common throughout the developing world, although many farmers combine it with shifting cultivation, irrigated cultivation, tree crops or cattle.[1] To understand the overall pattern of land use, we need to consider the demands each of these activities makes on the farmers' labour and capital constraints.

This volume discusses different types of technological change in PUC. These include adoption of high-yielding varieties, introduction of new crops, increased fertilizer application and pest control. Holden analyses the impact of a high-yielding maize variety, introduced in Zambia in the 1970s, accompanied by greater fertilizer use. This capital-intensive technological package discouraged extensive shifting cultivation, but depended on public support. Once the government reduced fertilizer subsidies and removed pan-territorial pricing as part of its structural adjustment policies, the process went into reverse. This provides an argument for renewed targeted support for intensive farming, although earlier policies put a heavy burden on government budgets.

Reardon and Barrett base their defence of 'sustainable agricultural intensification' (SAI) on roughly the same argument. They argue that, to produce more food without degrading the environment, farmers must use more capital, which they broadly define to include inorganic fertilizers, organic matter and land improvements. Reduced government support for farming, higher input prices and declining infrastructure investments encouraged farmers to follow an unsustainable path of intensification or to expand their activities further into the forest or other types of natural vegetation. Although they acknowledge that intensification *per se* will not necessarily reduce expansion, they argue that failure to move towards an SAI path will inevitably lead farmers to expand into the fragile margins. They also note that many quasi-fixed capital investments increase the productivity of the existing cultivated areas more than newly incorporated lands, which favours intensification over land expansion.

African agriculture takes place in a high-risk environment, related in part to widespread pests. Tsetse-flies transmit trypanosomosis in over 10 million km^2, causing morbidity and mortality in livestock and, to a much lesser extent, humans. Tsetse control makes animal traction possible, or at least more productive, and saves large amounts of human labour. While it obviously benefits farmers and animals, Reid *et al.* focus on the impact on forests, including woodlands. Trypanosomosis control encouraged agricultural expansion in their study area in Ethiopia, in part because households with fewer disease problems were able to plough more land.

Coxhead *et al.* also discuss the problems of pests and risk. Using household data from northern Mindanao, the Philippines, they explore how technological changes in vegetable production may affect farmers' demand for maize and vegetable land. Since vegetable production is more labour- and capital-intensive, one might expect a labour- and capital-constrained household that switches over to vegetables to occupy less land. However, Coxhead *et al.* find that technological changes that reduce yield variability and increase yields without affecting factor intensities will have small effects on farm area.

According to Pichon *et al.*, risk-minimizing strategies play a critical role in the land-use decisions of settlers in the north-eastern Ecuadorean Amazon. They conclude that two major reasons why these labour-constrained farmers grow coffee, despite the fact that it is a labour-intensive crop, are that it provides long-term income security and has a readily available market. This is another case where farmers proved willing to intensify even though they operate in a forest-abundant context. In this case, intensification seems to have reduced forest clearing. Those farmers whose production systems focus on coffee tend to maintain more than 50% of their plot in primary forest, even after several decades of working on their farm.

An important lesson emerges from these chapters. To predict the effect of technological change on total demand of farmland, one needs to adopt a whole-farm approach and take into account the interaction between different production systems within the farm. Each system fulfils household objectives, such as income generation, food security and risk prevention, to various degrees and has its own labour and capital requirements. Technological change that both increases yields and requires more labour has the potential to reduce overall farm demand for land, particularly in the short run.

2.5. Irrigated, intensive (lowland) agriculture

The Borlaug hypothesis suggests that new green-revolution technologies in lowland agriculture will save the forest by lowering food prices and thus making expansion less attractive and by increasing agricultural wages and thus making migration to frontiers less attractive. The chapters by Jayasuriya, Ruf and Shively and Martinez examine this issue in the Asian context. Rudel discusses the historical experience of the American South.

Shively and Martinez's study of Palawan in the Philippines provides evidence supporting the labour-market part of the Borlaug hypothesis. A project for improved small-scale irrigation systems raised the average cropping intensity (crops per year) and therefore labour demand. This resulted in a boom in the local labour market and pushed up wages. The availability of more and better-paid jobs in lowland agriculture in turn made it less attractive for the nearby upland population to expand their own agricultural activities, and forest clearing declined by almost 50%.

Green-revolution technologies, including irrigation, normally have a double effect on labour demand. Higher cropping intensity implies higher demand, but labour input per cropping season might decline, due to mechanization or the use of herbicides. The first effect dominated in Palawan. But Ruf presents a case from Sulawesi, Indonesia, where green-revolution technologies have saved labour and released it for use on the cocoa frontier.

Independently of its effects on labour markets, technological progress in intensive agriculture has output-market effects. Analysts normally use economy-wide models, such as the ones presented by Jayasuriya and Cattaneo, to study such effects. Generally, the price-reducing effect of higher agricultural supply resulting from the introduction of green-revolution technologies should favour forest conservation, particularly since food demand is generally inelastic. Nevertheless, there are important caveats. To significantly affect upland deforestation, upland and lowland crops should compete in the same domestic market. If one of them is traded internationally or they are not close substitutes in the domestic market, the impact is likely to be negligible. In addition, any positive output-market effect must outweigh effects pulling in the opposite direction. For example, if lowland technologies displace labour and induce migration to forest frontiers, this effect could potentially override the output-market effect. Further, as in other types of agriculture, technological progress in intensive agriculture can help farmers overcome capital constraints to expanding their agricultural areas by providing funds for investment, as evidenced in the cocoa study.

In the past, researchers have tended to overlook a point made by Jayasuriya – that technological progress in intensive agriculture (and elsewhere in the economy) can raise the demand for upland crops and therefore stimulate forest conversion to expand the area of these crops. In situations where upland products are not traded in the international market and have a high income elasticity, technological progress in lowland agriculture might increase pressure on forests. This may apply, for example, to vegetable or beef production.

2.6. Cattle ranching in Latin America

As White and his colleagues note, to farmers in tropical Latin America cattle represent status and high and stable incomes. To environmentalists they are a chewing and belching nemesis that destroys forests. Both views are correct. Cattle are often farmers' most profitable option. But cattle (not crops) are the main agent of deforestation in Latin America.

Pastures in the Latin American tropics take up a lot of land and often rapidly degrade. Therefore, many have argued that pasture intensification will reduce the need to chop down trees to create additional pastures. Moreover, if one can make the pastures more sustainable, farmers will not have to abandon their existing pastures and create new ones. The arguments have a certain logical appeal, but do they pass empirical tests?

Technologies to improve pastures and beef and milk production are available. Thus, as Vosti *et al.* note, there are three possible scenarios:

1. In the lose–lose scenario, farmers do not adopt land-saving technologies. As a result, their traditional system deteriorates, their income declines and deforestation continues or even accelerates.

2. In the win–win scenario, farmers adopt the new technologies and both income and forest cover increase.
3. In the win–lose scenario, farmers adopt the technology and income increases, but the new technology makes it more profitable to establish new pastures and deforestation increases.

Four chapters in this volume provide tentative answers regarding which scenario seems more likely. Vosti and his colleagues use a linear-programming farm model to show that smallholders in the western Amazon of Brazil are likely to adopt more intensive pasture and cattle production systems. These provide much higher income than traditional technologies; however, they increase pressure on forests. The technologies' greater profitability provides farmers with higher incomes, which relax their capital and labour constraints and permit them to increase their herd size to the privately optimal level. This implies that the win–lose scenario will dominate.

Roebeling and Ruben use a similar methodology to examine the impact of technological change in the Atlantic zone in Costa Rica. Their model predicts that a 20% increase in pasture productivity will lead to an almost 10% increase in pasture area and an almost 28% decline in forest area on the agricultural frontier. The increased pressure comes from large haciendas (> 50 ha) rather than small and medium-sized farms, whose main land uses are cash crops and forest plantations.

While Vosti *et al.* and Roebeling and Ruben both use partial equilibrium models, Cattaneo analyses the general equilibrium effects of improved pasture technologies. He argues that those who claim that improved pasture technologies reduce deforestation have not sufficiently considered the long-term effects. In the short term, the presence of labour and capital constraints makes many improvements in pasture technologies reduce deforestation. However, in the long term, when resources are more mobile, any improvement in the livestock sector will substantially increase deforestation. None the less, technological progress in the livestock sector is much more beneficial to farmers than new annual crop and tree technologies, again confirming the win–lose scenario.

Deforestation is not just a function of technological change. As White *et al.* suggest, the causal relation can also be reversed. They hypothesize that forest scarcity resulting from past deforestation promotes pasture intensification. This brings us back to Boserup (1965). If possible, farmers will expand before they intensify. What White *et al.* add to Boserup is the mechanism that generates this sequence. Forest scarcity leads to higher land prices. This makes it more attractive to expand beef and milk production through pasture intensification, rather than by purchasing additional land. The authors draw on field research in three locations with varying degrees of forest scarcity in Peru, Colombia and Costa Rica. In the Peruvian site, Pucallpa, which still has abundant forest, ranchers' optimal private choice is extensive cattle production and continued deforestation. Thus, we have a case of lose–lose. The Costa Rican site in the Central Pacific region has little remaining forest and high land prices. Farmers intensify to avoid pasture degradation, but with limited impact on forest, since most of it is already gone. The Colombian site, which has an intermediate level of forest cover, provides some reason for optimism. There the short-term effect of new technologies on forest cover appears positive, although the authors argue that to control deforestation in the long term will require other types of policy measures.

The above chapters join an increasing literature that questions whether more intensive pasture technologies will help conserve forests. Does this imply that one must choose between poverty and deforestation? Not necessarily. The general trade-off identified provides a potential entry point for policy-makers. In principle, at least, as Vosti *et al.* point out, policy-makers may be able to offer ranchers improved technologies in return for accepting other policies that limit their ability to expand their pastures. Thus, while pasture technologies are no panacea and, if adopted independently of other policy measures, may increase deforestation, they may form part of the overall solution.

3. Factors that Condition the Technology–Deforestation Link (Fig. 21.1)

In the introductory chapter, we identified several conditioning factors that determine whether technological progress in agriculture leads to more or less deforestation. The case-studies have explored the role of these different factors. Looking back at the seven main variables listed in Chapter 1 (section 3.2), we

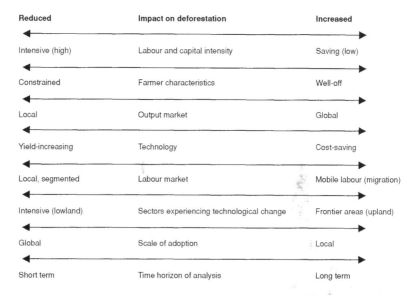

Fig. 21.1. Important dimensions in the technological change–deforestation link, with tentative impacts on deforestation.

have been able to amass a varying amount of empirical evidence on the role of the different factors. We have strong evidence to support the critical importance of the type of technology, the labour-market (migration) effects and the role of credit and higher income in relaxing capital constraints and stimulating farm investments. For all these variables, the empirical evidence corresponds well with the theory-based hypotheses of Chapter 2.

Many chapters have also discussed the role of farmer characteristics, output markets and agroecological conditions, but data appear less systematic and it is more difficult to draw general conclusions based on the evidence in the cases. Finally, the empirical evidence on the role of the institutional context, and the land-tenure regime in particular, is generally weak and calls for further comparative research.

3.1. The labour and capital intensity of the new technology

In Chapter 2, we classified technologies according to their factor intensities (labour per hectare, capital per hectare, etc.). Since most farmers are capital- and/or labour-constrained, how new technologies affect their total capital and labour requirements matters a great deal for how much land they can cultivate. This comes through clearly in the cases where the authors used linear-programming models (Holden, Roebeling and Ruben and Vosti *et al.*). In particular, when markets are imperfect, the households' endowments of labour and cash critically influence the outcome in regard to forest and it becomes much more likely that technological progress can promote forest conservation.

In situations where farmers are not capital- or labour-constrained, it is less important how labour- and capital-intensive new technologies are. Soybean is very capital-intensive, but subsidized credit and access to private credit removed a potential brake on expansion (Kaimowitz and Smith). Similarly, migration, both spontaneous and through government transmigration programmes, ensured a steady supply of labour to the cocoa frontier in Sulawesi (Ruf). How constrained farmers are does not only relate to the functioning of the labour and credit markets; the time horizon of the analysis also matters. In the long term, the technology's input intensities and farmers' constraints are less important (Cattaneo and Holden).

Farmers prefer to adopt technologies that enlarge their opportunities, rather than limiting them. Thus, for example, if farmers are labour- or capital-constrained, they put a high value on labour and are less likely to adopt labour- or capital-intensive technologies, respectively. Sometimes, however, they adopt these technologies anyway, provided they are very profitable or have other desirable characteristics, such as reducing risk or fitting in well with the farmers' seasonal labour requirements. Coffee adoption among smallholder settlers in Ecuador illustrates this point (Pichon *et al.*). The shifting-cultivation stories are mainly about farmers adopting labour-intensive technologies

(Holden, de Jong, and Yanggen and Reardon). But, again, intensification does not guarantee that deforestation stops or even slows down, particularly in the long term.

3.2. Farmer characteristics

Farmers range from poor, isolated and subsistence-orientated peasants to rich, commercially orientated landowners. Each type of farmer tends to specialize in different crops and production systems, making certain innovations relevant only for particular groups of farmers. The Roebeling and Ruben chapter on Costa Rica highlights this. In that case, the large haciendas only produce cattle, while small and medium farms are involved in a range of activities. Thus new pasture technologies mainly increase forest clearing from large farms, something which also has distributional implications.

Farmers respond differently to new technological innovations – in terms of both technology adoption and forest impact. Smallholders tend to be more cash-constrained. This might prevent them from using certain technological innovations, as illustrated by the soybean story from southern Brazil (Kaimowitz and Smith). In that case, only large commercial farmers adopted the technologies associated with large-scale deforestation. Capital-intensive technologies can therefore make poor farmers becoming losers in several ways: they cannot afford the new technologies, they might suffer from lower wages and output prices and deforestation reduces forest-based incomes and environmental services.

Not only are their constraints different, but smallholders tend to emphasize different objectives. For example, they generally emphasize food self-sufficiency and risk avoidance more than large farmers. Coxhead *et al.* find evidence that smallholders in their study area overuse natural resources as an insurance against yield (and price) risk. Thus, risk-reducing technologies could be appropriate for this group of farmers and might help conserve forests.

A low market value of their labour characterizes smallholders. Often they only participate in the low-paid, unskilled, rural labour market. If frontier land is open-access and allocated on a first-come, first-served basis, the low opportunity cost of smallholders' labour makes it attractive for them to migrate to the frontier. Once an active land market develops, poor smallholders may find it more difficult to purchase land, but large landowners may still take advantage of the low opportunity cost of their labour to hire them to engage in activities associated with forest clearing.

3.3. Output markets

The idea that technological progress increases supplies, which lowers output prices and sometimes even reduces farmer incomes, is often referred to as the 'treadmill effect'. Because the demand for food is generally inelastic, small changes in supply can lead to significant price changes, benefiting net consumers but making net producers lose.

The empirical question is how big this price effect is. The magnitude is the product of two factors: overall market demand elasticity and the relative supply increase. If only a small fraction of producers adopt the new yield-increasing technologies, the price effect will be small. This might apply to export crops in cases where each country has a small share of the global market. As a result, commodity booms tend to involve export crops, since world markets can absorb large supply increases with less effect on price. Sometimes, however, commodity booms are so large that they significantly influence global supply and hence depress global prices. This was the case in most of the commodity booms examined in this volume. Nevertheless, in several cases, the negative effect of increased supply on world prices was partially mitigated by rapid growth in consumer demand (Kaimowitz and Smith, and Wunder).

While the overall food demand might not respond much to price changes, this does not necessarily hold for particular food crops, as consumers can switch. Further, many agricultural products that are not foodstuffs, such as rubber and cotton, have synthetic substitutes, making their demand more elastic.

Technologies that producers widely adopt, such as new rice varieties, will typically have a large impact on market prices and this should put a brake on land expansion. Cattaneo, for example, in his economy-wide simulation model, discusses technological progress in the production of food crops (rice, manioc and beans) sold in the domestic market. Land availability does not constrain production and expansion goes on until lower prices make it unattractive to continue.

The link may be more complex in situations where governments heavily influence food prices. Rudel observes that, in the case of the USA, the government did not allow the market to work. Higher yield did not depress market prices as much as one might expect, because price-support programmes maintained high prices. In that case, the government used other policy measures to balance food markets and to convert marginal agricultural land to forest.

Finally, many types of technological change do not increase yields. They only reduce costs. Thus, they do not directly affect supply and output prices, although they may affect supply indirectly by increasing the profitability of production. For example, mechanization typically reduces labour inputs, often without increasing yields. A chain-saw can allow a farmer to clear four to five times more land, but does not necessarily improve yields. Thus an important dimension of new technologies is to what extent they increase yields or just lower costs.

Most of the case-studies do not report on significant price reductions due to technological progress, although Cattaneo's study of Brazil and Kaimowitz and Smith's study of soybeans in Brazil and Bolivia clearly point to downward pressure on prices resulting from supply increases. The fact that the cases do not identify more situations where such effects occurred is partly due to the farm-household (partial equilibrium) approach of most studies. But it might also reflect that many of the relevant technology changes taking place in frontier agriculture are location-specific and only result in a small change in overall market supply. More generally, it underscores the need to distinguish in the debate between development of new technologies that are applicable to large areas of intensive agriculture (à la Green-Revolution technologies) and the adoption of well-known technologies in frontier areas, which contribute a relatively small share of total production.

3.4. Labour markets and migration

In isolated forest-rich economies, one can expect labour-intensive technological change to have a positive or minimal impact on forests. Labour shortages and/or higher wages quickly constrain any expansion. On the other hand, where the regional and/or national labour markets function reasonably well and there is high labour mobility (migration), labour shortages are less likely to limit expansion. Several chapters (see particularly Cattaneo and Jayasuriya) note that the extent of interregional flows of labour and capital plays a crucial role in determining how much the agricultural sector expands, particularly over the long term.

When labour-intensive technological change takes place outside the frontier areas, active labour markets can help to curb deforestation. Employment opportunities outside the frontier will attract labour away from forest-clearing activities in the uplands, as illustrated by the Philippines irrigation study (Shively and Martinez). But, again, labour-saving technologies will foster greater migration to the frontier, as illustrated by the green-revolution technologies in Sulawesi (Ruf) and soybeans in Brazil (Kaimowitz and Smith).

Technological changes in agriculture that improve yields allow local agricultural production to feed more people. Although a higher carrying capacity might benefit the forest in the short term, in the long term it has several indirect effects that are likely to reduce forest cover. Higher populations are often associated with more public services and infrastructure, which attract additional migrants. Three examples in this volume are potatoes in 19th-century Switzerland (Mather), maize in Zambia (Holden) and bananas in Ecuador (Wunder). In all three cases, new crops sharply raised the carrying capacity of the region or country involved, but higher population densities were associated with forest loss.

3.5. Credit markets, farmer income and investment effects

Most farmers cannot borrow money freely and their cash holdings are limited. This sometimes makes them reluctant to adopt capital-intensive technologies, although there are definitely exceptions, cattle being the most prominent. When capital-constrained farmers do adopt capital-intensive technologies, their limited capital resources should make it much harder for them to expand their activities.

But, as noted earlier, new technologies not only influence farmers' demand for capital, they also affect how much capital they have access to. Technological progress increases farms' surplus and hence their ability to purchase farm inputs and make new investments (e.g. Ruf and Vosti *et al.*). Higher yields might also improve access to informal credit (Roebeling and Ruben). These factors make farmers able to increase their investments and buy more inputs, but the effect on deforestation is not obvious.

Consider the typical situation of a Latin American farmer who produces cattle, crops and forest products. Cattle offer less income per hectare but the highest rate of return on capital and labour. Capital constraints generally limit farmers' ability to expand their cattle herds. If a new crop technology boosts overall farm income and livestock are still farmers' most profitable alternative, the farmers may use their higher incomes from crop production to buy more cattle. Thus pasture may expand, rather than the cropping system that experienced the technological progress.

In contrast, a typical African farmer may combine one relatively high-yield–high-input ('land-intensive') system with a low-yield–low-input ('land-extensive') system. The former offers the highest returns to labour, but the farmer does not have enough capital or access to credit to specialize exclusively in that system, so she or he also uses the extensive system. In this case, greater access to credit or higher incomes from any source would allow the farmer to switch to the intensive system and reduce overall demand for land and deforestation (see Holden, and Reardon and Barrett).

Both cases boil down to the question of which type of investment gives the highest return: the land-extensive system, the land-intensive system or possibly some off-farm activity. Higher income due to technological progress would, *ceteris paribus*, reduce the demand for land in the last two cases but not in the first.

The question just raised has a bearing on the impact of higher wages and off-farm income, which have a dual effect on land demand. Higher opportunity costs of labour should make farmers work less in agriculture, thus reducing the demand for land. But more off-farm income relaxes the capital constraint. Farmers can now buy more seeds, fertilizer, machinery, cattle, labour, etc. As just argued, this latter effect could increase or decrease the demand for land, depending on whether land intensification or land extensification is more attractive. In their study of cattle in the western Amazon, Vosti *et al.* found the investment effect to dominate, leading to more pastures and less forest on the farm.

Once most natural vegetation has disappeared in an area, rising incomes resulting from agricultural productivity improvements can generate new demands for timber, fuel wood, fruits, nuts and other non-timber forest products. Growing markets for these products in turn often stimulate periurban households to plant trees as a commercial activity. In such contexts, forests cease to function as a residual land use and take on economic and social characteristics similar to those of agricultural land uses.

The role of capital (credit) constraints in determining farmers' demand for land and of technological change (and other factors) in relaxing these constraints provides very significant lessons from the cases of this book. Together with the migration effects, this appears key in any long-term analysis of the impact of new technologies on forest cover. It also challenges the conventional wisdom that poverty causes deforestation. Indeed, many studies in this book suggest the opposite – that poverty constrains deforestation.

3.6. Scale

Depending on what scale one focuses on, technological change may be seen to have rather different effects on deforestation. The Borlaug hypothesis appears more relevant at the national or global scale. At this scale, increased output and employment opportunities are likely to push down prices and discourage further forest conversion, although, if the technology is labour-saving and no alternative employment opportunities emerge, the opposite may occur. The counter-argument – that technological progress makes expansion more attractive – tends to assume that prices remain constant. This is a more plausible assumption when one looks at the household or village scale. Thus, situations that are win–lose at the local level may be win–win at the global level. This becomes less probable, however, once one takes into account the large areas of degraded lands, fallow and other land uses that fall under neither 'agricultural' land nor forest.

3.7. Short- and long-term effects

Technological changes can affect forest clearing in distinct ways and even opposite directions over time. In the short term, farmers take prices, wages, interest rates, labour and capital resources, government policies, transport, marketing and processing infrastructure and their own incomes as fixed and make their decisions accordingly. Over time, technological change can modify all of these things and this can lead to rather different outcomes. Particularly the chapters by Holden and Cattaneo illustrate that many of technological change's short-term positive effects on forest conservation disappear with time, as labour and capital move and relax the labour and capital constraints that played key roles in the short-term analysis.

Any technological change large enough to significantly affect land use stands a good chance of altering general economic development patterns and political power relations. Standard economic models do not easily capture these aspects. High agricultural growth rates spur economic development, urbanization and the growth of the industrial and service sectors. This, in turn, may induce farmers to abandon land with poor soils and topography and allow them to revert to forest, as happened in Europe (Mather) and North America (Rudel).

Any decision about what time period to focus on must take into account the issue of irreversibility. For many years, conventional wisdom had it that forest clearing in the tropics led to irreversible damage to forest ecosystems. New evidence suggests that these ecosystems might be more resilient than previously believed. The social processes involved in widespread deforestation may prove more difficult to reverse than the biological ones, or at least take just as long. The transformations in land use, demographics, political clout, incomes and wealth and patterns of demand resulting from technological change often linger long after the technology itself has run its course. The families that migrated from Ecuadorean highlands in the 1950s and 1960s to work on the banana plantations on the coast remained there for generations, even though the jobs that brought them there eventually disappeared (Wunder). New soybean technologies in Brazil helped powerful agroindustrial lobbies to emerge in the south and they became a permanent interest group vying for agricultural subsidies (Kaimowitz and Smith).

3.8. The policy context

Policies influence several conditioning factors. Governments can restrict or stimulate migration. They can subsidize or – more commonly – tax agricultural products. They can ban or encourage trade in agricultural commodities. They can favour certain crops or production systems. They fund agricultural research and extension services, which make new technologies accessible to farmers. Therefore sectoral and general macroeconomic policies help determine both technology adoption and the impact technological change has on the environment.

Unfortunately, as noted by Coxhead *et al.* and Reardon and Barrett, policy-makers only give secondary attention to the long-term environmental consequences of their policies. The impact on forests often appears as the unintended or unplanned side-effects of policies intended to boost supply, improve trade balance, etc. Moreover, in many cases, governments are not just determining the general economic environment in which farmers operate, but are actively promoting deforestation.

As experiences in the USA and Europe show, if the more general policy and economic contexts support both economic development and conservation, 'win–win' situations are more likely to occur and their positive impacts will be stronger. Low agricultural prices and labour-intensive agricultural systems resulting from technological change will probably reduce agricultural land use in forest-margin areas much more where agricultural wage rates are high and rural households have many non-farm employment options. Where these conditions do not apply, as in many modern-day developing countries, low agricultural prices may not deter poor households from moving to agricultural frontier areas, since the opportunity costs of their labour are low. Similarly, the market signals provided by technological change will have a much stronger effect if they are reinforced by other policy signals, such as effective regulations restricting farmers' encroachment on protected areas.

4. A Concluding Note

After all this, we are left in a world that defies simplistic explanations but requires clear and simple policies. It is a world in which agricultural innovation has provided huge benefits and yet poses real risks. The basic Borlaug hypothesis – that we must increase agricultural yields to meet growing global food demand if we want to avoid further encroachment by agriculture – still holds. Still, that by no means guarantees that specific agricultural technologies that farmers adopt will help conserve forests. The current trend towards more global product, capital and labour markets has probably heightened the potential dangers. Technologies that make agriculture on the forest frontier more profitable and that displace labour present particularly strong risks, while technologies that improve the productivity of traditional agricultural regions and are highly labour-intensive show the most promise.

Note

1   Following Ruthenberg (1980), we include systems with short and/or rare fallows in 'permanent upland cultivation'.

References

Boserup, E. (1965) *The Conditions for Agricultural Growth: the Economics of Agrarian Change under Population Pressure*. George Allen & Unwin, London, and Aldine, Chicago.

Ruthenberg, H. (1980) *Farming Systems in the Tropics*, 3rd edn. Clarendon Press, Oxford.

Policy Recommendations 22

David Kaimowitz and Arild Angelsen

1. Why Policy-makers Should Care

No policy-maker anywhere in the world makes decisions about agricultural research and technology transfer based solely on how those activities affect forests; nor should they. They usually think first about how to increase food production, earn more foreign exchange and raise farmers' incomes. If they paused for a moment to consider whether their efforts might have some bearing on deforestation, they might very well still go ahead with them even if they encouraged forest clearing. Indeed, most people would agree that sometimes crops and pasture should replace forests. We certainly would.

At the same time, many people also believe that the current rate of tropical deforestation exceeds reasonable limits; here again we include ourselves. Technological changes in agriculture can greatly influence whether that continues. While decision-makers must take into account a variety of potential impacts their policies may have, they should not ignore the effects on forests entirely. Radical changes, such as introducing a new crop or animal species, eradicating a major pest, shifting from slash-and-burn agriculture to sedentary systems and using machinery, chemical inputs or irrigation for the first time, can dramatically change land use. Policy-makers should consider this before promoting technologies with potentially negative effects, and might also include mitigating measures to avoid undesirable impacts on forests.

Another reason why policy-makers should understand how technological change affects forests is that research managers and development agencies increasingly seek to justify their budgets by claiming that their projects help conserve forests. As the world becomes increasingly urban and past scientific breakthroughs allow us to produce more food than markets demand, political support for agricultural research and technology transfer has declined. In contrast, public concern about the environment, and tropical forests in particular, has never been stronger. This has led many development agencies and research managers to 'repackage' their agricultural-technology work and market it as an activity that takes pressure off forests. Projects in agricultural frontier areas assert that, by helping small farmers produce more for longer periods on their existing fields, they can keep the farmers from abandoning their farms after several years and moving deeper into the forest. National and international research centres argue that, without the added production their new technologies make possible, farmers would inevitably have to clear additional forest to meet the rising demand for food.

Some policy-makers may take the stance that the only policy tools they need to conserve forests are protected areas and permanent forest estates. Within such areas, farmers should be kept out by strict regulation and everywhere else the government should leave markets to determine land use. Such views ignore the fact that public investments in agricultural research and technology transfer can powerfully influence land use, whether policy-makers mean them to or not. Besides, few developing countries have protected areas and permanent forest estates consolidated enough for them to rely solely on these approaches and ignore the potential impact of technological change.

As noted earlier, the findings presented in this book suggest that 'win–win' situations exist where new technologies can simultaneously improve both rural livelihoods and forest condition. In other instances, the different objectives conflict and policy-makers must decide how much forest they are willing to lose in return for higher agricultural production and/or farmer incomes ('win–lose'). Occasionally, one even comes across 'lose–lose' situations, where new technologies promote the conversion of forests to alternative land uses that provide little income or employment, cannot be sustained and/or are based on large direct or indirect subsidies.

2. Win–Win Outcomes

Our research has identified five main types of 'win–win' situations, where technological change can simultaneously meet both development and conservation objectives.

2.1. Agricultural technologies suited specifically for forest-poor areas

These technologies reduce pressure on forests and increase production and the incomes of farmers who adopt them. Some are specifically adapted to the natural environments of regions that have already lost most of their forest. Others require infrastructure, human capital or market access that farmers on the agricultural frontier do not possess. Prime examples of these types of technology include production systems involving highly perishable crops, irrigation investments in traditional lowland agricultural areas and crop varieties designed for regions that have been settled for many years. We can expect any increase in agricultural supply in already deforested regions to depress farm prices and hence discourage agricultural expansion in other areas.

The main caveat is that the technologies must not displace much labour, since people who lose their jobs may migrate to the agricultural frontier. Highly labour-intensive production systems in traditional agricultural regions, such as banana and tea plantations, and the cultivation of flowers, ornamental plants and vegetables can act as sponges for labour and discourage workers from migrating to forest-margin areas.

2.2. Labour-intensive technologies where labour is scarce and migration limited

Farmers in agricultural frontier areas are typically labour-constrained. To adopt a new technology that requires more labour per hectare, they have to stop cultivating some other area. This can reduce overall pressure on forests. However, these technologies will only simultaneously increase incomes and lower deforestation to the extent that they do not encourage in-migration from other regions.

The trick in making this win–win outcome work is to find labour-intensive technologies that farmers are willing to adopt and to avoid an inflow of migrants. In places where labour is scarce, farmers will prefer technologies that save labour, not labour-intensive technologies. Nevertheless, under certain circumstances, farmers will adopt labour-intensive technologies, even on the agricultural frontier. The most common examples involve high-value crops and dairy products whose production is intrinsically labour-intensive, such as bananas, cheese, coffee, coca leaves, pineapple and vegetables. The replacement of shifting cultivation by sedentary annual crop production is another example.

Besides helping to conserve forests, a good reason for policy-makers to promote labour-intensive technologies is that they benefit the poor more, since labour constitutes most poor households' main asset. In contrast, capital-intensive technologies that save labour have made the poor double losers. They cannot afford the new technology and the decline in labour demand depresses local wages.

Integrated conservation and development projects (ICDPs) typically seek to dissuade people living near protected areas from encroaching on those areas by helping them intensify agricultural production on their existing plots. To succeed in these efforts, the ICDPs must have viable labour-intensive alternatives to promote, similar to those mentioned above, and the households that would otherwise encroach upon the protected areas must be labour-constrained. The project must also have some means of keeping additional families and companies from moving in.

2.3. Promote intensive systems where farmers are also involved in low-yielding extensive farming practices

Developing-country farmers are typically involved in several production systems. Capital constraints might prevent them from engaging more in intensive farming, which can reduce overall farm demand for land. In this situation, government programmes might help the adoption of more intensive land uses, which might also be more sustainable.

Government fertilizer subsidies constitute a key policy issue in this regard. In recent years, many sub-Saharan African countries have removed fertilizer subsidies as part of their structural adjustment programmes (SAPs). This may encourage farmers to revert from sedentary agricultural systems to shifting cultivation. Standing forest constitutes a readily available cheap substitute for fertilizers, so they will only utilize the latter if they can obtain fertilizers at below market prices.

2.4. Agricultural technologies that substantially raise the aggregate supply of products with inelastic demand

Green-Revolution enthusiasts have long pointed to reduced pressure on forests as one of the main positive impacts of the widespread adoption of high-yielding varieties (HYVs). They argue correctly that, were it not for the spectacular increases in cereal production the Green Revolution made possible, developing-country food prices would have risen. This, in turn, would probably have encouraged agricultural expansion into marginal areas. The key elements here are that production rose enough to significantly affect prices and that lower cereal prices probably did not increase cereal consumption by very much. Research managers have made similar arguments in regard to livestock research in the Brazilian Cerrado. There, however, it appears doubtful that either of these two conditions applies.

2.5. Technologies that promote agricultural systems that provide environmental services similar to those of natural forests

Many 'agricultural' land uses provide reasonable levels of biodiversity, carbon sequestration, erosion control and other environmental services traditionally associated with forests. They can even serve as a source of 'forest' products, such as timber and fuel wood. While agricultural land uses will never eliminate the need to maintain certain areas in natural forests or plantations, agroforests and similar land uses may substitute for some forest functions. Agricultural research and technology transfer clearly have a role in trying to improve such systems and increase the likelihood that farmers will adopt them. Rather than seeking ways to create landscapes with highly intensive and artificial agricultural systems, on the one hand, and pristine forests, on the other, it might be better for policy-makers to encourage landscape mosaics with diverse multilayered cropping systems and forest fragments. As always, the solution depends on the specific objectives and the trade-off that exists between environmental services and agricultural production.

3. Win–Lose Outcomes

Despite what we would all like to believe, many of the impacts of agricultural technology are not win–win. Often higher incomes for farmers or lower prices for consumers come at the expense of forest cover and environmental services, creating a win–lose situation.

3.1. Agricultural technologies that encourage production systems that require little labour and/or displace labour

The prime examples here are technologies designed for mechanized cropping systems and extensive cattle ranches. By making these systems more profitable, technological innovations can provide incentives for farmers to devote more land to them. Since they do not require much labour, expanding these systems will not drive up labour costs and no feedback from the labour market will kick in to dampen the expansion. In the worst-case scenario, new technologies will actually displace labour and the displaced people will migrate to forest-margin areas and clear additional forest. In these situations, countries benefit from increased food production or foreign-exchange earnings but at the expense of environmental services and local livelihoods.

3.2. New agricultural products for sale in large markets in labour-abundant contexts

Many situations where rapid forest clearing occurs involve the introduction of some new crop for export or large domestic markets. More often than not, the new crops replace forests rather than pre-existing crops or unused degraded lands. The labour for these new activities may come from people who migrate from other regions, seasonally or permanently unemployed people within the region itself or people who abandon traditional activities to take up the new ones. At least in the first two situations, this implies a net increase in the amount of labour devoted to activities that involve forest clearing. The fact that production goes mostly to large markets outside the region often means that supply increases only modestly dampen prices. Typically, the economy booms, at least in the short term, but forests suffer. The major caveat here is that frequently the crops involved are tree crops, such as coffee, cocoa and rubber, which farmers grow in agroforest systems that provide substantial environmental services in their own right.

3.3. Eradication of diseases that limit agricultural expansion

Over the last century, the eradication of pests, such as the tsetse-fly, and diseases, such as malaria, have allowed farmers to occupy large new areas that had previously been off limits. Similarly, the control of foot-and-mouth disease in tropical Latin America may open large new markets to cattle ranchers and encourage them to expand their pasture area. While such disease-control efforts clearly have large benefits for both human health and farmers' incomes, they can also greatly intensify forest clearing.

3.4. Technological changes in forest margin areas with rapidly growing labour forces

Any improvement in the profitability of agriculture in places with remaining forest and abundant labour is likely to provoke greater deforestation. This applies both to situations with rapid spontaneous or directed colonization and to regions with high natural population growth. Technological changes have the greatest potential for fomenting inappropriate deforestation where other government policies, such as subsidized credit, price supports and infrastructure investments, effectively subsidize forest clearing. New technologies greatly magnify the effects of these distortions. Indeed, the combined effect of technological innovation and policy distortions may stimulate much more inappropriate forest clearing than the sum of the two individual effects.

4. Win–Lose + Lose–Win = Win–Win?

As noted previously, many technological changes that farmers are likely to adopt in forest-rich areas are win–lose. Farm income and agricultural production increase, but forest cover shrinks. Many regulatory conservation efforts are lose–win. They restrict farmers' opportunities, but – when enforced – help conserve forests. Perhaps, by creating a policy package that includes both elements one could construct a win–win outcome.

Governments play a central role in agricultural research and technology transfer and could potentially offer farmers subsidized technologies and inputs. In return, farmers might restrict their forest clearing. Access to specific farm programme benefits would be contingent on certain conservation practices. For this to work, however, would require the government to strictly enforce the agreement, which often proves quite difficult. Otherwise, farmers would have strong incentives to receive the subsidized technologies and encroach into forests. This has been a major problem in ICDPs. In principle, these are designed to create win–win packages but they have often been based on naïve assumptions about farmers' behaviour.

5. Forests or Environmental Services?

How policy-makers view the link between technology and forests depends partly on what environmental services they wish to preserve. For the sake of simplicity, this book's authors have tended to arbitrarily divide landscapes into forest and non-forest. Implicitly, this assumes that forest and non-forest are homogeneous categories. Real landscapes are more complex. They include various kinds of primary and secondary forests, fallow, plantations, agroforests, perennial crops, scrub vegetation, annual crops and pastures – to name but a few. Each offers different amounts of environmental services and (in some cases) forest products, such as types of biodiversity, carbon sequestration, recreational values, hydrological functions, marketable goods and products households consume directly. Policy-makers must think about which of these concerns them the most and why. To the degree decision-makers ultimately care more about these specific functions and not some arbitrary definition of forests, it may turn out that perennial crops or agroforests perform as well as or better than certain forests. For example, timber plantations may score lower in terms of biodiversity conservation and erosion control than scrub or fallow.

Many significant technological changes in agriculture involve tree crops, such as cocoa, coffee, oil-palm and rubber. Depending on whether one considers tree-crop plantations 'forested', 'deforested' or somewhere in the middle, one can draw quite distinct conclusions about how these technological changes affect forests. We believe tree crops often have a potential for win–win between farm income and environmental services, particularly when compared with the relevant alternatives and not the status quo situation (which might not be a realistic alternative).

6. Economic Liberalization, Market Integration and Globalization

Agricultural markets are increasingly global. The process is partly technologically driven and partly politically driven. Improvements in processing and transport technology have made it possible for farmers to sell their products far away. Export-orientated development strategies, currency devaluations associated with SAPs and trade liberalization have removed barriers to trade and actively promoted it.

The globalization of agricultural markets makes it much harder for localized agricultural productivity gains to feed through into lower prices and slower growth of cropland and pasture. Global markets are simply too large for most productivity increases to significantly affect prices. Perhaps more importantly, fluctuations in agricultural production in traditional agricultural regions tend to swamp the price effects arising from technological change. As a result, trade liberalization and SAPs greatly increase the likelihood that technological changes in agriculture will have negative or negligible impacts on forests. While agricultural production historically has been closely linked to local population growth, global market demand now increasingly determines local land use.

7. Poverty, Economic Growth and Forests

Many people claim that technological change in agriculture will discourage deforestation by reducing poverty either at the household or at the national level or both. Poor people and countries are excessively concerned with the short term and this leads them to deplete their forest resources too quickly. These analysts imply that, if technological change increased these households' and countries' incomes, that would allow them to take a more long-term view. Others emphasize that the demand for environmental services, such as the recreational benefits associated with forests, generally increases as income rises, while the demand for fuel wood and bush meat declines. For example, higher urban incomes often stimulate tree planting in the nearby periurban surroundings. Technological change also leads to higher economic growth, which may push up wages and discourage people from migrating to marginal agricultural frontier areas or devoting their time to clearing inaccessible forests with poor soils. At the national level, higher per capita incomes may contribute to the governments' capacity to formulate and implement environmental policies. All this suggests that technological change may help families and countries simply to grow their way out of their environmental problems – what we referred to as the economic development hypothesis in Chapter 1.

On the other hand, technological change can also fuel agricultural expansion by providing the capital farmers require for that purpose. If capital markets were perfect, farmers could simply borrow the money they need to enlarge their farms, but in many cases they are not. This forces farmers to finance at least part of their investments involving land clearing with savings, some of which can come from higher productivity and lower costs. Higher incomes also generate additional demands for agricultural products. This pushes up prices and stimulates farmers to enlarge their farms. Economic development provides new sources of capital to invest in infrastructure projects that allow farmers to move into previously inaccessible forests.

We still know surprisingly little about the net effect of these different processes. Some analysts posit the existence of an 'environmental Kuznets curve' for forests: at lower income levels, the additional income will raise deforestation, but subsequent increases will reverse that trend. The econometric evidence to support this idea remains weak. And, even if such a curve exists at the national level, there are still many aspects we do not understand. For example, we still know little about the relative contributions of each factor, the level of income beyond which deforestation begins to decrease or how the question plays out at the household level. For the moment, no one can guarantee that economic development – whether agriculturally driven or not – will lead to a forest transition and an end to inappropriate deforestation. Informed proactive policies will have to do that.

Index

acidification 14, 203, 215, 253, 260, 338
agroecological conditions 1, 7, 97, 272, 283–284, 304, 395
agroforestry 4, 16, 246, 407–409
alang alang *see Imperata*
alley cropping 265
Alternatives to Slash-and-Burn (ASB) 4
　see also slash-and-burn
Amazon 11, 12, 13, 54, 66, 69–89, 91, 94–96, 99, 104, 113–130, 153–165, 188, 190, 195, 202, 205, 206, 213–228, 312, 388, 391, 393, 399
　Brazilian 11, 13, 69–89, 94–96, 113–130, 195, 202, 205, 206, 214, 225, 312, 388, 393, 399
　Colombian 99
　Ecuadorean 12, 153–165, 391
　Peruvian 94, 105, 213–228
Andean Common Market 196, 197, 207, 209
animal traction 14, 240, 241, 244, 283, 284, 285, 390
annual crops 71, 76–79, 80, 81, 83, 84, 85, 87–88, 117, 118, 127, 128, 199, 203, 204, 206, 208, 215, 216, 217, 218, 222, 226, 227, 246, 296, 349, 351, 360, 388, 393, 405, 409

bananas 13, 15, 137, 138, 139, 149, 167–192, 302, 386–387, 398, 401
　Ecuador 13, 167–190, 386, 387, 398, 401
　Panama disease 169, 172, 175
　Rwanda 244
　varieties 13, 170–184, 187
　　Cavendish 13, 174–184, 191, 387
　　Gros Michel 170–176, 387
　see also commodity booms
beans 253, 260
　see also soybeans
benefit cost analysis 101–102, 196, 209, 213
biodiversity 2, 115, 154, 208, 209, 231, 246, 247, 302, 347, 406, 409
bioeconomic models 118–128, 138–149
　see also models
Bolivia 13, 195–209, 386, 387, 397

413

booms *see* commodity booms
Borlaug (hypothesis) 3–4, 53–64, 391, 400, 402
Borneo 16, 304, 367–379, 389
 Kalimantan (Indonesia) 367–379, 389
 Sarawak (Malaysia) 367–369, 389
Boserup 6, 8, 22, 36, 91–92, 163, 163, 224, 254, 292, 305, 388, 393
Brachiaria 100–104, 215
Brazil 11, 12, 13, 53, 69–89, 94–96, 104, 113–130, 195–209, 214, 225, 291, 296, 312, 386, 387, 388, 393, 396, 397, 398, 399, 401, 406
 Brazilian Amazon 11, 13, 69–89, 94–96, 113–130, 195, 202, 205, 206, 214, 225, 312, 388, 393, 399
 Brazilian Cerrado 13, 195–209, 406
Burkina Faso 240, 244, 274–275, 283–285, 313

Cameroon 245–246
capital
 capital/land ratios 121, 122, 123, 124, 125, 129, 180, 197, 198, 208, 321
 constraints 4, 5, 9 10, 11, 12, 16, 19, 23, 25, 26, 27, 28, 82, 94, 101, 102, 103, 123, 124, 146, 147, 148, 150, 162, 182, 188, 190, 191, 207, 215, 226, 246, 390, 392, 393, 395, 399, 400, 406
 costs 13, 23, 122, 140, 141, 227, 387
 goods 6, 20, 215
 human 55, 59, 62, 63, 404
 inputs 6, 21, 22, 121, 139, 146, 222, 232, 387
 intensive 26–28, 128, 391–392
 investment 15, 102, 104, 215, 226, 232, 234, 236
 markets 32, 79, 86, 129, 140, 402, 410
 neutral 21, 28
 quasi-fixed 232, 236, 237, 238, 239, 241, 390
 saving 21, 26, 28, 135, 137, 142, 146, 147, 150
carrying capacity 39, 95, 118, 128, 129, 251–267, 388, 398
cassava 14, 79, 128, 251–267, 370, 388–389, 397
cattle 10, 11–13, 39, 44, 69–150, 271–286, 308, 392, 399
 see also dairy farming/systems; haciendas; livestock; ranching
Cavendish 13, 174–184, 191, 387
 see also bananas, varieties
CGE model 29–32, 72–87, 317–332
 see also general equilibrium
chain-saws 155, 292, 296, 300, 305, 345, 397
chemicals 180, 195, 202, 233, 291, 335, 339, 344, 355, 357, 363, 388, 403
chitemene 14, 252–267, 388
CIAT (Centro International de Agricultura Tropical) 94, 207
coca 214, 405
cocoa 15, 31, 167, 169, 170, 173, 178, 182, 188, 245–246, 291–313, 386, 387, 388, 389, 391, 392, 395, 408, 409
 Cameroon 245–246
 Côte d'Ivoire 15, 273, 291–313, 386, 387
 Sulawesi 15, 291–313, 386, 387, 389, 391, 395, 398
cocoyam 245–246
coffee 12–13, 154–165
 Borneo 371
 Brazil 116, 196, 198, 199
 Côte d'Ivoire 297, 300, 302
 Ecuador 154–165, 188, 391, 395
 export taxes 352
 Philippines 348, 349, 352, 360
commercial farming 42, 54, 71, 205, 207, 208, 240, 241, 326, 348, 349, 359, 371, 385, 388, 396, 400
 see also dairy farming/systems; haciendas; ranching

commodity booms 9, 13, 15, 16, 30,
        167–192, 195–209, 291–313,
        383, 386–389, 391, 397, 398,
        408, 409
    bananas 13, 15, 167–192,
        386–389
    cocoa 15, 31, 170, 291–313,
        386–389
    coffee 199
    oil 174, 178, 179, 184
    soybean 13, 195–209, 386–389
    see also bananas; cocoa; coffee;
        soybeans
communal forests 39, 45, 375
conservation 26, 54, 55, 66, 106, 226,
        232, 237, 240, 242, 401, 404,
        408, 409
    areas 154
    biodiversity 115, 154, 246, 409
    forests 1, 2, 4, 5, 9, 10, 26, 30, 33,
        40, 88, 113, 379, 388, 392,
        395, 400
    integrated conservation and
        development programmes (ICDP)
        4, 405
    soil 207, 226, 240, 241, 246, 359
cost benefit analysis see benefit cost
        analysis
Costa Rica 12, 72, 90–107, 135–151,
        228, 393, 396
Côte d'Ivoire 15, 273, 291–313, 386,
        387
credit
    access/availability 13, 135, 137,
        140, 142, 149, 155, 157, 162,
        163, 164, 224, 260, 261, 285,
        338, 358
    constraints 10, 16, 25, 162, 242,
        253, 254, 266, 354, 358,
        398–400
    markets 7, 116, 148, 241, 253,
        254, 266, 267, 285, 395,
        398–400
    policies 14, 106, 137, 148, 150,
        241
    subsidies 54, 93, 115, 129, 169,
        190, 197–198, 199, 202, 203,
        204, 205, 207, 208–209, 214,
        238, 240, 242, 260, 266, 387,
        395, 408
    supply 262
        formal 140, 147, 148, 149, 150,
            399
        informal 137, 139, 144, 146,
            147, 148, 207, 209, 387, 395
cut-and-carry systems 100

dairy farming/systems 41, 94, 97, 103,
        118–129, 405
    see also cattle
debt crisis 205
demand elasticities see elasticities
demographic factors/changes 46, 136,
        187, 191, 251, 317, 378, 401
Denmark 11, 36–38, 47–50, 385
discounting 5, 75, 126, 140, 141, 233,
        253, 254, 266
diseases 14, 15, 167, 169, 172, 174,
        176, 184, 187, 199, 207, 252,
        271–286, 292, 295, 301, 302,
        352, 353, 354, 359, 386, 390,
        408
    control 14, 15, 271–286, 408
    malaria 279, 283, 408
    Panama disease 169, 172, 176
        see also bananas
    trypanosomosis (tsetse) 14, 15,
        271–286, 390, 408
distribution
    income 2, 8, 71, 84, 150, 238
    land 66, 242, 279
drudgery aversion 251, 255

ecological
    benefits 389
    bioecological 292–293
    conditions 165, 292, 303
    crisis 37
    effects 278
    niches 301
    zones 271
econometric models 138, 222–225,
        342–344, 356–358
    see also models

economic liberalization 14, 16, 231,
    237, 238, 239, 240, 245, 247,
    248, 409–410
economics of scale 20, 185, 197, 198,
    208, 240, 387
Ecuador 12, 13, 153–191, 291, 386,
    387, 391, 395, 398, 401
    Ecuadorean Amazon 12, 153–165,
        391
education 4, 37, 40, 99, 155, 221, 224,
    242, 293
elasticities
    labour supply 9, 25–26, 30–32, 73,
        234, 255, 262, 292, 363, 395
    output supply 9, 31, 221,
        292–294, 319–331, 397, 406,
        408
    output demand
        income 15, 236, 238, 322, 328,
            330, 331, 392
        prices 7, 9, 15, 31, 236, 238,
            241, 254, 322, 323, 328, 330,
            331, 392, 397, 406
    substitution 73, 74, 76, 321, 327
employment 335–345
    off-farm 138, 139, 165, 237, 253,
        262, 376, 402
    unemployment 115, 351, 387, 401
environmental services 4, 8, 27, 227,
    406–407, 408, 409
erosion 8, 43, 47, 142, 199, 202, 208,
    209, 231, 233, 244, 258, 304,
    309, 406, 409
Ethiopia 14, 242, 271–286, 390
exchange rates
    overvalued 174, 179, 183, 186,
        199, 204, 205
    policies 240, 352
    real 69, 85, 86
exports
    bananas/Ecuador 167, 169, 174,
        175, 176, 179, 183, 186
    cocoa 167
    coffee 352
    crops 9, 31, 86, 135, 174, 179,
        186, 205, 246, 322, 323, 397,
        407
    cycles 191
    rice 328

rubber/Borneo 368, 369, 370,
    373
soybeans
    Bolivia 196, 197, 207
    Brazil 195, 196, 199, 205, 206
    USA 199
extensive (farming) systems 26–28,
    216, 226, 265, 284, 390, 399
    see also low-input systems
extensive margin 29–30, 32, 354, 359,
    388

fallow 3, 14, 16, 32, 43, 117, 127, 128,
    156, 213–228, 245, 252–266,
    292–313, 347–355, 367–376,
    388–389, 400, 409
FAO (the Food and Agriculture
    Organization) 2, 107, 136
farm models *see* household models
farming systems *see* agroforestry;
    chitemene; dairy farming/
    systems; extensive systems;
    fallow; grassmound; haciendas;
    intensive systems; kudzu;
    sharecropping; shifting
    cultivation; slash-and-burn
fertilizers 60, 240, 265, 311
    maize-fertilizer 14, 251, 265
    subsidies 14, 55, 240, 244, 406
finger millet 240, 252, 253, 256, 259,
    260, 261, 388
fires 104, 252, 300, 301
firewood *see* fuel
fiscal
    balance 239
    fiscally sustainable 241–242, 243,
        247
    incentives 71, 199, 205
    revenues 154, 207
fish 286, 309, 310, 387
food security 9, 11, 12, 16, 72, 83, 87,
    88, 245, 247, 253, 256, 391
foreign exchange 154, 178, 209, 403,
    407
    see also exchange rates
forest
    legislation 35, 37, 40, 41, 45, 49,
        378, 385, 389

primary forest 4, 178, 218,
    296–297, 300–301, 303, 305
products 106, 126, 146, 155, 208,
    220, 292, 321, 322, 330, 349,
    368, 399, 409
rent 30, 292–296, 305, 308, 309,
    383, 386
secondary forest 8, 14, 215, 216,
    218, 220, 222, 225, 226, 227,
    296, 297, 300–301, 302, 303,
    305, 341, 349, 367, 370, 371,
    372, 389, 409
transition 5, 11, 35–50, 66, 203,
    385, 411
France 11, 42–46, 47–50, 385
fuel
    coal 11, 35, 41, 43, 44, 340, 344
    fuel wood (firewood) 11, 35, 37,
        39, 41, 44, 45, 47, 48, 49, 271,
        330, 344, 385, 399, 406, 410
    oil 115, 155, 168, 174, 178, 179,
        186, 214
fundikila *see* grassmound
fungicides 179
    see also herbicides; insecticides;
        pesticides

gender 27, 265
general equilibrium
    analysis 11, 29–32, 69–89,
        317–334
    CGE models 69–89
        see also models
    effects 19–20, 29–32, 196, 202,
        204, 209, 238, 241, 393
Germany 37, 179
Ghana 242, 291, 296
GIS *see* remote sensing
grassmound (fundikila) 253–260
Green Revolution 3, 15, 53–66, 96,
    238, 292, 296, 308, 319, 320,
    327, 328, 329, 332, 391, 392,
    398, 406
greenhouse-gases 115
Gros Michel 170–176, 370
    see also bananas, varieties
groundnuts 246, 256, 259, 276

haciendas 135–151, 170, 172, 178,
    393, 396
Hecksher–Ohlin model 319–322
    see also models
herbicides 21, 55, 142, 155, 164, 180,
    200, 292, 296, 302, 308–311,
    312, 313, 387, 391
    see also fungicides; insecticides;
        pesticides
Hicks neutral 21
high-yielding varieties (HYV) 2, 327,
    390, 406
    see also reference for the particular
        crops
horticulture 227
    see also vegetables
household (farm) models 23–28,
    118–128, 138–149, 216–225,
    253–262, 336–344, 353–358
    see also models

ICRAF (International Centre for Research
    on Agroforestry) 4
IFPRI (International Food Policy
    Research Institute) 72, 73,
    107
ILRI (International Livestock Research
    Institute) 107, 277
Imperata 309, 310
imports
    dependence 351
    food 245, 327–328, 352
    free trade 321
    fuel 37, 41
    inputs 239
    restrictions 352, 359, 360
    substitution 199, 329, 352
income
    distribution 2, 71, 77, 84, 150
    effects 12, 26, 31, 88, 144, 149,
        236, 319, 323, 326, 328
    elasticities 15, 236, 238, 322, 328,
        331, 392
    farmer 202, 398–400
    maximization 23, 139–140, 154,
        255, 261
indigenous people 115, 368

Indonesia 4, 15, 16, 291–313, 317,
    323, 328, 331, 367–379, 389,
    391
  Kalimantan 367–379, 389
  Sulawesi 15, 291–313, 386, 387,
    389, 391, 395, 398
industrialization 2, 11, 35, 41, 44, 47,
    54, 64, 360, 385
infrastructure 5, 8, 69, 70, 85, 87, 130,
    136, 147, 172, 173, 182, 191,
    196, 197, 204, 221, 224, 231,
    232, 234, 242, 245, 253, 260,
    349, 378, 387, 390, 398, 400,
    404, 408, 410
  irrigation 243, 335–345
  market 165, 238, 240, 247, 252,
    352
  transportation 10, 13, 115, 137,
    149, 169, 183, 205, 295
insecticides 180, 276–277
  see also fungicides; herbicides;
    pesticides
integrated conservation and
  development projects (ICDP) 4,
  405, 409
intensive (farming) systems 10, 12, 14,
    26–28, 79, 96, 113, 118–125,
    127, 128, 129, 265, 266, 267,
    284, 391–392, 399, 405–406
  see also capital intensive; cattle;
    dairy farming/systems
interest rates 7, 75, 102, 136, 139, 140,
    204, 205, 240, 241, 262, 400
  subsidies 136
investment effects 23, 28, 387,
    398–400
irrigation 15, 62, 180, 186, 187, 241,
    243, 308, 318, 331, 335–345,
    391–392, 398, 405

Jones specific-factor model 322–327
  see also models

Kalimantan (Indonesia) 367–379, 389
Kenya 273
kudzu 14, 118, 213–228, 388, 389
Kuznets 5, 411

labour
  constraints 19, 25, 30, 123, 150,
    157, 161, 162, 163, 220, 225,
    226, 227, 285, 300, 308, 348,
    355, 356, 358, 364, 390, 393
  intensive crops/technologies 9, 10,
    11, 12, 15, 21, 26, 28, 78, 79,
    84, 88, 144, 146, 147, 163,
    172, 191, 197, 265, 296, 342,
    351, 390, 391, 395, 398,
    405–406
  labour/land ratios 242, 294, 329
  markets 4, 7, 24, 65, 122, 129,
    196, 197, 202, 204, 234, 237,
    238, 244, 246, 253, 254,
    265–266, 267, 285, 326, 331,
    336, 341, 345, 391, 392, 395,
    396, 398, 402, 407
  neutral technologies 21, 28, 312,
    320, 322, 323, 326, 327, 329
  saving crops/technologies 6, 8, 9,
    20, 21, 28, 137, 142, 146, 147,
    150, 155, 186, 224, 225, 237,
    256, 266, 277, 285, 296, 300,
    305, 308, 309, 312, 335, 344,
    388, 389, 399, 400
  supply 9, 25–26, 30–32, 73, 115,
    125, 234, 236, 255, 262, 292,
    295, 323, 326, 363, 395
land
  capital/land ratio 197
  degradation 5, 6, 8, 11, 32, 72, 75,
    78, 105, 106, 107, 400, 407
  landless 115, 262
  policies 242–243, 297, 313
  prices 12, 78, 92, 97, 99, 100,
    103, 105, 141, 202, 242, 318,
    393, 394
  rights 115, 170, 190
    see also property rights
legumes 101, 102, 118, 119, 121, 122,
    124, 141, 259, 276
liberalization see economic liberalization
linear programming (LP models)
    118–128, 138–149, 253–265
  see also models
livestock
  disease control 271–286
  markets 93–94

systems 113–130
  see also cattle; dairy farming/
  systems; haciendas; ranching
  technology 82–84, 92
loans 204, 242, 338
  see also credit
logging 49, 71, 72, 75, 86, 137, 141, 155, 295, 297, 302, 317, 327
low-input systems 120, 232, 233, 252, 399
  see also extensive systems
lowland agriculture 2, 15, 16, 29, 30, 36, 62, 135, 168, 170, 190, 207, 214, 225, 245, 319, 323, 326, 327–332, 335–345, 348, 383, 391–392, 405
LP-models see linear programming

maize 244, 246, 265
malaria 279, 283, 408
  see also diseases
Malaysia 16, 317, 329, 367–379, 389
Mali 239, 243–244
Malthusian 36
manioc see cassava
manure 233, 240, 357, 363
marketing margins 75, 140, 144, 146, 147, 148
mechanization 179–183, 195–209
migration 29–30, 69–89, 292–296, 300–301, 398, 405–406
millet see finger millet
Mindanao 348–360, 390
  see also Philippines
miombo 251–267
  see also woodlands
models
  bioeconomic 118–128, 138–149
  CGE 29–32, 72–87, 317–332
  econometric 138, 222–225, 342–344, 356–358
  remote sensing (GIS) 273–278
  Hecksher–Ohlin 319–322
  household (farm) 23–28, 118–128, 138–149, 216–225, 253–262, 336–344, 353–358
  Jones specific-factor 322–327

  linear programming (LP) 118–128, 138–149, 253–265
  Markov 72, 278–286
  von Thünen 23–28
monocultures 170, 259, 260

national parks 106, 313, 348
NGOs 242, 245, 247
non-timber forest products (NTFPs) see forest, products
Norway 37

oil see fuel
oil palm 15, 214, 317, 319, 323, 332, 409
open access 75, 254, 396
  see also property regimes/rights
organic
  fertilizers 232
  matter 206, 231, 232, 233, 240, 244, 246, 252, 253, 292, 390
overgrazing 273

paddy see rice
pan-territorial pricing 260, 265, 390
parastatals 240, 241, 242, 247, 260, 261
pastures 91–107, 113–130
  improved 94
  investment 101–103
  policies 93
  technologies 94–96, 97–101, 103–105
perennial crops 11, 70, 72, 79–82, 84–85, 88, 99, 117, 127, 128, 156, 162, 164, 215, 216, 222, 227, 233, 244, 349, 359, 409
Peru 12, 14, 91–107, 213–228, 388, 393
pesticides 3, 142, 215, 308, 339, 343, 345
  see also fungicides; herbicides; insecticides
pests 2, 95, 96, 101, 196, 252, 255, 292, 295, 301, 302, 352, 353, 354, 359, 386, 390, 403, 408

pests *continued*
    pest control 2, 196, 390
Philippines 15, 16, 323, 335–345,
    347–360, 390, 391, 398
    Mindanao 348–360, 390
    Palawan 335–345
plant breeding 200, 203
plantains 135–150, 173, 214, 245,
    246, 302
    see also bananas
plantations 13, 15, 37, 44, 136, 137,
    138, 139, 141, 148, 149, 150,
    151, 155, 167–192, 215, 227,
    291–313, 317, 367, 369,
    386–387, 393, 401, 405, 407,
    409
    see also bananas; cocoa; coffee
planting material 10, 291, 301, 308,
    352
political lobbying 209, 386
potatoes 39, 42, 189, 349, 352, 359,
    398
poverty 2, 4, 5, 62, 63, 115, 116, 227,
    246, 349, 410
price subsidies 135, 147, 149, 150, 260
primary forest *see* forest
productivity *see* total factor productivity
profit maximization 25, 27, 75, 125,
    127, 140, 162, 220, 221, 234,
    337, 342, 343, 344
    see also income maximization;
    utility, maximization
property regimes/rights 7, 15, 37, 54,
    75, 93, 106, 266, 313, 317, 318,
    319, 326, 327, 329, 330, 332
public goods 106, 203, 242, 247, 277,
    283, 284, 286, 398
Pucallpa *see* Peru

ranching 91–107, 113–130, 135–150,
    392–394
    see also cattle; dairy farming
    /systems; haciendas; livestock
rate of exchange *see* exchange rates
reforestation 11, 34, 35–50, 55, 59, 60,
    61, 62, 63, 64, 66, 104, 106,
    291–313, 368, 373, 374, 378,
    385

regional development 72, 115, 173
remote sensing 14, 273–278
rent
    economic rent 76
    forest rents 30, 292–293,
        295–296, 305, 308, 309, 383,
        386
    land rents 23, 27, 75
    rent gradient 23
resettlement 258, 282
    see also migration; transmigration
Ricardo 292, 323
rice (paddy) 243, 245, 296, 300, 302,
    303, 308, 309, 312, 317–332,
    375
risk 347–360
    averse 88, 233, 251, 337, 348,
        353, 355, 358, 359
    neutral 355
roads *see* infrastructure
rubber 15, 16, 31, 116, 215, 317, 319,
    323, 332, 349, 367–379, 388,
    389, 397, 408, 409
rural development 135, 136, 260
Rwanda 233, 240, 244

Sarawak (Malaysia) 367–379, 389
secondary forest *see* forest
seeds
    improved 6, 20, 137, 258
    parastatals 240, 241
    policies 240
    prices 240, 262, 371
    subsidies 238, 240, 244, 246
    true potato seeds 352
settlers 12, 61, 95, 99, 136, 137, 149,
    153–164, 186, 195, 391, 395
    see also migration; resettlement;
        transmigration
sharecropping 54, 155, 172, 198, 199,
    308
shifting cultivation 5, 14, 236, 252,
    253, 254, 255, 265, 267, 303,
    305, 349, 383, 388–389, 390,
    395, 405, 406
    see also chitemene; fallow; kudzu;
        slash-and-burn
silviculture 42, 47, 48, 49

silvopastoral 106
slash-and-burn 4, 213–228, 246, 370, 403
soil erosion *see* erosion
soils 222, 224, 252, 265, 283, 284, 312
soybean 13, 61, 62, 85, 86, 189, 195–209, 304, 386, 387, 388, 395, 396, 398, 401
structural adjustment programmes/policies 14, 137, 207, 238, 242, 251, 390, 406
subsidies *see* credit; fertilizers; fiscal; interest rates; price subsidies; seeds
subsistence
    crops 15, 31, 240, 319
    farming 60, 62, 65, 155, 156, 178, 285, 318
    hypothesis 3–4, 26
    requirements 70, 74, 255, 257
substitution effects 26, 31, 144, 150, 355
substitution elasticities *see* elasticities
Sulawesi 15, 291–313, 386, 387, 389, 391, 395, 398
supply elasticities *see* elasticities
sustainability 2, 5, 6, 8–9, 40, 43, 76, 79, 94, 105, 115, 116, 129, 139, 142, 153, 164, 218, 231–247, 254, 266, 390, 392, 406
    sustainable agricultural intensification 8–9 14, 231–247, 390
Switzerland 11, 36–50, 385, 398

Tanzania 240, 242, 252, 259, 272
taxes 45, 93, 106, 141, 168, 205, 207, 239, 295, 352, 401
technology transfer 206, 207, 403, 404, 407, 408
tembawang 374–376, 378
Thailand 317, 328
timber
    demand 399
    prices 327

    production 37, 45, 48, 76, 126
    supply 2, 37
total factor productivity 6, 11, 20, 21, 22, 77–85, 232
tractors 55, 62, 163, 202, 204, 241, 276, 312, 339, 345
trade
    balance 205, 401
    free trade 24, 321, 360, 410
    international 49, 71, 319, 323, 330, 392
    policies 199
    restrictions 178, 351–352, 371
    terms of 77, 79, 247
    UNCTAD 178
    WTO 352
transmigration 308, 310, 330, 331, 387, 395
    see also migration; resettlement
treadmill effect 31, 397
tree crops 8, 22, 27, 117, 160, 291–313, 367, 370, 378, 389, 390, 408, 409
trypanosomosis (tsetse) 14, 15, 271–286, 390, 408
    see also diseases

unemployment 13, 115, 407
upland agriculture 2, 15, 16, 30, 39, 44, 61, 254, 279, 280, 297, 317, 318, 319, 322–327, 328, 329, 330, 331, 332, 335–345, 347, 348, 349, 351, 352, 359, 360, 368, 371, 375, 378, 383, 390–391, 392
urbanization 49, 65, 199, 234, 317, 401
    see also migration; resettlement; transmigration
USA 2, 53–66, 90, 106, 169, 199, 200, 205, 206, 216, 385, 397, 401
utility
    function 138, 139
    maximization 25–26, 138, 139, 234, 255, 261, 353, 363, 364
    see also income, maximization; profit maximization

vegetables 15, 16, 317, 319, 326, 328, 347–360, 364, 390, 391, 392, 405
  see also horticulture
von Thünen 23, 27

wages 9, 10, 23, 31, 70, 84, 139, 169, 174, 186, 197, 204, 237, 246, 255, 309, 323–331, 336–337, 341–342, 391, 396–400
watershed 347–360
weeds
  control 207, 216, 252, 297, 300, 309, 313, 335, 344
  herbaceous 213, 215, 218
  invasions 5, 95, 96, 118, 219, 259, 296, 305, 308, 309, 371, 387

weeding 21, 119, 121, 122, 137, 140, 142, 160, 215, 218, 220, 227, 232, 245, 254, 259, 277, 373, 389
wheat 3, 47, 188, 199, 202, 276, 292
'win–win' situations/policies 1, 2, 9, 10, 12, 15, 16, 113, 227, 393, 400, 401, 404–407, 408–409
woodlands 37, 43, 49, 196, 203, 206, 251–267, 271, 279, 280, 281, 283, 385, 390
World Bank 3, 207

Zambia 14, 246, 251–267, 272, 388, 390, 398
Zimbabwe 233, 240, 246, 273, 275, 283, 285